Towards Principled Oceans Governance

Problems such as the effects of global warming, overharvesting of fish stocks, the impacts of marine pollution, ineffective maritime regulation and enforcement, and security concerns are common throughout the world's oceans, seas and marine regions. Such crises are calling into question the ocean management paradigm as reflected in the 1982 United Nations Convention on the Law of the Sea. As a result countries are beginning to voyage beyond traditional fixations on sovereign *rights* to exploit marine resources, to a new emphasis on social and environmental *responsibilities* and increasing calls for more "principled" ocean governance. International legal principles, such as precaution and ecosystem-based management, have emerged but the principles only set general directions. A sea of conceptual and practical challenges face countries and citizens as they navigate towards implementation strategies and measures.

Australia and Canada have been at the forefront of efforts to operationalize integrated oceans and coastal management. Throughout the 1990s both countries devoted considerable effort to developing strategies to give effect to international ocean management obligations. *Towards Principled Oceans Governance* brings together papers authored by leading Australian and Canadian policy-makers and scholars in ocean law and policy, to consider principled oceans governance from the perspective of two large coastal states that have actively engaged in domestic, regional and international ocean governance agendas. The book assesses the obligations, compliance, implementation and trends in international ocean law and their impact on Australia and Canada, and explores how these countries have responded to the challenges of ocean governance, looking at the legal, policy and political responses that the countries have adopted. While Australia and Canada alone cannot set the international ocean governance agenda, this volume considers the ways in which they have attempted to influence that agenda and the "lessons learned" from their respective successes and failures, lessons that could be applied to nearly all coastal states seeking to engage in an oceans governance policy.

Donald R. Rothwell, Ph.D., is Challis Professor of International Law, University of Sydney, Australia, and Director of the Sydney Centre for International and Global Law. He is a past president of the Australian New Zealand Society of International Law (ANZSIL), and member of the Commission on Environmental Law, World Conservation Union (IUCN). **David L. VanderZwaag**, Ph.D., is Canada Research Chair in Ocean Law and Governance, Dalhousie University, Canada. He is a member of the Commission on Environmental Law (CEL), World Conservation Union (IUCN) and presently chairs the IUCN Specialist Group on Oceans, Coasts and Coral Reefs.

Routledge advances in maritime research

Edited by H.D. Smith
Cardiff University, UK.

The oceans and seas of the world are at a critical juncture in their history. The pressures of development brought about by the globalisation of the world economy continue to intensify through the major sectors of ocean use. In parallel marine management and policy issues become larger, more numerous, and more urgent. The response of this series is to provide in-depth analysis of ocean development, management and policy from a multidisciplinary perspective, encompassing a wide range of aspects of interrelationships between the oceans and seas on the one hand, and maritime human activities on the other. Several strands run through the series.

- Studies of the development and management of major ocean industries and uses including shipping and ports; strategic uses; mineral and energy resources; fisheries and aquaculture; the leisure industries; waste disposal and pollution; science and education; and conservation.
- Inter- and multidisciplinary perspectives provided by the natural sciences; geography, economics, sociology, politics, law and history.
- Responses to the need to devise integrated ocean policies and management measures which cover the deep oceans, the bordering seas, and coastal zones.
- Regional studies at a variety of geographical scales from large ocean regions to regional seas.

The series is of interest to all concerned professionally with the oceans and seas, ranging from scientists and engineers to surveyors, planners, lawyers and policy makers working in the public, private and voluntary sectors. It is also of wider public interest to all those interested in or having a stake in the world ocean and its bordering seas.

1 **Development and Social Change in the Pacific Islands**
Edited by A.D. Couper

2 **Marine Management in Disputed Areas**
The case of the Barents Sea
Robin Churchill and Geir Ulfstein

3 **Marine Mineral Resources**
Fillmore C.F. Earney

4 **The Development of Integrated Sea-Use Management**
Edited by Hance D. Smith and Adalberto Vallega

5 **Advances in the Science and Technology of Ocean Management**
Edited by Hance D. Smith

6 **World Fishery Resources**
James R. Coull

7 **International Law and Ocean Use Management**
Lawrence Juda

8 **Sustainable Ocean Governance**
Adalberto Vallega

9 **Shipping and Ports in the Twenty-First Century**
Globalization, technological change and the environment
Edited by David Pinder and Brian Slack

10 **Managing Britain's Marine and Coastal Environment**
Towards a sustainable future
Edited by Hance D. Smith and Jonathan Potts

11 **International Maritime Transport**
Perspectives
Edited by James McConville, Alfonso Morvillo and Heather Leggate

12 **Towards Principled Oceans Governance**
Australian and Canadian approaches and challenges
Edited by Donald R. Rothwell and David L. VanderZwaag

Towards Principled Oceans Governance

Australian and Canadian approaches and challenges

Edited by Donald R. Rothwell and David L. VanderZwaag

LONDON AND NEW YORK

First published 2006
by Routledge
2 Park Square, Milton Park, Abingdon, Oxon OX14 4RN

Simultaneously published in the USA and Canada
by Routledge
711 Third Avenue, New York, NY 10017

Routledge is an imprint of the Taylor & Francis Group, an informa business

First issued in paperback 2012

Typeset in Garamond by Wearset Ltd, Boldon, Tyne and Wear

British Library Cataloguing in Publication Data
A catalogue record for this book is available from the British Library

Library of Congress Cataloging in Publication Data
A catalog record for this book has been requested

ISBN13: 978-0-415-38378-3 (hbk)

ISBN13: 978-0-415-51239-8 (pbk)

Contents

Contributors

François N. Bailet, Ph.D., is Honorary Adviser to the President of the International Ocean Institute Network (IOI) and Adjunct Professor in Dalhousie Law School and Marine Affairs Programme, Dalhousie University, in Halifax, Nova Scotia, Canada. His current research interests are ocean governance and the implementation of the Law of the Sea Convention and the UNCED process, with a focus on maritime security, regionalization and capacity building.

Sam Bateman, Ph.D., is a former officer in the Royal Australian Navy and now a Professorial Research Fellow at the Centre for Maritime Policy at the University of Wollongong, Wollongong, Australia and a Senior Fellow at the Institute of Defence and Strategic Studies, Nanyang Technological University, Singapore. He is a former member of the National Oceans Advisory Group (NOAG) established by the Australian Government to advise on the implementation of *Australia's Oceans Policy*.

Richard J. Beamish, O.B.C., C.M., Ph.D., F.R.S.C., is the Senior Scientist and former Director at the Pacific Biological Station, Fisheries and Oceans Canada, in Nanaimo, British Columbia, Canada. He is a member of the Order of Canada and Order of British Columbia and a Fellow of the Royal Society of Canada. He has published over 150 scientific articles on a range of topics from climate impacts on fish, to new species of fish. His current research examines the factors that regulate the abundance of Pacific salmon in the Pacific Ocean.

Anthony Bergin is an Associate Professor of Politics at the University College, University of New South Wales, Australian Defence Force Academy (ADFA), Canberra, Australia. He has published widely on Australian and Asia Pacific oceans policy. From 1991–2003 he was the Director of the Australian Defence Studies Centre, a security think tank based at ADFA.

Marian Binkley, Ph.D., is Dean of Arts and Social Sciences and Professor of Anthropology in the Department of Sociology and Social Anthropology at Dalhousie University in Halifax, Nova Scotia, Canada. Her research interests include anthropology of coastal communities, occupational health and safety of fishers, and tourism. Her current studies focus on the restructuring of the town of Lunenburg, Nova Scotia, from a fishing-based economy to one of tourism, and the impact of the Atlantic Canadian fishery crisis on the health and safety of Newfoundland fishers.

Roger H. Bradbury, Ph.D., is a marine scientist with interests in coastal ecosystems, particularly coral reefs, and an international reputation in advanced modelling and analysis. Educated in Australia and Canada, he is a Visiting Fellow at the Centre for Resource and Environmental Studies at the Australian National University, Canberra, Australia, and a Fellow with the Resource Futures Program in CSIRO. Roger is the principal of the consulting firm Tjurunga, specializing in the science of complexity.

Aldo Chircop, JSD, is a faculty member at Dalhousie Law School, Faculty of Graduate Studies (Marine Affairs Programme) and Faculty of Arts and Social Sciences (International Development Studies Programme) in Halifax, Nova Scotia, Canada. Dr. Chircop held the Canadian Chair in Marine Environmental Protection at the World Maritime University, Malmö, Sweden. His disciplinary and interdisciplinary research interests are in comparative ocean law and policy, law of the sea, international marine environmental law, maritime law and integrated coastal management. He is co-editor of the *Ocean Yearbook* (University of Chicago Press) and co-author of *Maritime Law* (Toronto: Irwin Law, 2003).

Geoff Clark is Professor of Law, School of Law, James Cook University, Cairns, Australia. He holds or has held a number of judicial and tribunal appointments at the federal and state level, including as a member of the National Native Title Tribunal (five years), a member of the Land Court and as Deputy Chairperson of the Queensland Aboriginal Land Tribunal. His research interests are in native title and in the field of alternative dispute resolution with particular emphasis on the design of mediation systems to cater for cross-cultural issues and decision making.

Janna Cumming is a graduate of the Faculty of Law, University of Victoria, Victoria, British Columbia, Canada, and a member of the British Columbia Bar. She is currently practicing in Vancouver as a staff lawyer for Legal Services Society of British Columbia. Her research interests include a broad range of international public law and policy issues.

Rodney Dillon lives in Hobart, Tasmania, Australia and worked at the Pasminco zinc works for 23 years before his election to the Aboriginal and Torres Strait Islander Commission Board in 1999. A Regional Councillor since 1993, he is also the former chair of the South East Tasmanian Aboriginal Corporation. Mr. Dillon is an advocate of indigenous peoples' land and sea rights, including associated resource, hunting and fishing rights, and an active international campaigner for the repatriation of indigenous human remains.

Liza D. Fallon is an Environmental Consultant and completing her doctoral research at the University of Tasmania, Hobart, Tasmania, Australia. She is currently focusing on issues of global governance, sustainability and the Southern Ocean fisheries. Her industry experience includes environmental management, environmental impact assessment and four voyages to Antarctica with the Australian National Antarctic Research Expeditions. She has published on environmental management and environmental impact assessment, coastal and ocean management, land-based water management, sustainable tourism and visitor center planning.

Susanna D. Fuller is a doctoral candidate in the Biology Department at Dalhousie University in Halifax, Nova Scotia, Canada. Her thesis focuses on marine sponge populations in the Northwest Atlantic and the effects of industrial fishing practices on these populations. She is particularly interested in how science informs policy and legislation within environmental and fisheries contexts.

Alison Gill, Ph.D., is a Professor at Simon Fraser University, Burnaby, British Columbia, Canada, with a joint appointment in the Department of Geography and the School of Resource and Environmental Management. She is currently Associate Dean of Arts and Social Sciences. Her research interests are in the processes of community change especially as they relate to tourism. She is a principal investigator in the Linking Science and Local Knowledge Node of the Ocean Management Research Network.

Diana Ginn is an Associate Professor in the Faculty of Law at Dalhousie University in Halifax, Nova Scotia, Canada. She teaches and writes in a variety of areas, including aboriginal title, administrative law, property law, and gender and the law.

Bruce G. Hatcher, Ph.D., is a generalist who studies the relationships between ecosystem structure, function and human activities in marine environments. Educated in Canada and Australia, he currently holds the Chair of Marine Ecosystem Research at Cape Breton University, Sydney, Nova Scotia, Canada, and is the Director of the Bras d'Or Institute there. Certified as a Senior Ecologist by the Ecological Society of America, Dr. Hatcher did much of the work for this chapter as the Professor and Director of the interdisciplinary Marine Affairs Program at Dalhousie University in Halifax, Nova Scotia, Canada.

Larry Hildebrand is Adjunct Professor of Marine Affairs and Environmental Studies at Dalhousie University in Halifax, Nova Scotia, Canada, and is the Manager of Sustainable Communities and Ecosystems for Environment Canada in Halifax. He serves on the editorial board for the journal *Ocean & Coastal Management*, is the co-founder and current Vice President-Liaison of the Coastal Zone Canada Association, and is a Board of Directors member of the Canadian Coastal Science and Engineering Association and The Coastal Society in the U.S. He is the 2002 recipient of the H.B. Nicholls Award for leadership in the field of coastal zone management in Canada.

Douglas M. Johnston, Ph.D., is Emeritus Professor of Law at the University of Victoria, Victoria, British Columbia, Canada, where he held the Chair in Asia-Pacific Legal Relations between 1987 and his retirement in 1995. His previous appointments were at Dalhousie University (1972–87), University of Toronto (1969–72), and Harvard University (1966–69). He has taught and written extensively in international and comparative law, marine and environmental law and policy, international relations and modern Chinese studies. He is currently active as a member of the Board of Governors of the Maritime Awards Society of Canada (MASC).

Russ Jones or Xuya K'aadangaas ("Smart Raven") belongs to the Haida First Nation and resides in Skidegate, Haida Gwaii, British Columbia, Canada. He is a fisheries consultant with a fisheries and engineering background who works mainly for First Nations clients in British Columbia, Canada. His work focuses on fisheries management and policy issues including topics such as co-management and marine protected areas.

Lorne K. Kriwoken, Ph.D., is a Senior Lecturer at the Centre for Environmental Studies and a Research Associate at the Institute of Antarctic & Southern Ocean Studies, University of Tasmania, Hobart, Tasmania, Australia. Dr. Kriwoken has conducted research and published in the fields of ocean policy and law, coastal zone management, marine protected areas, World Heritage areas, sustainable polar tourism and environmental impact assessment. He is a member of the World Commission on Protected Areas (IUCN), a member of the Environment Institute of Australia and New Zealand and on the editorial board of *Polar Record*.

Ted L. McDorman is Professor, Faculty of Law, University of Victoria, Victoria, British Columbia, Canada. Professor McDorman has written extensively on ocean law, policy and management issues. Since 2000, he has been "editor-in-chief" of *Ocean Development and International Law: The Journal of Marine Affairs*.

Ransom A. Myers, Ph.D., holds the Killam Chair of Ocean Studies at Dalhousie University in Halifax, Nova Scotia, Canada. Dr. Myers current major research is on the meta-analysis of data from many populations and communities, and a global assessment of sharks.

Chrys-Ellen M. Neville, B.Sc., is a research biologist at the Pacific Biological Station, Fisheries and Oceans Canada, in Nanaimo, British Columbia, Canada. She has over 15 years' experience working in fisheries research and has co-authored over 30 scientific articles and reports including several award winning publications. Ms. Neville has worked on a variety of projects including factors impacting the early marine survival of Pacific salmon and the impact of climate on fish stocks. Her recent interests include the modeling of ecosystems including Bowie Seamount.

Donald R. Rothwell, Ph.D., is Challis Professor of International Law, Faculty of Law, University of Sydney, Sydney, Australia, and Director of the Sydney Centre for International and Global Law. His major research interest is international law with a specific focus on law of the sea, law of the polar regions, dispute resolution and international law in Australia. He is currently working on a project reviewing the regime of navigation under the law of the sea, and is the current President of the Australian and New Zealand Society of International Law (ANZSIL).

Veronica Sakell is former Director, Australia's National Oceans Office, Hobart, Tasmania, Australia. Under her leadership, the National Oceans Office was established and the South-east Regional Marine Plan development commenced. Ms. Sakell previously held senior executive positions in private industry and in the Department of Premier and Cabinet in Tasmania.

Phillip Saunders is Dean of Dalhousie Law School in Halifax, Nova Scotia, Canada. Professor Saunders' teaching and research interests are in marine and environmental law, maritime boundary delimitation, offshore oil and gas and aquaculture. He was formerly with the International Centre for Ocean Development, a Canadian development agency, as Senior Policy Advisor and as Field Representative, South Pacific.

Martin Tsamenyi, Ph.D., is Professor of Law and Director of the Centre for Maritime Policy at the University of Wollongong, Wollongong, Australia. Dr. Tsamenyi has written extensively on ocean policy making and developing legal frameworks to implement the Law of the Sea Convention and has undertaken consultancy for several governments and international organizations. He is active in the development and implementation of *Australia's Oceans Policy* and served on the Steering Committee charged with the implementation of *Australia's Oceans Policy* in the southeast region.

David L. VanderZwaag, Ph.D., is Professor of Law and Canada Research Chair in Ocean Law and Governance at Dalhousie University in Halifax, Nova Scotia, Canada. He was a co-founder of Dalhousie's interdisciplinary program in Marine Affairs. He is a member of the Commission on Environmental Law, World Conservation Union (IUCN) and Chair of the Specialist Group on Oceans, Coasts and Coral Reefs.

Geoff Wescott, Ph.D., is Associate Professor of Environment, Melbourne Campus of Deakin University, Australia. Dr. Wescott has been actively involved in coastal and park management for 25 years and has published extensively in these fields. He was the Chair of the National Reference Group (NRG) of the Marine and Coastal Community Network until 2003. He continues as a member of the NRG and represents the Network on Australia's National Oceans Advisory Group. He is also Deputy Chair of the Board of Directors of Parks Victoria (Australia) and a former member of the Victorian Coastal Council.

Derek Woolner is a Visiting Fellow at the Strategic and Defence Studies Centre at the Australian National University, Canberra, Australia. He was Director of the Foreign Affairs, Defence and Trade Group in the Research Service of the Parliament of the Commonwealth of Australian until 2002. Mr. Woolner has written numerous papers on the structure and management of Australia's maritime border protection regimes and his contribution to this chapter was made whilst working in the Australian Defence Studies Centre of the University of New South Wales, at the Australian Defence Force Academy.

Foreword

Donna M. Petrachenko[1]

In September 2003, the World Parks Congress was held in the beachside location of Durban, South Africa. Over 3,000 delegates met to debate issues facing natural resource management. Since the latter part of the 1800s when the national park system began, our focus has been on the terrestrial world. At the end of the last century, when 1998 was designated as the International Year of the Oceans, many believed that at last, worldwide attention would be given to the planet's greatest natural resource, the oceans. Leaders at the 2002 World Summit on Sustainable Development identified conservation and management of marine and coastal areas as one of the pressing challenges for countries domestically and internationally. Detailed targets for action were set out in many areas including destructive fishing practices, marine pollution, high seas biodiversity conservation, and marine protected areas.

In recent years, articles have begun to appear in the public domain highlighting the threats that human activities are posing to the marine environment. While understanding of the need to do something may be increasing, tangible results or even small steps appear few and far between. At Durban, while accolades were given for the great expansion of the coverage of protected areas to over 12 percent of the world's surface, little mention was made of the fact that less than one percent of oceans, coasts and seas are protected and even less is truly managed. Perhaps the inability to amass support to face the threats head-on comes from the fact that in the ocean environment, the credo of "think globally, act locally" which has benefited sustainable development, *is* very difficult to apply. Who are the custodians of the oceans? How can oceans effectively be governed to ensure that the principles of ecosystem and integrated management, intergenerational equity, and the precautionary approach are implemented? These questions are for scientists, policy- and decision-makers alike. This book is a definite step forward to inform public debate and raise awareness of the many complexities that exist.

While oceans are of importance to every part of the globe, the two countries, Australia and Canada, which are the focus of this book, have vast coastlines and histories of indigenous cultures and aspects of national development linked to the oceans. Both perform leadership roles in their regional spheres of influence with concomitant results on the international stage. The two countries have ratified the United Nations Convention on Law of the Sea, Australia in October 1994, and Canada in November 2003. The way of life in many coastal communities is tied to the health of the marine environment and the sustainability of living and non-living resources. Both countries have delved into ocean governance issues. Australia has a Commonwealth *Oceans Policy* and Canada has an *Oceans Act*. Tackling the issues associated with moving from sector specific management (whether it be oil and gas exploration and development, wild fisheries, aquaculture, marine transportation, or tourism) to a holistic integrated approach to governance is at the heart of strategic oceans' directions in both countries.

At the same time there are a number of areas where the countries diverge. Jurisdiction in marine areas is shared in Australia among the Commonwealth and state governments, whilst in Canada, the federal government has almost exclusive jurisdiction. Approaches to indigenous issues generally and those associated with jurisdiction differ as well. The beach culture in Australia has a profound impact upon the national psyche. In Canada, the three oceans have their own individual characteristics that are reflected in the associated patterns of human use. Australia has made a priority of marine protected areas since the establishment of the Great Barrier Reef Marine Park in 1975, while Canada's first federal marine protected area to protect biodiversity, the Endeavour Hot Vents Marine Protected Area was designated only in 2003. It is the exploration of these similarities and differences that serves as the foundation for the Australian Canadian Oceans Research Network (ACORN).

I am honored to have been involved with ACORN and was particularly struck by the range of disciplines and depth of experience of participants at the June 2002 Canberra Challenge Workshop which forms the basis for this book. I would also like to express my appreciation to the Asia-Pacific Economic Cooperation (APEC) Marine Resources Conservation Working Group for having the insight to initiate work on oceans governance in the region and support the work of ACORN. This type of leadership fosters informed debate, identifies gaps in current knowledge and practice, and contributes to the emergence of strategic approaches to oceans conservation and management.

As Chairperson of the Canberra Workshop, I found it to be a really interesting experience. The Workshop was designed around each research team group, of which there were 11, presenting their research results to not only their own peers but primarily to a challenge team of six distinguished Australian and Canadian oceans governance opinion leaders.[2] The challenge team provided the presenters with valuable debate and feedback that was instrumental in helping research teams finalize their papers, which are presented in this book.

It is critical at this time in our evolution that we address the lack of integrated frameworks for oceans governance both domestically and internationally. Over the years our terrestrial experience has influenced our attitudes to oceans. Moving from coastal, to three nautical miles, then to 200 nautical miles, then beyond to the outer limits of the continental shelves we are now beginning to debate the future of the common heritage of mankind – the high seas. Our ability to explore the depths is increasing as is access, development and utilization of deep-sea resources. These areas, previously thought of as too remote and available to all who wished to venture there, are now being recognized as needing urgent action and attention. The question is whether we can take steps in time before precious and unique ecosystems are irreparably harmed.

Applying precaution, using a multidisciplinary integrated approach and having foresight to take innovative actions such as involving more that just resource users and governments in the determination of the future, would certainly be a positive indicator of things to come. There must be informed public discussion as effective governance of our oceans needs to reflect the progression of societal values. If this can occur at domestic and regional levels, then perhaps we can begin to see precaution and other governance principles as they truly are, not as defined by a specific discipline, whether lawyers, policy-makers or scientists. Our success will be measured by future generations and we can hope that responses to the challenges posed by governance of the oceans are not left to be defined only after a significant event forces us to take action in hindsight. Hopefully the steps we are taking now will be viewed as leading to positive results and not just as waves of rationality in a perpetual sea of vagueness.

Notes

1 Assistant Deputy Minister – Special Envoy to Asia-Pacific Fisheries and Oceans Canada; Visiting Deputy Secretary Department of Environment and Heritage Commonwealth Government of Australia; Visiting Professorial Fellow Centre for Maritime Policy, Faculty of Law, University of Wollongong; and Lead Shepherd APEC Marine Resources Conservation Working Group.

2 Ms. Donna Petrachenko – Chairperson; Dr. Arthur Hanson – Canada's Oceans Ambassador; Dr. Ian McPhail – Former Chair of the Great Barrier Marine Parks Authority; Dr. Conall O'Connell – Deputy Secretary, Environment Australia; Dr. David VanderZwaag – Canada Research Chair in Oceans Law and Governance; Dr. Donald Rothwell – Faculty of Law, University of Sydney; and Ms. Veronica Sakell – Former Director, Australia's National Oceans Office.

Preface

Just leaving port. That image describes countries around the globe, including Australia and Canada, as they try to voyage beyond traditional fixations on sovereign *rights* to exploit marine resources to a new emphasis on social and environmental *responsibilities* and increasing calls for more "principled" ocean governance. International legal principles, such as precaution and ecosystem-based management, have emerged from the 1992 United Nations Conference on Environment and Development and the 2002 World Summit on Sustainable Development, but the principles only set general directions. A sea of conceptual and practical challenges face countries and citizens as they navigate towards implementation strategies and measures.

This book is the outcome of a comparative research program in ocean law, policy and management focusing on Australian and Canadian approaches towards principled oceans governance. The book draws on papers authored by leading Australian and Canadian policy-makers and scholars in ocean law and policy and presented at workshops hosted by the Australian Canadian Oceans Research Network (ACORN) in Vancouver (2000) and Canberra (2002). It highlights national challenges in implementing key international oceans and environmental law principles in oceans and coastal management, with a particular focus on the principles of integrated coastal and ocean management, integrated maritime enforcement, precaution, ecosystem-based management, indigenous rights, and community-based management.

ACORN was formed in 1993 at the initiative of oceans law and policy scholars based at Dalhousie University and the University of Tasmania. During the first phase of ACORN's work, attention was given to general developments in Australian and Canadian oceans law and policy following the adoption and entry into force of the 1982 United Nations Convention on the Law of the Sea and the 1992 United Nations Conference on Environment and Development (UNCED). The outcome of that work was published as *Oceans Law and Policy in the Post-UNCED Era: Australian and Canadian Perspectives* (1996), Kriwoken, Haward, VanderZwaag and Davis (eds).

Phase 2 of ACORN's research agenda has focused on Australian and

Canadian initiatives to give effect to evolving international normative principles within complex federal, legal, policy and political frameworks. A feature of the current work is therefore a focus on national and local responses in an attempt to provide something of a "report card" on Australian and Canadian progress in this field, but also to highlight the challenges facing all coastal states as they come to grips with ocean policy development and implementation in the new global oceans governance environment.

The present volume consists of core overview papers emanating from the project. Additional sectoral papers in the areas of shipping, oil and gas, fisheries and tourism have been published in a special volume 26(1) of the *Dalhousie Law Journal*.

An additional feature of ACORN Phase 2 has been the strong support given to the project by Environment Australia and Fisheries and Oceans Canada. Without the financial and in-kind support of both agencies, this book would not have been possible. To that end, particular thanks must be extended to Phil Burgess, Dick Carson and Donna Petrachenko who all made significant contributions to the project at various stages. Sam Baird, Chief, Oceans Policy, Fisheries and Oceans Canada – Pacific Oceans Directorate, has been an indefatigable supporter of this project over the past five years and his logistical management skills and personal energy were especially integral to the success of the Vancouver and Canberra workshops. ACORN has also been assisted by secretariat assistance provided by Andrew Brooke (Canberra) and Shelley Schnurr (Vancouver).

Various other entities and persons are also acknowledged. A strategic grant from the Social Sciences and Humanities Research Council of Canada (SSHRC) supported Canadian participation in ACORN 2, and AquaNet, a Centres of Excellence Network for Aquaculture funded by the Natural Sciences and Engineering Research Council of Canada and SSHRC through Industry Canada, provided budgetary support for the comparative indigenous rights analyses, in particular. The Canadian High Commission (Canberra) and Mr. Gaston Barban, Deputy High Commissioner provided assistance during the Canberra 2002 workshop. Art Hanson, Ian McPhail, Conall O'Connell and Donna Petrachenko were members of the "Challenge Team" that critiqued the papers at the 2002 Canberra workshop.

Other members of the ACORN team who contributed in various ways to the finalization of this project include Anthony Charles, Rod Dobell, Sue Dobson, Keith Evans, Nathan Evans, Liz Foster, Craig Forrest, Stuart Gilby, Marcus Haward, Richard Herr, Jane Hutchinson, Stuart Kaye, Hugh Kindred, David Leary, Ted McDorman, Cameron Moore, Rosemary Rafuse, Dawn Russell, Douglas Sanders, Krista Singleton-Cambage, Ian Townsend-Gault, Cheryl Webb and Michael White.

In the production of this book, particular thanks are extended to the library staff of Dalhousie University, the University of Sydney and the University of Victoria, and to Susan Rolston, Seawinds Consulting Services, who undertook considerable editing responsibilities. Molly Ross was endlessly patient with her word processing. The research assistance of Dalhousie law students, Annie Makrigiannis, Emma Butt and Tricia Warrender is also recognized.

Donald R. Rothwell, Sydney
David VanderZwaag, Halifax
April 2005

Part I

Introduction

1 The sea change towards principled oceans governance

Donald R. Rothwell and David L. VanderZwaag

Introduction

Crises in ocean and coastal management are facing states throughout the world. The effects of global warming,[1] overharvesting of fish stocks, the development consequences of aquaculture,[2] the impacts of multiple forms of marine pollution, competition over maritime space between both international and local actors, ineffective maritime regulation and enforcement, and security concerns are common throughout the world's oceans, seas and marine regions.[3]

Such crises are calling into question the ocean management paradigm that has dominated since the adoption of the four 1958 Geneva Conventions on the Law of the Sea[4] and their subsequent replacement by the 1982 United Nations Law of the Sea Convention (LOS Convention).[5] That paradigm has emphasized state entitlements and general responsibilities over ocean areas[6] and has encouraged a top-down, bureaucratic and technical approach to managing ocean uses where scientific knowledge is viewed as the main litmus in rational decision making.[7]

In this chapter, a brief review will be undertaken of the development of the global oceans legal framework, before an assessment is undertaken of the challenges confronting principled oceans governance. The state of oceans governance in Australia and Canada will also be briefly outlined.

The global legal framework

The global agenda for oceans management can be seen in three phases. First, the developments up till 1958 in the international law of the sea and some initial efforts at marine environmental management and protection.[8] These initiatives resulted in the four 1958 Geneva Conventions on the Law of the Sea, all of which entered into force throughout the 1960s and had ongoing influence well into the 1980s.

Second, the period 1959–82, which saw a rapid development in the international law concerning the oceans culminating in the finalization of the LOS Convention.[9] At this time there were parallel developments in international environmental management highlighted by the 1972 Stockholm Declaration on the Human Environment,[10] and new international instruments regulating pollution from ships,[11] ocean dumping,[12] land-based pollution[13] and regional fisheries management.[14] New institutions such as the United Nations Environment Programme (UNEP) with a strong oceans management focus were also created.[15] Scientific expertise in marine environmental protection was also given prominence through the formation of the Group of Experts on the Scientific Aspects of Marine Environmental Protection (GESAMP).[16]

Third, the period 1982 to the present. During this time, with the new law of the sea framework established by the LOS Convention essentially settled, adjustments have primarily been made to the law of the sea by way of supplementary agreements regarding the deep seabed[17] and straddling and highly migratory fish stocks.[18] New environmental initiatives, however, have transformed international environmental law and management. The outcomes of the 1992 United Nations Conference on Environment and Development (UNCED) have been particularly influential, including the Rio Declaration,[19] Agenda 21,[20] and the Convention on Biological Diversity.[21] With these instruments have come further development of overarching environmental norms such as the precautionary principle, intergenerational equity[22] and sustainable development.[23]

In 2002, the World Summit on Sustainable Development (WSSD) provided an opportunity to assess progress in meeting the objectives and goals of Agenda 21 and the state of global environmental protection.[24] WSSD demonstrated that while there has been considerable progress, challenges remain, particularly with respect to fleshing out and implementing the principles of sustainable development.[25] While the record of negotiation and, for the most part, implementation of the new legal instruments for the oceans has been impressive, compliance and enforcement with the new norms has been variable. This is particularly evident in the contrast between the willingness of coastal states to assert sovereign rights over the oceans,[26] and their reluctance to accept responsibility for ocean management of not only areas within their own national jurisdiction and control but also in the case of areas beyond national jurisdiction.[27]

Oceans governance

Against the background of the development of the legal framework for managing the oceans, parallel developments have been taking place in the concept of ocean governance around the world. However, the concept is not easy to pin down because of multiple dimensions.[28] The term captures the ongoing evolution towards more "participatory decision making" involving not just government agencies and departments but a broader range of participants including the private sector, scientists, community groups, non-governmental organizations (NGOs), academics, First Nations and others. Ocean governance also suggests a wider range of approaches to influencing human behavior towards the oceans and coastal areas beyond the traditional command–control icon where governments establish environmental or marine conservation standards backed-up with sanctions, such as fines and imprisonment. Those approaches include economic incentives and disincentives, voluntary programs, community-based management, co-management and integrated ocean/coastal planning.[29]

Being an open-textured concept, ocean governance casts a wide discursive net. It allows discussion of the complex array of contributing factors to the degradation of marine resources and areas, such as poverty, population growth, urban sprawls along the coast, over-consumption, consumerism, limited financial and technical resources, and lack of political will. It opens the door to interdisciplinary reflection on "good governance." For example, what legal, institutional, societal, economic and ethical changes are needed to ensure ocean ecosystems capable of supporting human needs without over-compromising marine biodiversity?

Various trends stand out in ocean governance, reflective of the post-modern era. Those trends include increasing reliance on "soft law" instruments, such as conference declarations and codes of conduct; a growing influence of intergovernmental institutions such as the World Trade Organization (WTO), the European Union and the North American Free Trade Agreement; a fast-changing and fragmented proliferation of rules and regulations; the expanding role of non-legal disciplines such as ethics, sociology, ecology and economics in public policy formation; and an increasing skepticism towards the primacy of science and expert opinions in reaching decisions. Competing social interests (such as full employment and a clean environment) and conflicting social values (such as deep and shallow ecology) place a difficult burden on legislators, administrators and the courts.[30]

Various legal principles have emerged through international agreements, declarations and codes of conduct, and they are acting like engines of change in the normative transition.[31] Those principles include, among others, integration, precaution, the ecosystem approach, community-based management and indigenous rights.[32] However, each of the principles tends to be confusing because of multiple meanings and still evolving interpretive discussions.[33] The integration principle carries at least nine differing connotations[34] with the two most common uses in the oceans governance context being the calls for integrated coastal/ocean management[35] and integrated maritime compliance and enforcement.[36] A topic of considerable controversy is whether integration of economic, social and environmental goals, the core dimension of sustainable development, should be simply a procedural balancing act or a substantive destination, for example, where ecological integrity is not compromised.[37]

The precautionary principle/approach continues to be elusive and very controversial.[38] It is heralded by some as an "environmental savior," holding off the evils of ever-expanding globalization and commercialization pressures through fundamental regulatory shifts such as reversing the legal burden of proof to proponents of change.[39] Others perceive the precautionary principle/approach as the potential doomsday for human technological and scientific innovations.[40]

The ecosystem approach, often used interchangeably with the term ecosystem-based management[41] and being embraced under the auspices of the Food and Agriculture Organization (FAO)[42] and the Convention on Biological Diversity,[43] also raises implementing challenges. While setting the general directions of governing human uses based on ecosystem rather than political boundaries and moving from species protection to broader biodiversity considerations,[44] the approach has not settled the tensions between ecocentric and anthropocentric perspectives.[45] How far ecosystem processes and functions should be protected in light of socio-economic demands is left unanswered.[46] How the approach can be actualized in light of the limited understanding of marine ecosystems is another critical issue.[47]

The ecosystem approach also displays considerable overlap with other principles, which may engender a sense of intellectual frustration over where the concept begins and ends. For example, the precautionary approach, public participation and integrated coastal/ocean planning are all viewed as important supportive principles.[48]

A mix of measures has been suggested to put the ecosystem approach to fisheries into practice. They include improving the selectivity of fishing gear; lessening of effects of "ghost fishing" by lost nets through biodegradable materials and active "sweeping" campaigns; prohibiting destructive fishing practices such as trawling, at least in some sensitive areas; encouraging less-impact fishing methods such as trapping and longlining; establishing marine protected areas; substantially reducing fleet sizes; restoring fish habitats; applying environmental impact assessment review to at least new proposed fisheries; and promoting the use of eco-labeling.[49]

Perhaps the largest impediment to the ecosystem approach to fisheries is in the area of compliance and enforcement. Even traditional fisheries management, focusing on single stocks, has been plagued by the problems of illegal fisheries, misreporting and failures by flag states to control their fishing vessels, especially on the high seas.[50]

The Plan of Implementation, adopted at WSSD in 2002, has placed further elaboration and implementation of the ecosystem approach on the political agenda. Paragraph 30(d) encourages the application by 2010 of the ecosystem approach.[51]

The principle of community-based management, recognized in Principle 22 of the Rio Declaration on Environment and Development,[52] is also fraught with implementation issues. How to define a community, for example, on the basis of local place or a collective interest, is problematic.[53] The principle is sometimes given an expanded interpretation to include co-management where government shares decision-making authority with communities.[54] The principle has the tendency to "blend" with indigenous rights, as aboriginal communities are one of the major avenues for local empowerment.[55] The extent of management rights is also open to debate with a spectrum of rights possible, running from the narrow right to share in the access to fisheries, to a broader bundle of collective rights such as the right to habitat protection and to be involved in regional planning and international allocation agreements.[56]

Indigenous rights to offshore areas and resources are also subject to considerable uncertainties. Whether aboriginal title, recognized for terrestrial areas,[57] can extend to ocean areas is unresolved.[58] The extent of aboriginal rights beyond fishing for food and a moderate commercial livelihood[59] remains to be defined with various rights "waiting in the wings" including the right to steward or manage fisheries[60] and the right to environmental integrity as a basis for ensuring cultural survival.[61] The nature of the fiduciary duties of the Crown to Aboriginal peoples also needs to be clarified including the exact parameters of the duty to consult[62] and the obligation to support community-based and clean production-oriented fisheries.[63]

Australian/Canadian responses

This volume explores how two of the world's largest coastal states[64] – Australia and Canada – are navigating in the transition towards principled oceans governance. While Canada has established a legislative foundation for integrated planning through a new *Oceans Act*,[65] Australia has facilitated planning pursuant to a *National Oceans Policy*.[66] How effective the planning processes will be remains to be seen for both countries. Numerous common constraints include federal–provincial jurisdictional tension, lack of certainty over how management plans will be given legal force, and limitation of planning powers to marine areas.[67]

Integration in maritime compliance and enforcement is also complicated in Australia and Canada by the multiple agencies that have a role in offshore enforcement. In addition, there are numerous activities that need to be monitored and controlled including fishing, vessel-source pollution, offshore hydrocarbon exploration and exploitation, drug smuggling, illegal immigration, maritime terrorism and safety at sea. Both countries also face significant challenges in responding to maritime compliance and enforcement because of the variable environmental conditions in which such operations are undertaken, ranging from polar to tropical extremes.[68]

Both Australia and Canada have also experienced the challenges of implementing the precautionary principle/approach. While both countries have incorporated precaution into various pieces of federal legislation and have already seen courts struggling to interpret legal implications, Australia stands out in front of Canada through broader incorporation of the principle/approach at the state/territorial level and through application of precautionary environmental impact assessments to marine fisheries proposals.[69]

A related challenge is dealing with ecosystem-based management. While Canada has talked about and even legislated for marine ecosystem-based management,[70] Australia has led in implementation partly through its well-funded Collaborative Research Centres (CRCs) such as the CRC for the Sustainable Management of the Great Barrier Reef in Townsville and the CRC for Antarctic Research in Hobart. It is clear that more needs to be done in this area to foster bureaucratic change so as to recognize public concern over the need for a healthy marine ecosystem. To that end, proposals for an "ecosystem bill of rights" are worthy of greater consideration.[71]

An important aspect of integrated marine and coastal management is the role of community involvement. A range of options is available in devolution of management to the community level. However, to date there has been limited consideration in the academic literature of what legal and institutional provisions may be necessary to support community-level governance.[72] In Canada, attempts have been made to invoke co-management in the Atlantic fisheries and joint project agreements and "community quotas" have been issued mainly to sectors of the commercial fishing industry and not to geographically defined local communities. In Australia, Coastcare, a community-based grants scheme, has had a positive impact in its promotion of community projects and education. However, the Australian experience also suggests that some communities have not been seriously involved in coastal decision making and have been used as an unpaid workforce.

Perhaps one of the most significant challenges facing principled oceans governance for Australia and Canada is the full recognition of and integration of indigenous rights in governance structures. Because of their differing historical, legal and political backgrounds, these issues have been addressed differently in the two countries. In Canada, the extent of Aboriginal title over the ocean seabed and adjacent waters remains unresolved.[73] First Nations have had a clear history of ocean use and interest off the Pacific Northwest coast, and recently Haida, Tsimshian and Nuu-chah-nulth First Nations have filed claims to sea title and fishing rights uses. Fish and wildlife co-management arrangements have also been incorporated into modern land claims such as the Nunavut Land Claims Agreement in the Arctic. Concerns remain, however, with many First Nations in British Columbia over possible offshore oil and gas activities and the threats of salmon farming to wild salmon stocks.[74]

The Australian approach to resolving native title and rights claims is based on the provisions of the *Native Title Act 1993* (Cth), which established the National Native Title Tribunal to mediate indigenous land claims including marine claims. Following recent court decisions, which have partly resolved the status of offshore native title in Australia, there has been a surge of interest in this area to such an extent that as of 2001 there were approximately 120 native title claimant applications in sea areas around Australia.[75] However, the Australian government's narrow focus on Aboriginal title and rights to marine resources has been subject to criticism and a wider vision has been urged where Aboriginal peoples' involvement in resource management is considered in the context of human rights and social justice. In particular, it has been argued that there is more scope for actual empowerment of indigenous peoples in the management of marine resources and in the fishing industry.[76]

Conclusion

What then is the future of oceans governance? Recently, there has been increasing debate over the institutional, ethical and theoretical dilemmas confronting global society in trying to ensure sustainable seas.[77] Key institutional hurdles are evident in the shift to oceans governance such as conflicts with state sovereignty in the move towards transboundary cooperation, differences among states in their capacities to govern ocean affairs, and limited and fragmented funding to promote effective and equitable ocean management.[78] Some of the difficult ethical issues surrounding ocean governance include the appropriate roles for scientific and professional expertise in the move towards participatory democracy; the conciliatory challenges raised by weak and strong sustainability perspectives and differing moral philosophies towards nature; the controversial nature of ocean equity, for example, the accommodation of traditional and indigenous fishing community interests with industrial fishing; and the unresolved interpretive viewpoints relating to the precautionary principle. There are also major conceptual problems such as the competing orders of rationality – biological, social, political and legal.

Against this background, there has been an increasing focus on integration in oceans management and governance.[79] This has taken place not only through initial efforts in the LOS Convention to give greater priority to marine environmental protection,[80] but as a result of the development of environmental norms and principles especially following UNCED,[81] and new international governance initiatives focused particularly on the marine environment and the high seas.[82] This raises questions concerning future directions for regime building in ocean governance. Three sub-global levels of regime building are thought to hold particular promise: the large marine ecosystem (LME),[83] the regional-sea level and the sub-regional sea scale. Within the Pacific Ocean, the Asia-Pacific Economic Cooperation (APEC) forum in particular may be an appropriate mechanism for promoting more effective ocean governance.[84]

The need for effective oceans governance has now become a matter of global concern. The United Nations, a body with a long-standing interest in ocean affairs, is actively engaged in the process of monitoring developments. The United Nations Secretary-General has noted:

> The most effective approach to ensure the protection of vulnerable ecosystems is through the adoption of integrated, multidisciplinary and

> multisectoral coastal and ocean management at the national level, as recommended in chapter 17 of Agenda 21 and by the Plan of Implementation of the World Summit on Sustainable Development, as well as by the Consultative Process. It is also necessary to adopt an ecosystem-based management [*sic*], providing a more holistic approach, which would focus on managing marine ecosystems as a whole rather than specific individual elements within them and to enable the development of a longer-term sustainable strategy.[85]

The result of all these developments is that the health of the marine environment has probably never before been the subject of such sustained international attention. While for the most part the tools – both legal and political – for addressing these issues are relatively well developed, there remains significant variance in implementation of the agreed standards. Given the truly global nature of the world's oceans and their interconnected ecosystems, failure to address problems in one ocean or one sector of the ecosystem will have implications elsewhere. World attention is increasingly therefore turning to the "implementation gap."[86] The practices of two significant maritime states in addressing some of these issues is therefore of some importance in terms of defining how states will meet the challenges ahead.

Notes

1 There is now strong scientific evidence of the impacts of global warming upon the world's oceans which is acting as an additional catalyst for global oceans action, see United Nations Secretary-General "Oceans and the law of the sea: Report of the Secretary-General" (3 March 2003) UN Doc. A/58/65, para 196.

2 See the discussion in "A new way to feed the world" *The Economist* 9 August 2003 at 9.

3 Many of these "threats" to the world's oceans are discussed in J. Temple Swing "What future for the oceans?" (September/October 2003) 82(5) *Foreign Affairs* 139–52.

4 These Conventions are as follows: Convention on Fishing and Conservation of the Living Resources of the High Seas 559 UNTS 285; Convention on the Continental Shelf 499 UNTS 311; Convention on the High Seas 450 UNTS 82; and the Convention on the Territorial Sea and Contiguous Zone 516 UNTS 205.

5 1833 UNTS 396.

6 See David VanderZwaag, Bruce Davis, Marcus Haward and Lorne K. Kriwoken "The Evolving Oceans Agenda: From Maritime Rights to Ecosystem Responsibilities" in Lorne K. Kriwoken, Marcus Haward, David VanderZwaag and Bruce Davis (eds) *Oceans Law and Policy in the Post-UNCED Era: Australian and Canadian Perspectives* (Kluwer Law International, London: 1996) 1–9.

7 For a recent critique on the limits of science in resolving disputes where clashes in culture and values are at stake, see Holly Deremus and A. Dan Tarlock "Fish, farms, and the clash of cultures in the Klamath Basin" (2003) 3 *Ecology Law Quarterly* 279–350.

8 Included amongst these initiatives were the 1946 International Convention for the Regulation of Whaling 161 UNTS 74, and the 1954 International Convention for the Prevention of Pollution of the Sea by Oil 327 UNTS 3 (OILPOL).

9 This period was also characterized by several pivotal decisions by the International Court of Justice on the law of the sea, such as the *North Sea Continental Shelf Cases* (*Federal Republic of Germany* v. *Denmark*; *Federal Republic of Germany* v. *The Netherlands*) [1969] ICJ Reports 3; and the *Fisheries Jurisdiction (Merits) Case* (*U.K.* v. *Iceland*) [1974] ICJ Rep 3.
10 (1972) 11 *International Legal Materials* 1416.
11 1973 International Convention for the Prevention of Pollution from Ships, as amended by the 1978 Protocol 1340 UNTS 61 (MARPOL 73/78).
12 1972 Convention for the Prevention of Marine Pollution by Dumping from Ships and Aircraft 932 UNTS 3 (London Dumping Convention).
13 See, e.g. 1974 Convention on the Prevention on Marine Pollution from Land-Based Sources (1974) 13 *International Legal Materials* 352 (Paris Convention).
14 See, e.g. 1980 Convention on the Conservation of Antarctic Marine Living Resources 1329 UNTS 47 (CCAMLR).
15 UNEP was responsible for the development of a regional seas program that resulted in the adoption of several regional seas conventions aimed at pollution control and environmental management, as discussed in P.H. Sand *Marine Environmental Law in the United Nations Environment Programme: An Emergent Eco-regime* (Tycooly Publishing, London: 1988).
16 GESAMP was established in 1967 by a number of UN agencies to address marine pollution problems. In 1993, the sponsoring organizations agreed to extend the role to cover all scientific aspects of the prevention, reduction and control of degradation of the marine environment. "GESAMP Introduction." Online at: www.gesamp.imo.org/gesamp.htm (accessed 12 April 2005).
17 1994 Agreement Relating to the Implementation of Part XI of the United Nations Convention on the Law of the Sea of 10 December 1982 (1994) 33 *International Legal Materials* 1309.
18 1995 Agreement for the Implementation of the Provisions of the United Nations Convention on the Law of the Sea of 10 December 1982 Relating to the Conservation and Management of Straddling Fish Stocks and Highly Migratory Fish Stocks (1995) 34 *International Legal Materials* 1542.
19 (1992) 31 *International Legal Materials* 876.
20 Reproduced in N.A. Robinson (ed.) *Agenda 21 and the UNCED Proceedings* Vol IV (Oceana, New York: 1993).
21 (1992) 31 *International Legal Materials* 818.
22 For a recent discussion of implementation in the Canadian context, see Jerry V. DeMarco "Law for future generations: The theory of intergenerational equity in Canadian environmental law" (2004) 15 *Journal of Environmental Law and Practice* 1–46.
23 These principles have been subject to comment by international courts and tribunals, see especially *Case Concerning the Gabčíkovo-Nagymaros Project* (*Hungary* v. *Slovakia*) (Judgment) [1997] ICJ Reports 7; *Southern Bluefin Tuna Cases* (*Australia* v. *Japan*; *New Zealand* v. *Japan*) (Provisional Measures), ITLOS Cases No 3 & 4, 17 August 1999 (1999) 38 *International Legal Materials* 1624.
24 See United Nations *Report of the World Summit in Sustainable Development* (26 August–4 September 2002) UN Doc. A/CONF/199/20.
25 See comments in A.J. Hanson "Measuring progress towards sustainable development" (2003) 46 *Ocean & Coastal Management* 381–390.
26 Most particularly evidenced by claims to 200 nautical mile exclusive economic zones and continental shelves, including in some cases extended continental shelves beyond 200 nautical miles. For comment, see E. Mann Borgese "Sovereignty and the Law of the Sea" in T.A. Mensah (ed.) *Ocean Governance: Strategies and Approaches for the 21st Century* (The Law of the Sea Institute, Honolulu: 1996) 35–8.

27 See, e.g. Montserrat Gorina-Ysern "World ocean public trust: High seas fisheries after Grotius – towards a new ocean ethos?" (2004) 34 *Golden Gate University Law Review* 645–714; Lee A. Kimball "Deep-sea fisheries of the high seas: The management impasse" (2004) 19 *International Journal of Marine and Coastal Law* 259–87.

28 Gilles Paquet and Kevin Wilkins *Ocean Governance: An Inquiry into Stakeholding* (Centre on Governance, University of Ottawa, Ottawa: 2002). The following discussion on ocean governance is drawn from David VanderZwaag, Sean LeRoy and Rod Dobell "Ocean Governance" Workshop Backgrounder for 2003 Ocean Management Research Network National Conference (1 November 2003). Online at: www.maritimeawards.ca (accessed 12 December 2003).

29 The broader array of approaches may be viewed as part of a shift towards "communicative governance." See J.W. Van der Schans *Governance of Marine Resources: Conceptual Clarifications and Two Case Studies* (Ebaran, Delft: 2001) 226–40; Jurgen Habermas *Between Facts and Norms: Contributions to a Discourse Theory of Law and Democracy* (MIT Press, Cambridge, MA: 1996); I.M. Young *Democracy and Inclusion* (Oxford University Press, Oxford: 2001).

30 Nicolas de Sadaleer *Environmental Principles: From Political Slogans to Legal Rules* (Oxford University Press, Oxford: 2002) 245–61.

31 See, e.g. Jon M. Van Dyke "The Rio principles and our responsibilities of ocean stewardship" (1996) 31 *Ocean & Coastal Management* 1–23.

32 Other principles include pollution prevention, polluter pays, public participation, intergenerational equity and intragenerational equity. For a further discussion on the role of principles see Philippe Sands "International Law in the Field of Sustainable Development: Emerging Legal Principles" in Winfried Lang (ed.) *Sustainable Development and International Law* (Graham & Trotman/Martinus Nijhoff, London: 1995) 53–72.

33 See, e.g. Jaye Ellis "The Straddling Stocks Agreement and the precautionary principle as interpretive device and rule of law" (2001) 32 *Ocean Development & International Law* 289–311.

34 Variations in meaning include external integration (ensuring integration of social, economic and environmental dimensions); internal integration (reducing fragmentation in pollution control through a unified permitting approach); vertical integration (enhancing cooperation among levels of government); horizontal integration (coordinating the efforts of sectoral government departments and agencies); multidisciplinary integration (promoting the use of natural science and insights of social sciences and humanities into decision making); integrated culturing (supporting ecological engineering where production systems are integrated with ecosystem functions); and international integration (incorporating international obligations and responsibilities into domestic laws and policies). See David VanderZwaag, Gloria Chao and Mark Covan "Canadian aquaculture and the principles of sustainable development: Gauging the law and policy tides and charting a course" (2002) 28 *Queen's Law Journal* 279–334 at 286–7.

35 See Chircop and Hildebrand in Chapter 2 and Sakell in Chapter 3 in this volume. For a discussion of the further uncertainties surrounding the integrated coastal/ocean management approach, see Karen Nichols "Integrated coastal management: Problems of meaning and method in a new arena of resource regulation" (1999) 51 *Professional Geographer* 388–99.

36 See Bailet, Cumming and McDorman in Chapter 4 and Bateman, Bergin, Tsamenyi and Woolner in Chapter 5 in this volume.

37 For discussion of the ongoing interpretive tension, see John C. Dernback "Sustainable development: Now more than ever" (2002) 32 *Environmental Law*

Reporter 10003; and Holly Doremus "The rhetoric and reality of nature protection: Toward a new discourse" (2000) 57 *Washington & Lee Law Review* 11–73. For the view that ecological integrity should be protected through adoption of biodiversity reserves while ecosystem health should be ensured through the human uses that are ecologically sustainable, see J. Baird Callicut and Karen Mumford "Ecological Sustainability as a Conservation Concept" in John Lemons, Laura Westra and Robert Goodland (eds) *Ecological Sustainability and Integrity: Concepts and Approaches* (Kluwer Academic Publishers, Dordrecht, 1998).

38 See, e.g. David Freestone and Ellen Hey (eds) *The Precautionary Principle and International Law: The Challenge of Implementation* (Kluwer Law International, The Hague: 1996); Arie Trouwborst *Evolution and Status of the Precautionary Principle in International Law* (Kluwer Law International, The Hague: 2002).

39 For a good discussion of the complexities involved in assigning burdens of proof and setting standards of proof, see Carl F. Cranor "Asymetric Information, The Precautionary Principle, and Burdens of Proof" in Carolyn Raffensperger and Joel A. Tickner (eds) *Protecting Public Health and the Environment: Implementing the Precautionary Principle* (Island Press, Washington, DC: 1999) 74–99.

40 See, e.g. Julian Morris "Defining the Precautionary Principle" in Julian Morris (ed.) *Rethinking Risk and the Precautionary Principle* (Butterworths-Heinemann, Oxford: 2000) 1–21.

41 See, e.g. Trevor Ward and Eddie Hegerl *Marine Protected Areas in Ecosystem-based Management of Fisheries* (A Report for the Department of the Environment and Heritage, Commonwealth of Australia, Canberra: 2003); Hanling Wang "Ecosystem management and its application to large marine ecosystems: Science, law, and politics" (2004) 35 *Ocean Development & International Law* 41–74.

42 See FAO Fisheries Department "The Ecosystem Approach to Fisheries" *FAO Technical Guidelines for Responsible Fisheries* No. 4, Suppl. 2 (FAO, Rome: 2003). The FAO has preferred the term ecosystem approach for various reasons, including consistency with the term "precautionary approach." See S.M. Garcia, A. Zerbi, C. Aliaume, T. Do Chi and G. Lassere "The ecosystem approach to fisheries: Issues, terminology, principles, institutional foundations, implementation and outlook" *FAO Fisheries Technical Paper* No. 443 (FAO, Rome: 2003) at 6.

43 Decision V/6, adopted at the 5th Conference of the Parties (COP) in 2000, called upon parties and international organizations to apply the ecosystem approach, recommended the application of 12 principles and endorsed five operational guidelines. Decision VII/11, adopted at the 7th COP in 2004, suggested annotations and implementation guidelines for the 12 principles.

44 See Lawrence Juda "Considerations in developing a functional approach to the governance of large marine ecosystems" (1999) 30 *Ocean Development & International Law* 89–125.

45 See Volkmar Hartje, Axel Klaphake and Rainer Schliep *The Intenational Debate on the Ecosystem Approach: Critical Review, International Actors, Obstacles and Challenges* (Federal Agency for Nature Conservation, Bonn: 2003) 12; Sudhir Chopra and Craig Hansen "Deep ecology and the Antarctic marine living resources: Lessons for other regimes" (1997) 3 *Ocean and Coastal Law Journal* 117–48; W.M. Von Zharen "An ecopolicy perspective for sustaining living marine species" (1999) 30 *Ocean Development & International Law* 1–41; Bruce Pardy "Changing nature: The myth of the inevitability of ecosystem management" (2003) 20 *Pace Environmental Law Review* 675–92.

46 See, e.g. Charles R. Malone "The Federal Ecosystem Management Initiative in the United States" in Lemons, Westra and Goodland, note 37.

47 For example, the limited understanding of critical habitats for endangered/threatened aquatic species has been identified in the Canadian context. See G.S. Jamieson (ed.) *Proceedings of the DFO Pacific Region Critical Habitat Workshop* (2003) *DFO Canadian Science Advisory Secretariat Proceedings Series* 2003/010; R.G. Randall, J.B. Dempson, C.K. Minas, H. Zowles and J.D. Reist (eds) *Proceedings of the National DFO Workshop on Quantifying Critical Habitat for Aquatic Species at Risk* (2003) *DFO Canadian Science Advisory Secretariat Proceedings Series* 2003/012.
48 FAO Fisheries Department, note 42 at 17–18.
49 Ibid., at 29–42.
50 Ibid., at 71. See also Rosemary Gail Rafuse *Non-Flag State Enforcement in High Seas Fisheries* (Martinus Nijhoff Publishers, Leiden: 2004).
51 *Report of the World Summit on Sustainable Development*, Johannesburg, South Africa, 26 August–4 September 2002, A/CONF. 199/20, Chapter I, Resolution 2. Online at: www.johannesburgsummit.org (accessed 18 December 2003).
52 (1992) 31 *International Legal Materials* 876. Principle 22 states: "Indigenous people and their communities, and other local communities, have a vital role in environmental management and development because of their knowledge and traditional practices. States should recognize and duly support their identity, culture and interests and enable their effective participation in the achievement of sustainable development."
53 Evelyn Pinkerton "Community-based Management & Co-management" Workshop Backgrounder for the 2003 Ocean Management Research Network National Conference (7 November 2003). Online at: www.maritimeawards.ca (accessed 16 December 2002).
54 National Round Table on the Environment and the Economy *Sustainable Strategies for Oceans: A Co-Management Guide* (Renouf Publishing, Ottawa: 1998) 13.
55 David VanderZwaag *Canada and Marine Environmental Protection: Charting a Legal Course Towards Sustainable Development* (Kluwer Law International, London: 1995) 38–40.
56 Pinkerton, note 53.
57 See *Delgamuukw* v. *British Columbia*, [1997] 3 S.C.R. 1010, 153 D.L.R. (4th) 193.
58 See the study of proprietary rights in sea spaces where Canadian jurisprudence is compared in relation to Australia, the United States and New Zealand. C. Rebecca Brown and James I. Reynolds "Aboriginal title to sea spaces: A comparative study" (2004) 37 *U.B.C. Law Review* 449–93.
59 Such a right was recognized in *R.* v. *Marshall* based upon a treaty right, [1997] 3 S.C.R. 456, 177 D.L.R. (4th) 513.
60 Emily Walter, R. Michael McGonigle and Celeste McKay "Fishing around the law: The Pacific Salmon Management System as a 'structural infringement' of aboriginal rights" (2000) 45 *McGill Law Journal* 263–314.
61 See David VanderZwaag "The precautionary principle and marine environmental protection: Slippery shores, rough seas, and rising normative tides" (2002) 33 *Ocean Development & International Law* 165–88 at 174.
62 See Richard F. Devlin and Ronalda Murphy "Contextualizing the duty to consult: Clarification or transformation?" (2002) 14 *National Journal of Constitutional Law* 167–216.
63 Walter *et al.*, note 60 at 309–12.
64 Each country manages maritime zones on three oceans. Canada has the world's longest coastline of about 244,000 kilometers. "Canadian Geography Facts." Online at: www.dfo-mpo.gc.ca/canwaters-eauzcan/facts-faits/cat_list_e.asp?Catid=5 (accessed 12 April 2005). Australia has a coastline 69,630 kilometers long (including nearby islands). National Oceans Office "Background Paper 1 – The Wild Sea." Online at: www.oceans.gov.au/background_paper_1/page_003.jsp (accessed 12 April 2005).

65 S.C. 1996, c. 31. Introduced as Bill C-98 in 1995, it was re-introduced as Bill C-26 and adopted in the 1995–96 legislative session. The *Oceans Act* was assented to on 18 December 1996 and came into force on 31 January 1997.
66 Commonwealth of Australia *Australia's Oceans Policy* (Environment Australia, Canberra: 1998).
67 See the discussion in Chapter 2 by Chircop and Hildebrand and Chapter 3 by Sakell in this volume.
68 See the discussion in Chapter 4 by Bailet *et al.* and Chapter 5 by Bateman *et al.* in this volume.
69 See the discussion in by VanderZwaag, Fuller and Myers in Chapter 6, and Kriwoken, Fallon and Rothwell in Chapter 7 in this volume.
70 See the discussion by Hatcher and Bradbury in Chapter 8 in this volume.
71 See the discussion by Beamish and Neville in Chapter 9 in this volume.
72 See the discussion by Binkley *et al.* in Chapter 10 in this volume.
73 See the discussion by Ginn in Chapter 11 in this volume.
74 See the discussion by Jones in Chapter 12 in this volume.
75 See the discussion by Clark in Chapter 13 in this volume.
76 See the comments by Dillon in Chapter 14 in this volume.
77 See the discussion by Johnston in Chapter 15 in this volume. For a further discussion, see Lee A. Kimball *International Ocean Governance: Using International Law and Organizations to Manage Marine Resources Sustainably* (IUCN – The World Conservation Union, Gland: 2001).
78 For the suggestion that the implementation gap be addressed through the development and strengthening of regional indigenous organizations see Lennox Hinds "Oceans governance and the implementation gap" (2003) 27 *Marine Policy* 349–56.
79 For an early discussion on this topic, see A. Underdal "Integrated marine policy: What? Why? How?" (1980) 4 *Marine Policy* 159–69. For a more contemporary analysis, see D.M. Johnston and D.L. VanderZwaag "The ocean and international environmental law: Swimming, sinking and treading water at the millennium" (2000) 43 *Ocean & Coastal Management* 141–61.
80 See especially LOS Convention, Part XII "Protection and Preservation of the Marine Environment."
81 As discussed in P. Sands *Principles of International Environmental Law* I (Manchester University Press, Manchester: 1995); and P.W. Birnie and A.E. Boyle *International Law and the Environment* 2nd edition (Oxford University Press, Oxford: 2002).
82 This has been an area of particular interest for the United Nations Open-ended Informal Consultative Process on Oceans and the Law of the Sea which, at its fourth meeting in 2003, focused on protecting vulnerable marine ecosystems. See United Nations Secretary-General "Oceans and the law of the sea: Report of the Secretary-General – Addendum" (29 August 2003) UN Doc. A/58/65/Add.1, para. 136.
83 For a further discussion of the LME concept and governance challenges, see Lawrence Juda and Timothy Hennessey "Governance profiles and the management of the uses of large marine ecosystems" (2001) 32 *Ocean Development & International Law* 43–69.
84 Besides the Fisheries Working Group, the Marine Resources Conservation Working Group has been active on various fronts including overseeing implementation of the *Action Plan on Sustainability of the Marine Environment*. See "Marine Resource Conservation Group." Online at: www.apec.org/apec-groups/working_groups/ marine_resource_conservation.html (accessed 12 April 2005).
85 United Nations Secretary-General, note 1, para 219.
86 See discussion in L. Hinds, note 78.

Part II

Integration

2 Beyond the buzzwords

A perspective on integrated coastal and ocean management in Canada*

Aldo Chircop and Larry Hildebrand

Introduction

It is now more than eight years since the *Oceans Act* came into force as Canada's modern legal framework for integrated coastal and ocean management (ICOM).[1] Although there have been several integrated management initiatives at the national, regional and provincial levels, the assessment of the record to date is not a simple matter. This difficulty is well illustrated by the 2001 parliamentary review of the *Oceans Act* and the federal government's response.[2] The House of Commons' Standing Committee on Fisheries and Oceans (Standing Committee) concluded that

> [T]he Oceans Act is fundamentally sound and [the Committee] does not recommend any major amendments to the Act at this time. Nevertheless, the Committee has some concerns over the administration of certain aspects of the Act. Certain principles and programs that were key elements of the Act do not appear to have been as fully implemented as they could or should have been. In addition, a number of more specific concerns were raised particularly with respect to the creation of Marine Protected Areas and Integrated Management (Part II, Oceans Management Strategy) and marine services (Part III, Powers, Duties and Functions of the Minister) that the Committee believes should be given due consideration.[3]

The Standing Committee held hearings across Canada and received inputs from diverse interests groups. These inputs are visible in the 16 recommendations proposed by the Standing Committee.

In its response to the Standing Committee's recommendations, the federal government is of the view that although there is still much to be done, much has been accomplished.[4] The federal government expressed pleasure that the Act is seen as fundamentally sound. It listed many coastal and ocean initiatives as part of the record of the administration of the Act. The specific responses provided to each recommendation are more reserved. Through guarded language the federal government disagreed with[5] or offered explanation or clarification of the basis[6] of many of the recommendations. Where the government tended to agree it showed a willingness to consider or agree with in part[7] or simply confirm that the recommendation was already being followed in whole or in part.[8] Only one recommendation was agreed to without comment.[9]

This exchange is evidence of the intertwining political and bureaucratic agendas and processes over the key legislation prescribing integration in Canada. It illustrates a major challenge for scholar and practitioner alike: how to assess the record on integration in the context of conflicting claims. The assessment of ICOM initiatives is difficult for several reasons including complexity, lack of well-established and documented baselines, unclear or insufficient indicators, lack of systematic project monitoring and time frame of review to capture stated short-, medium- to long-term goals. The identification of indicators can be particularly difficult as it may not always be possible to quantify results, and important as qualitative assessments may be, they are necessarily prone to highly subjective interpretation moderated only by political justifications and bureaucratic constraints. This complex task is further accentuated in the context of states with complex systems of government operating in situations of geographical, ecosystemic, political, socio-economic and cultural diversity. Federal states are a case in point. Understanding how well a particular initiative is doing may depend on individual observation, motive, context and point in time.

Much is at stake: Canada's ocean activities account for an estimated CDN$20 billion in annual domestic economic activity, and this figure does not represent the total value of the country's seaborne trade.[10] In addition, the fate of coastal and marine ecosystems, and the well-being of innumerable aboriginal and coastal communities depend on the management of the marine environment.

This chapter attempts to rationalize an approach to assessing federal policy, planning or management initiatives by developing a theoretical framework drawing from Canadian ICOM practices. It then proceeds to consider specific experiences in the context of this framework in Canada. The experiences are a mixture of "old" and "new" federal initiatives at the national and regional levels. The Atlantic Coastal Action Program (ACAP) can be considered a "mature" initiative because of its longevity (13 years). *Canada's Oceans Strategy: Our Oceans, Our Future* (*Oceans Strategy*) was adopted in 2002 but was preceded by a five-year gestation period. The Eastern Scotian Shelf Integrated Management Initiative (ESSIM), commenced in 1999, is still at a gestation phase. Despite the "immaturity" of the latter two examples, both provide useful insights into the concept and practice of integration, and offer a useful comparison with ACAP. All three have grappled or are grappling with integration in their own individual way, but in a common constitutional context. The *Oceans Strategy* is different from the other two in that it is national in character. ACAP and ESSIM are regional, both in the Atlantic, but are led by two different lead agencies, respectively Environment Canada and DFO. All three are challenging to assess.

The authors identify questions to be asked and factors to be weighed in relation to both the development process of an initiative, the decision as manifested in text, and actual results where these are ascertained. The assessment of the integration record requires a sifting of buzzwords. The analysis leads to a qualitative assessment and conclusions on what could or should have been achieved, or is likely to be achieved, given intended objectives and the influence of relevant factors. This approach produces "relative judgments."

Because of the experimental nature of this study, the authors do not embark on in-depth influence analysis, but rather propose what they hope will be a useful systematic approach for more in-depth analysis of case studies. Although applied in a federal context, the analytical framework would be equally useful to the study of provincial initiatives.

Context

ICOM initiatives in Canada occur in theaters of biogeophysical, socio-economic and cultural diversity. Canada borders on the Arctic, Atlantic and Pacific Oceans. The differences between these three marine environments are further accentuated by intra-regional ecosystemic diversity.[11] In addition, the Great Lakes constitute significant hydrospace that is subject to many of the interests and activities of marine areas. Canada's extensive river systems also have an intricate relationship to the marine environment, particularly for the definition of management areas that include watersheds.

The socio-economic and cultural differences are also significant. New Brunswick is the only officially bilingual province, while Newfoundland and Labrador has the highest persistent unemployment rates in Canada. The level of wealth across the country is very variable, with the poorest provinces and highest rates of unemployment being in the Atlantic region. The Arctic region presents a totally different scene with a fragile environment, low population density, dominant Aboriginal peoples presence, and growing political consciousness and aspirations.

Questions and factors

Triggers

The drivers of the policy-making, planning or management process are a first consideration. Is the process a result of foresight or simply a reaction to an unforeseen problem, event or emergency? The challenge for the decision-maker is to remain ahead of events so as to avoid substituting reactive for proactive approaches. Reactive approaches may lead to inefficient responses and defensive posturing, possibly characterized by optics more than content, in the light of uneasiness of political masters and public critique.

Sectoral and integrated coastal and ocean initiatives in Canada generally have been the result of triggers or pressures. The triggers have tended to be a singular or series of events leading to a crisis that has prodded the government into action. These events have not necessarily been unforeseeable, yet decision-makers remained unprepared. The 1999 Supreme Court of Canada's decision in *R.* v. *Marshall* triggered a series of important developments in aboriginal rights in coastal and ocean resource development, but in reality constitutional recognition of aboriginal rights was foreseeable as a result of preceding case law.[12] Since the *Constitution Act, 1982*, a pattern in the constitutional recognition of aboriginal rights in Canada has emerged.[13] However, the Department of Fisheries and Oceans (DFO) was unprepared for the assertion of fishing and other rights by First Nation bands in areas licensed to other local resource users and the conflict this generated. The collapse of the northern Atlantic cod stocks in the early 1990s was the result of longstanding overfishing. The collapse led to a series of haphazard political, management and fiscal responses by the federal government, e.g. The Atlantic Groundfish Strategy (TAGS).[14]

Pressures have tended to influence the development of initiatives as a result of osmosis. Pressures are influences resulting directly from larger policy initiatives or changes in the governance environment. For example, in the 1990s the privatization drive across a range of government services affected maritime administration services. The introduction of marine service fees and privatization of oil spill response were part of this phenomenon. Somewhat similarly, the Atlantic Coastal Action Program (ACAP) was influenced by the trend for inclusive participation in resource management. ACAP thus employs a stakeholder-based approach under the influence of a growing trend in community-based co-management.[15]

Ideally, the decision-maker should rationalize ICOM decisions on the basis of projected goals and objectives based on foresight, in addition to responding to crises. ICOM is much more than crisis management. Sustainability requires ongoing costs/benefits assessment with reference to an ecosystem's ability to produce the intended goods and services.

Problem-response and baselines

There are three tasks to address in this factor. The first addresses to what extent, if at all, is a particular initiative problem-oriented, and at what scale? Is it responding to a problem as it arises, or is it anticipating it? These questions are relevant for assessment of stated goals and objectives. While clarity is highly desirable, goals and objectives must be flexible enough to respond to unforeseen issues as they arise. Ideally, an ICOM initiative should manifest a long-term vision and an ability to respond to issues as they arise while maintaining a steady course.

The second task is to ascertain the existence of an integrated approach and then to assess how it is formulated. The concept of integration is widely recognized as a basic principle of ICOM.[16] Integration is a response to the sectoralization of the environment and, in a marine context, the piecemeal approach to ocean development that remains so pervasive throughout the world. Sectoralization frequently results in multiple use conflicts and adverse consequences on the marine environment because the activities of other users and the cumulative environmental impacts are not anticipated. The integrated approach requires a holistic approach, where local action must take place within the context of the "big picture." Thus, the manner in which a problem is identified and formulated must take into consideration its context. The integrated approach also has implications for management area and institutional responses, as discussed below.

The third task concerns baselines. The function of baselines is to enable performance measurement against a starting point in fact and time. Baselines are required in any evaluation process to enable a quantitative and qualitative assessment of outputs, outcomes and impacts.

Management area

In planning and management initiatives, provision for the definition of the operational area or the actual definition of such area is closely related to the problem addressed. Does the problem define the area, or does the area define the problem? There are two approaches: functionalist and administrative.

In a functionalist approach, area definition is premised by the full extent of the problem, its impacts and the response needs. Of the two, theoretically this is more consistent with the integrated approach and facilitates an ecosystem-based approach, as long as the management area coincides with the relevant ecosystem. Size *per se* is not necessarily an issue. There could be management and administrative disadvantages, such as how far the logic of integration might take problem definition and identification of a relevant ecosystem, and the involvement of more institutional actors with different mandates. Ecosystem boundaries might not be easy to define and in any case are not likely to be permanent. A functionalist approach suggests that boundaries might need to be reviewed in the light of changing scientific evidence and understanding.

The administrative approach, whereby a jurisdictional area is defined in advance, follows identification of problems in an area addressed pursuant to an institutional mandate. There is convenience, simplicity and clarity in the applicability of a mandate-based approach. These benefits might occur at the cost of relevance and effectiveness. A major difficulty is that the problem might not be dealt with holistically and that an ecosystem-based approach might not be possible.

Knowledge base

Several questions arise when we consider the knowledge base that a decision draws upon. First, what knowledge of the problem and management area is available? A lack of knowledge could result in an initiative that is more politically than knowledge driven.

How is that knowledge created and is it accessible? The integrated approach necessitates a multidisciplinary knowledge of a problem, and this in turn might result in an interdisciplinary response. Also relevant here is the extent to which the approach to knowledge building is inclusive of sources other than natural science so as to incorporate other disciplines, traditional ecological knowledge and local user or community knowledge. An inclusive approach suggests that the knowledge is not elitist, but canvasses all available sources. This could be very significant in the eyes of participating actors. It is also useful to ascertain the extent to which a government initiative draws on non-governmental research and knowledge capabilities (e.g. universities, private sector). Is there an epistemic community behind the decision? Are government experts networking with non-governmental experts? Are decision-makers drawing knowledge directly from stakeholders (i.e. value of consultation but also the lobbying this entails) or hiring knowledge (consultants) or are they simply using in-house expertise? Is there an opportunity for epistemic communities and the public at large to question or peer review the science made available?

From a pragmatic perspective, it is important to ask how knowledge is made available to the decision-maker. It is difficult for decision-makers to deal with scientific uncertainty, information shrouded in jargon, or information presented in an unusable manner. Scientists may be specialists, but managers are generalists. Also, the relationship between science and management is not an easy one, as science may not always produce the definitive answers that administrators and their political masters seek. In turn, scientists are fiercely independent and tend to object to administrative controls.

Policy

What should be considered here is a policy decision and the policy framework within which it occurs. There are various policy factors that may facilitate or constrain ICOM both at the development stage (policy-making process) and in substantive content (the policy decision). Miles defined policy as "a purposive course of action," suggesting a rationalized decision in view of achieving stated ends.[17] Policy must clearly convey its purpose and the action foreseen to avoid ambiguity of expectations. Clarity, consistency, predictability and equity are important criteria. The content must withstand at least a *prima facie* probing analysis. There should be benefits ensuing that are justified by the costs (socio-economic, ecosystemic) incurred. There is no objective standard. Rather the decision must be justifiable according to an identifiable set of values, interests or policy promises, and those that decide must be held accountable. Relevant questions to consider relate to the degree of politicization of a decision, the extent to which a policy reflects diversity, the extent of integration or sectoralization, targets of the decision, the relationship of the policy decision to other policies (coordination, complementarity or conflict), resource commitments, intended effect (e.g. allocation, distribution, organization, etc.) and accountability.

The policy rationalization process is influenced by values or beliefs held by the decision-maker and interests that are actively pursued. In assessing ICOM initiatives, it is useful to enquire whose interests are driving the policy development process and in whose interests the final decision is made. The integrated approach implies an inclusive approach, and the policy decision made is necessarily rationalized on the basis of the diversity of interests. A policy decision that is particular rather than general can be expected to be more exclusive than inclusive in the interests captured. Thus where the context is characterized by diversity, the interests of diversity cannot be served and integration cannot be achieved through an exclusive approach. This is not to say that sectorally-based or -oriented decisions are necessarily problematic, rather that decisions that purport to be integrated ought to be looked at differently from those that are not. Therefore what are the underlying values and interests and whether or not a particular initiative is inclusive or exclusive are pertinent questions.

Efficiency is a further criterion for assessing a policy decision.[18] Policy making in a federal system necessarily occurs at different levels of government as well as in different sectors, frequently in a parallel manner. When policy making at different levels occurs without cross-referencing, a lack of efficiency and possibly also a lack of effectiveness can be expected. This is so because resources, especially limited resources (whether human or material), cannot be said to be used efficiently if duplication occurs and objectives are reached at a higher overall cost, irrespective of whether duplication occurs as a matter of right, principle or simple competition.

ICOM policy making occurs within the larger governance and socio-economic framework. Accordingly, it is to be expected that there will be a relationship with other policy-making processes, frequently elbowing for attention and resources, at times complementing and at other times competing with other processes. The ICOM process can be negatively or positively influenced by extraneous factors, and likewise affects these other processes. The presence and degree of influence or spill-over of other decision-making processes, such as trade and energy, can be vital to explain why a particular ICOM process is driven by shipping and/or offshore development. This poses a challenge for the integrated approach as it may well be that it is the consequence, rather than the cause of a coastal and ocean management challenge, that may have to be addressed, and in itself this poses limits to integration.

Finally, the policy-making process occurs in context and rarely is this static. Thus fisheries management in Canada in the 1990s faced massive stock collapses, loss of livelihoods and displacement of coastal communities, and the ensuing decisions had to reflect the ecological, political, economic and social crises. Coastal and marine resource allocation must take into consideration the context of aboriginal rights. In the 1990s, the privatization drive resulted in significant change in the institutional framework for shipping, and ostensibly what was supposed to result in integration of ocean management functions resulted in fragmentation of maritime administration functions. Contextual pressures significantly influenced policy making. The lessons of the 1990s and into this millennium in Canada suggest that ICOM prospects may be significantly shaped by context before they are even initiated.

Legal framework

Legal factors that influence ICOM in Canada draw on Canada's federal character, the historical division of powers and its international obligations. The legal factors considered here are proposed as related classes of issues, namely property and jurisdiction, aboriginal rights, statutory schemes (federal and provincial) and applicable international law.

Federal and provincial property and jurisdiction

The first factor concerns Canadian maritime zones and related authority exercised over ocean areas through the international law of the sea. This is a facilitating factor for ICOM because it produces some degree of certainty for Canada's maritime claims in the international community. Although Canada only became a party to the United Nations Convention on the Law of the Sea (LOS Convention) in November 2003, it had in effect legislated through Part I of the *Oceans Act* and earlier statutes almost all maritime zone entitlements under that treaty.[19] Canada has an extensive system of straight baselines in the Atlantic, Arctic and Pacific oceans that captures extensive inshore waters as Canadian territory.[20] Canada has also claimed many bays on all three oceans as historic bays, mostly without protest from other maritime powers.[21] Modern maritime zone claims include a 12 nautical mile territorial sea, a 24 nautical mile contiguous zone, a 200 nautical mile exclusive economic zone (EEZ) and a continental shelf.[22] Although the full seaward limits of the first three have been determined, the outer limits of the continental shelf have not yet been determined. Given the broad margin character of Canada's potential claim in the Atlantic and Arctic oceans, the lack of a seaward limit at this time could constrain the full exercise of continental shelf rights and responsibilities in relation to non-living resources and sedentary species.

The extent of Canadian authority over the various maritime zones is highly variable. Internal waters and the territorial sea are subject to sovereignty, in effect entailing the exercise of the totality of jurisdictions and powers that may be exercised on land, subject to the constraint of the international right of innocent passage. The contiguous zone permits the exercise of enforcement jurisdiction for customs, fiscal, immigration and sanitary purposes. For example, Canada has the right to turn away or apprehend ships that may carry illegal immigrants. In reality this power is constrained by humanitarian considerations and the frequent lack of seaworthiness of rogue ships.[23] The EEZ provides sovereign rights over natural resources, exclusive rights over other economic activities and jurisdiction for environment, marine science, artificial islands and installations purposes. The overall constraint in this maritime zone relates to the specificity of the existing rights. In practice, however, Canada has provided for the application of federal and provincial laws over offshore activities.[24] The continental shelf within 200 nautical miles (i.e. co-extensively with the EEZ) poses no special issues. It is outside 200 nautical miles and up to the as yet undefined outer limits of the continental margin that Canada has a potential constraint to its sovereign rights over natural resources (including sedentary species). In this "outer" continental shelf area, offshore mineral activities will potentially be subject to an international tax payable to the International Seabed Authority.[25] At the same time, the exercise of rights over sedentary species (the only living resources tied to the continental shelf regime) may enable Canada to protect the seabed and subsoil habitats of such species outside the 200 nautical mile limit.[26]

At a sub-national level, there is the ongoing constraint of provincial claims to maritime property rights as distinct from federal jurisdiction. For the most part, the federal government (as "Canada") exercises the rights and duties of a coastal state in the law of the sea.[27] Maritime areas are generally deemed to be "extra-territorial" and therefore *prima facie* are subject to this national level of authority. However, this has not discouraged some provinces from testing their claims over various maritime areas. British Columbia does not enjoy a territorial sea but has property over the waters, seabed and subsoil of the area enclosed between Vancouver Island and the mainland.[28] Newfoundland probably has a territorial sea of three nautical miles, but not a continental shelf.[29] New Brunswick and Nova Scotia have not judicially tested their long-standing and pre-Confederation claims in the Bay of Fundy.[30] Nova Scotia has on occasion reminded the federal government that Sable Island is part of the province. Nova Scotia has strong grounds for a legal claim on historic grounds to maritime areas off its Atlantic shores.[31] On different occasions in the past, provincial courts have exercised jurisdiction over causes of actions in bays.[32]

The constitutional law of Canada and case law do not effectively settle property and jurisdictional issues. The *Constitution Act, 1867* allocated extra-territorial matters, fisheries, navigation and shipping to the federal government, and property and civil rights to provincial governments.[33] At the same time, the property boundaries of Nova Scotia and New Brunswick were protected as at the time of Confederation, suggesting that whatever these provinces brought into Confederation by way of property is still protected today.[34]

In practice, although provincial perceptions have tended to constrain federal initiatives in ICOM, both levels of government have approached their differences in a pragmatic and functional approach and situation-by-situation manner. This has enabled ocean development to proceed while provincial claims remained unaffected. Two examples of this concern offshore development and aquaculture. Following the *Newfoundland Offshore Reference*,[35] the federal, Newfoundland and Nova Scotia governments entered into political offshore accords that were legislated concurrently at both federal and provincial levels.[36] In aquaculture, the federal government concluded agreements with several provincial governments that in effect recognize the provinces' lead role in the licensing and management of this marine use.[37]

Aboriginal government

As seen earlier, aboriginal rights are increasingly finding constitutional protection. While these rights can be seen as a type of encumbrance on the Crown, their full extent remains uncertain. One view is that such rights attach to Crown title wherever that title may be asserted.[38] The potential outcome of differences between the federal government's regulatory conservation authority and aboriginal groups that claim a right to manage a resource as part of their aboriginal rights is also unclear. However, the 2002 agreement between the Department of Indian Affairs and Northern Development and Quebec's Inuit indicates the federal government's disposition to addressing the right of Aboriginal peoples to share benefits in non-living and living resources.[39]

Statutory schemes (federal and provincial)

Federal and provincial statutory schemes may also facilitate or constrain ICOM. The most important federal statute in support of ICOM is clearly the *Oceans Act*. Divided into three parts, this Act defines the maritime zones of Canada, provides for integrated management, and allocates powers, duties and functions to DFO. The Act gives DFO general authority to lead and facilitate ICOM. The Act is a type of "constitution" for Canada's ocean space. As a result, much of it is declaratory, organizational and norm setting at a level of generality.

Despite its comprehensiveness, the Act does not cover all relevant factors for ICOM. Arguably, what is *not* covered by the Act is as important as what is covered. A significant potential constraint to integrated coastal management is the exclusion of rivers and lakes, and by implication watersheds.[40] The full extent of application to terrestrial areas is also questionable, despite a reference to integrated management plans for "all activities or measures in *or affecting* estuaries, coastal waters and marine waters" (emphasis added).[41] For integrated coastal management purposes, this is an obvious contradiction in the legislation. Federal initiatives under this Act would have to find creative ways to apply integrated management plans to the "land" component of the land–sea interface. The application of such initiatives to rivers and lakes would have to be orchestrated under the authority of other legislation.[42]

The Act is primarily framework legislation and to date contains no subsidiary legislation other than what was imported from the statutes that it now supersedes. Stakeholders have perceived this absence of new regulation as a weakness, and the Standing Committee has in fact recommended the adoption of regulations.[43]

Beyond the *Oceans Act* and federal and provincial environmental protection acts lies a myriad of federal and provincial sectoral legislation. This legislation establishes mandates for coasts and oceans-related concerns that interact with DFO's lead role in ICOM. Occasionally, there is at least implicit, if not explicit, conflict or lack of complementarity between sectorally allocated powers and the integrationist role of the lead agency. This has the potential of constraining ICOM. For instance, DFO, the Department of the Environment (Canadian Wildlife Service) and the Parks Canada Agency have mandates to establish protected areas, although under different names.[44] The Department of Transport is the maritime administration of Canada, but the Canadian Coast Guard (CCG) is part of DFO.[45] Culturally and functionally (because of the shipping and navigation concerns), the CCG is naturally closer to the Department of Transport (which still hosts marine institutions such as the Marine Safety and Ports and Harbours) than to its current institutional home.[46] The Nova Scotia and Newfoundland offshore petroleum boards have separate federal and provincial statutory authority to grant offshore licenses and to set conditions for the conduct of exploration and development activities. This can result in actual overlaps between offshore uses and other licensed or serviced uses, and protected areas established by other departments under other statutory authority. The shipping legislation does not fully apply to offshore activities.[47] Likewise, water quality criteria for discharges into the marine environment are different for offshore activities and shipping.

As a result, there are some overlaps of statutory mandates and inconsistencies in regulatory standards for different users of the marine environment, even in the same area. Potentially, these legislative factors will constrain ICOM.

Interdepartmental conflict resolution mechanisms are not necessarily legislated. Interdepartmental overlaps in mandates and consequent turfing may be addressed through memoranda of agreement or joint committees designed to harmonize or dovetail efforts, for example, the agreement to address overlapping mandates to establish protected areas.[48]

Applicable international law and policy

Increasingly, international law plays a significant role in informing and guiding Canada's domestic legal system. Canada is party to numerous ICOM-relevant treaties and these are implemented through federal statutes.[49] Canada has also implemented treaties that it generally supports, but is not a formal party to.[50] There is also international customary law that applies in Canada without necessarily being legislated through a statutory scheme.[51] The relevance of international law lies in the existence of international standards which Canadian courts have invoked in interpreting and applying law consistently with Canada's international obligations.[52] The factor to be weighed here is the existence or otherwise of international law that should inform and guide a particular ICOM initiative and whether such an initiative is consistent with or contrary to an international commitment.

Institutional framework

Institutional actors

A variety of institutional actors are involved in ICOM and these can be discussed in terms of extent of authority, function performed and interests represented. The principal concern here is the extensive range of government actors: cabinets, ministries, departments, agencies and inter-ministerial/departmental committees. Authoritative actors in Canada are located in federal, provincial and municipal[53] levels of government, aboriginal government, and in cases of delegation (e.g. through boards, tribunals or crown corporations) or privatization (private sector bodies). Parliamentary and provincial legislature committees may also have an indirect role to play, such as the Standing Committee for Fisheries and Oceans, which conducts periodic reviews of the *Oceans Act* and its implementation. The court of law, by facilitating dispute settlement or clarifying the import of a particular law, can have far-reaching influence on ICOM, e.g. the *Marshall* decision's impact on aboriginal resource rights and the Supreme Court's jurisprudence on Canadian maritime law that has significantly curtailed a widespread practice of judicial application of provincial private law in a maritime setting.[54] Key questions to be asked in relation to authoritative decision-makers is who is driving, leading, facilitating or constraining an ICOM initiative, and why?

Authoritative actors have a legal mandate to perform, are subject to political and bureaucratic pressures, and can be targets for criticism or perceived as sources of benefits. The manner through which they react to these pressures can facilitate or constrain their ability to perform their mandates. For instance, significant criticism of the first version of the *Oceans Act* as a bill forced some reconsideration and re-introduction as an improved bill.[55] Similarly, public criticism of the *Oceans Strategy* discussion paper probably led to its five-year "freeze" until the actual strategy was released in 2002.[56]

Although not possessing authoritative decision-making power, non-governmental organizations (NGOs) may screen decision making in the interests of accountability, project particular interests, disseminate information and educate the public alone or in partnership with decision-makers.

Communities, whether working through NGOs or through an incorporated body, may also share local authoritative decision making with a level of government through co-management initiatives. The contributions of such organizations in shaping a particular ICOM initiative, peer reviewing it or in promoting accountability should be identified.

As institutional actors, industry stakeholders act to influence, pressure or lobby authoritative decision-makers in pursuit of particular interests. For example, when the federal government acted to privatize contingency planning response services to ships, over 30 objections were registered. In turn these objections led to the establishment of a federal commission to enquire into the basis of fees for such services.[57]

Nature and clarity of ICOM mandates

Controversial or unclear ICOM mandates can give rise to resistance or conflicting expectations between lead and other actors. Mandates can be either formal, an authoritatively assigned power, or informal, where in the absence of a specific allocation, a power may be assumed or expected to be assumed by a concerned actor.

Perhaps the most important role belongs to DFO, which is designated as the lead agency for ICOM in Canada. DFO also has the power to assume non-designated responsibilities over any other ocean matter within federal jurisdiction that is not assigned to another minister.[58] It must be emphasized that this lead role is with reference to the *Oceans Strategy*, integrated planning and management, and marine protected areas (MPAs). Other departments have their own separate *de jure* lead roles in their respective sectors. Hence the recommendation of the Standing Committee that "the government affirm that the Minister for Fisheries and Oceans has the primary responsibility for all matters relating to the management of Canada's oceans" could only be met with the inevitable government response: "[B]oth the Department of Fisheries and Oceans and the *Oceans Act* fully respect the existing mandates, responsibilities and authorities of other federal departments and agencies. This is important because nearly every federal department or agency has some level of responsibility related to Canada's oceans, and therefore has a legitimate and necessary role to play in the future of oceans management."[59]

DFO (through its minister) "shall lead and facilitate the development and implementation of a national strategy for the management of estuarine, coastal and marine ecosystems."[60] This broad function in relation to ecosystems is accompanied by a complementary function with reference to different types of waters, i.e. "the development and implementation of plans for the integrated management of all activities or measures in or affecting estuaries, coastal waters and marine waters."[61] The dual role of leader and facilitator in both "strategizing" and "planning" is a significant combination of powers that enables DFO to embark on its own initiatives and at the same time assist with the initiatives of other departments.

The *Oceans Act* mandate does not provide DFO with a *carte blanche* for the exercise of its powers under the Act. First, the powers are actually legal duties, meaning that lack of leadership or facilitation by DFO would be at odds with the Act. The extent to which inaction is legally actionable or simply a matter of political accountability is unclear. The political undertone of inaction was well-captured by the Standing Committee in recommending that the Minister for Fisheries and Oceans "exercise his role as the minister with overall responsibility for the management of Canada's oceans more proactively."[62]

The Act requires that strategies and plans be based on principles of sustainable development, integrated management and precaution.[63] The ability of these principles to facilitate or constrain ministerial activity depends on their definition, although, in general terms, sustainable development and precaution are defined.[64] Precaution also benefits from further development and application in other statutes and case law. Integration, on the other hand, is not defined in any manner and could be problematic in its application.[65] The absence or flexibility of definitions is arguably useful for the federal government in launching initiatives that are guided only by general norms.

A more significant constraint is DFO's duty to cooperate with "other ministers, boards and agencies of the Government of Canada, with provincial and territorial governments and with affected aboriginal organizations, coastal communities and other persons and bodies, including those bodies established under land claims agreements."[66] The duty here is arguably more than a duty to consult, and is probably a duty to "collaborate" in the exercise of powers of leadership and facilitation. The diversity of actors to be consulted necessarily involves a high degree of complexity in communications and interactions leading to decision making. In fact, ICOM initiatives should also be screened to ascertain to what extent, if at all, they address interdepartmental coordination and cooperation in view of a harmonized approach.

The extent to which mandates facilitate or constrain ICOM also has to be considered with reference to the institutional "heritage" of the body concerned. For example, a long-standing criticism of DFO has been the heavy emphasis on fisheries. Even following the entry into force of the *Oceans Act*, where DFO received explicit responsibilities and powers for "oceans," the institutional fisheries stigma remained, and fisheries constituencies still lobby for a higher profile for fisheries, interests.[67] It is suggested that in the eyes of non-fisheries stakeholders, this stigma may constrain ICOM initiatives as it might suggest bias. In this respect, as lead agency for oceans, DFO needs to balance the needs and demands of integrated and multi-sectoral management with the sectoral aspect of its mandate, i.e. fisheries. Failure to create this balance could potentially create a conflict of interest in the department's dual mandate (i.e. oceans/multi-sectoral leadership v. fisheries/sectoral leadership), and possibly undermine integrated management.

Institutional behavior

The ICOM inquiry should also extend to the behavior of relevant actors. There are those who are inclined towards cooperative or competitive behavior or possibly non-involvement. Cooperation might result from normative expectations, such as the expectation of collaboration from the Minister of Fisheries and Oceans. Competition is likely the result of inter-governmental and bureaucratic turfing. Non-involvement may simply be passivity, possibly as a result of disinterest, perceptions of lack of relevance, or lack of resources to commit.

How decisions are made is also important. The institutional culture will determine how meaningful overtures of cooperation might turn out to be. For instance, if the public expects consultation before an initiative is launched or a decision is made and this does not occur, resistance and non-compliance can be expected. The decision to proceed with a consensus-based approach at the First ESSIM Forum Workshop should be perceived as a new way of doing business in the Maritimes, possibly resulting in better constituency reception of this initiative than others. At the same time, it has raised expectations on how DFO should proceed in the future.

Participatory processes

Central to governance is a class of factors loosely referred to as "participatory processes". They may be expressions of participatory rights and expectations and potentially play a significant role in legitimizing or constraining ICOM knowledge building, decision making, implementation and compliance. Aspiration for good governance is increasingly enhancing ways to facilitating stakeholder and public participation, beyond mere information and consultation. Federal and provincial environmental assessment legislation now provides for public hearings or other types of participative processes.[68] The intensity of these "participatory processes" is even more visible in relation to aboriginal communities, local communities and affected individuals who may demand inclusion as a matter of right.

The traditional protection of individual rights (including property rights) through principles of natural justice has now evolved into a more far-reaching requirement of procedural fairness in most administrative decision-making bodies.[69] In particular, there may be legitimate expectations that a particular procedure or process be followed because of the expectations arising from a statutory scheme, government representations or treaty membership, and ultimately, the credibility of government.[70] In oceans and environmental contexts, this is particularly relevant with reference to DFO's duty to cooperate and in the conduct of environmental assessment hearings.

Integration, because of the implied diversity, suggests an inclusive approach. The quantity or quality and timeliness of inclusion raise questions of equity, or fairness. Administrative decision making that affects individuals or groups in a fundamental manner is bound to observe participatory entitlements as a matter of procedural fairness.[71] Participatory rights have become very important in the administrative state. In an ICOM context, where administrative decisions may grant or take away a license to hunt or fish or pollute, issue, confirm or deny maritime documents, permit a reduction in goods and services provided by an ecosystem, or facilitate the urbanization of a coastline, among others, it is to be expected that those who are affected will want to be consulted. It is difficult to envisage compliance if the inequity of a decision-making process provokes resistance and grievance, in turn forcing adoption of costly enforcement or conflict management measures. Consequently, it is legitimate to ask the extent to which the policy-making process is equitable in the eyes of both the decision-maker and decision-receiver.[72] This might help explain the degree of cooperation or otherwise in an ICOM initiative.

A participatory framework can satisfy these equity concerns in various ways, for example, co-management, public hearings or consultations and discussion papers accompanied by workshops. There is no limit to the possibilities. However, the context of the participatory process, the range and intensity of participation, and the degree of satisfaction or non-objection of stakeholders should favor some possibilities over others. The key question is whether there is good process under the circumstances. This is a relative test. An ICOM initiative should be screened for such participatory processes.

The diversity implied by the integrated approach and the consequent inclusive participation provide an opportunity for competing interests to influence a decision. Truly inclusive participation may avoid many potential conflicts simply by ensuring access and exchange of information to avoid misunderstandings. However, in some situations differences grounded on values and entrenched interests could mature into open conflicts. An ICOM initiative should anticipate this and include conflict management mechanisms.

Resources

ICOM initiatives entail costs. A development proposal may consume ecosystem goods and services. A proposal may also allocate benefits to some, and decrease benefits to others. There might be a cost for non-action, or a higher cost associated with one option over another. There could be opportunity costs. Government may need to appropriate funding to support an initiative. Government may need to levy taxes to fund an initiative, or donations or other voluntary allocations solicited. In all these instances, costs are incurred because an ICOM initiative needs to be resourced. In successful ICOM initiatives, the original investment may be multiplied as a result of leveraging other resources. This could be evidence of buy-in or ownership by stakeholders who recognize the value of the initiative and commit to its continuity by allocating more resources. Indeed, this could be an indicator of sustainability. The inverse of this is when there is no resource allocation to an initiative, suggesting a lack of genuine commitment. The absence or insufficiency of resources may stultify an initiative. This could be an indicator of lack of sustainability.

Evaluation

An ICOM initiative should also be screened to determine if its design includes a monitoring and evaluation process to enable it to measure progress or the lack thereof, and to adjust to lack of results, change and unforeseen circumstances.[73] As indicated earlier, an effective evaluation process needs to start with a reliable set of baselines, performance indicators, critical assumptions and clear objectives with targets to be achieved.

Results should be measured in terms of outputs (immediate products), outcomes (short- to medium-term) and impacts (long-term). These should be both quantitative and qualitative indicators. Ultimately, ICOM should achieve effectiveness in terms of (1) behavioral change (actual or potential; incentives) and (2) impact on the environment, economy, health, etc. Costs will be incurred too; accordingly, there should be the possibility of measuring the benefits against the costs, and this will indicate the level of efficiency of an initiative. As a matter of good governance, there should be transparency in the evaluation process. Ultimately, constituencies have to be satisfied with both process and results.

Assessing specific ICOM initiatives

At this point the discussion moves from the theoretical framework to specific integrated management initiatives in Canada. How have the factors presented above facilitated or constrained specific federal ICOM initiatives in Canada?

Canada's Oceans Strategy: Our Oceans, Our Future

The Canadian federal government released the long-awaited *Canada's Oceans Strategy* (*Oceans Strategy*) in mid-summer 2002.[74] The *Oceans Strategy* constitutes the policy framework for Canada's vision for the management of its ocean space and is likely second in importance only to the *Oceans Act* for ICOM in Canada. The Strategy goes to some length to assert the importance of ocean governance, specifically in terms of inter- and intra-governmental collaboration, shared responsibility and an inclusive approach to decision making. Because of its novelty, this instrument can only be assessed with reference to the process that generated it and the actual content.

A reading of the *Oceans Strategy* and the accompanying communication documents would lead the public to assume that interest in and preparatory work for ICOM in Canada began only in the mid-1990s. There is no reference at all to the significant level of cooperative effort that was undertaken from the late 1980s through the early 1990s under an interdepartmental federal initiative known as Marine Environmental Quality (MEQ).

The 1987 *Oceans Policy for Canada*[75] recognized that many federal agencies share the responsibility for and must cooperate in the maintenance and enhancement of the quality and sustainability of the marine environment. In support of this need, an Interdepartmental Committee on Oceans (ICO) was established to coordinate and guide marine programs and policies at the federal level. ICO recognized that coordination at the federal level would be essential and that an overarching framework for marine environmental quality would be necessary. Thus in 1989, ICO established a Director-General level sub-committee to oversee the preparation of a federal MEQ framework and action plan. A working group of this DG Sub-committee, co-chaired by Environment Canada-Atlantic Region and DFO-Ottawa and composed of federal departments and agencies with a stake in the marine environment, was established to lead this process.

In 1992, 17 federal deputy ministers/presidents endorsed a document entitled "Framework for the Management of Marine Environmental Quality within the Federal Government."[76] Following this endorsement, the ICO Sub-committee further directed the Working Group to prepare a federal MEQ Action Plan that identified interdepartmental activities related to the marine environment and provided for the overall coordination of related policies and programs of the federal government. Once completed, a national MEQ framework and action plan would be prepared, as a cooperative effort involving the provinces, territories, First Nations, industry, universities and the public.

The federal "Framework" and MEQ Action Plan set out a strategy for the management of marine environmental quality in Canada.[77] It consists of an overall objective,[78] a set of guiding principles[79] and a series of specific goals and related actions.[80] The MEQ Working Group was active through early 1995 when proposed interdepartmental transfer of resources all but eliminated discussion and precluded further collaboration. The MEQ initiative was relegated to oblivion and the proposed national MEQ framework and action plan were never pursued.

Even a cursory examination of the MEQ initiative's principles, objectives, goals and proposed actions demonstrate that much of the current ICOM thinking as espoused by DFO was anticipated. However, the *Oceans Strategy* provides no acknowledgment of this important earlier cooperative work before DFO was designated lead federal department for coasts and oceans by the *Oceans Act*. Two observations have to be offered in this regard: first, there is loss of corporate memory, suggesting inefficiency and possibly leading to the proverbial "re-invention of the wheel," and second, this lack of acknowledgment of other institutions' contributions may not bode well for future interdepartmental cooperation that will be so essential to ICOM.

Trigger

The responsibility of DFO to develop a national oceans strategy is conferred by the *Oceans Act*.[81] Accordingly, this particular initiative cannot be said to have been triggered by an event, but rather it constitutes the fulfillment of a legal mandate. Nor can it be said, in the view of these authors, that any particular crisis triggered the federal government to exercise its own mandate. However, it can be said that the eventual release and content were influenced by ongoing public pressure and perceptions of appropriate political timing. The release of the *Oceans Strategy* on the eve of the impending Rio + 10 UN conference in Johannesburg, South Africa, is no coincidence, and suggests an opportune forum to showcase Canadian oceans expertise.

Problems and baselines

The legal mandate provided to DFO is with reference to a national strategy for the management of Canadian estuarine, coastal and marine ecosystems. Because of its generality as a management framework at a national scale, the Strategy does not, and could not, purport to address in depth any one problem or class of problems. Rather, it provides a framework to address any coastal and ocean problem or issue in existence or that might arise through integrated management plans, which are also mandated under the *Oceans Act*.

Consequently, the *Oceans Strategy* does not identify any baselines, and could not possibly do so. This will have to be addressed at the level of integrated management plans. At the same time however, the federal government has indicated in the *Oceans Strategy* that evaluation will be based on identified performance indicators, possibly to enable results-based management. Results-based management requires the identification of objectives in the form of outputs, outcomes and impacts (change over the short, medium and long term), the achievement of which would be measured against performance indicators, and against a situation or conditions at a particular moment in time (physical and temporal reference points).

The overall performance of the Strategy will be difficult to assess without identified baselines. However, the performance of individual integrated management plans and MPAs established under the impetus of the *Oceans Strategy* might be deemed to constitute the performance of the Strategy itself.

Management area

The *Oceans Strategy* addresses management area concerns at various levels. On a macro policy level, the Strategy applies to all Canadian coastal and marine environments within national jurisdiction. Under certain conditions, it also purports to apply to areas outside national jurisdiction. On an operational level, the Strategy provides for the identification of two types of management areas: large marine ecosystem (large ocean management area, or LOMA) and coastal management area. Both would be defined in ecosystemic terms, and the coastal management area would be related to the larger ocean management area.

The Strategy emphasizes that the ecosystemic approach may well produce management areas that cut across different jurisdictional zones. What the Strategy does not anticipate at this stage, but should be anticipated in integrated management plans, is that ecosystemic "boundaries" are not necessarily permanent, may vary in the case of overlapping ecosystems, and may fluctuate over time. This seems to have been anticipated in the South-East Marine Management Plan in Australia.[82] In effect, the conscious decision to steer away from formalistic jurisdictional boundaries for management areas signifies the arrival of a functionalist approach to ocean zoning. There are merits in this approach, but it will live side-by-side, rather than displace zoning for sectoral purposes, such as offshore oil and gas exploration, development and production licenses. How the two types of zoning, the first for integrated management purposes and the second for specific sectoral purposes, interrelate remains to be seen.

There may be limitations to this approach to management area definition. The Strategy is careful to stipulate an ecosystemic approach or an ecosystem-based approach, and not an ecosystem management approach. This is not necessarily a problem-oriented approach, and in fact other than the need to identify priorities, there is no hint in the Strategy that ICOM in Canada will be problem oriented. The advantage of a problem-oriented approach is that the ecosystemic definition of the management area would be more directly related to the area of influence of the problem (which might cut across different systems). What should logically result from the approach in the Strategy is that generalized ecosystems will be identified, and then problems therein will be targeted for planning and management action, and not the other way round.

Knowledge base

The *Oceans Strategy* is purportedly a knowledge-based instrument drawing from national and international experience in ocean management. Knowledge is understood as solid multidisciplinary science subject to peer review. The emphasis on marine science is accompanied by user knowledge (industry, fisherfolk, local communities) and the traditional ecological knowledge of aboriginal communities.[83] Perhaps this is where the *Oceans Strategy* misses the point on integration, in terms of the need for a better understanding of the relationship between communities/users and the marine environment. The understanding of human behavior in relation to complex systems will require more than marine science. A notable under-emphasis is the role of academic and research institutions operating outside the governmental framework. This suggests that the traditional uneasiness of civil servants with the academic establishment continues and defeats the expressed intent on integrating knowledge. Moreover, there seems to be an implicit assumption that government bodies in Canada have the necessary knowledge and expertise to undertake the complex integration required in ICOM. Government would provide capacity to communities and to the international community, but little attention seems to be paid to the need to build capacity within the various levels of government in Canada.

On the positive side, there is an important commitment in the *Oceans Strategy* to the dissemination of knowledge and access to information by stakeholders and participants in integrated management plans. This will be essential to leveling the playing field among the various actors and facilitate informed and meaningful participation.

Policy factors: decision-making process and content

As noted earlier, a 1997 discussion paper initiated the process leading to the release of the *Oceans Strategy*. Subsequent federal government hearings invited public and stakeholder feedback. The critical reactions reflected the great diversity among ocean interests across the country and the high expectations on the federal government for this initiative. A major weakness at that time and until the actual release of the *Oceans Strategy* was that there was no clearly stated integrated national ocean policy or public release of a sequel to the discussion paper. As a result, the discussion paper was perceived as identifying potential issues that could be addressed in an oceans management strategy in the absence of well-defined objectives. This major weakness has now been addressed: the *Oceans Strategy* is the integrated national ocean policy of Canada.

During this interim period the initiative for regional integrated management planning was well under way, and it was in fact thought that the learning-by-doing approach at the regional level would eventually lead to the development of a national strategy (bottom-up approach). This period also saw three successive ministers for DFO and the emergence of several issues that derailed DFO's attention (e.g. *Marshall* decision and the Burnt Church crisis in New Brunswick). By 2002, the federal decision to develop the Strategy internally and release it was effectively made without further public consultation. To remedy this shortfall in follow-up public consultations, the Strategy is very careful in advocating an inclusive approach to integrated management decision making.

On the policy content side, the fundamental goal of the Strategy is "to ensure healthy, safe and prosperous oceans for the benefit of current and future generations of Canadians."[84] This lofty goal is supported by three major objectives: (1) marine environment protection, (2) promotion of sustainable economic opportunities, and (3) the exercise of international leadership.[85] The pursuit of the goal and objectives is advanced as a principled approach.[86] The principles advanced are drawn from the *Oceans Act*, namely sustainable development, integrated management and precaution. Although these fall short of the full range of principles stated in the Rio Declaration on Environment and Development, several other principles in the latter are identified in the Strategy.[87] Thus, ecosystem-based approaches, indigenous knowledge and coastal communities have a place and role in the Strategy.

Sustainable development is used more as a buzzword than a principle that can have operational significance. Nothing new is offered in relation to precaution, which is advanced both as principle and approach. However, of considerable significance are the federal government's commitment to the "wide application of the precautionary approach to the conservation, management and exploitation of marine resources in order to protect these resources and preserve the marine environment," the promotion of an ecosystem-based approach to management, application of conservation measures and establishment of MPAs, investing in knowledge building, and maintaining ecosystem integrity.[88] The *Oceans Strategy* falls short of stipulating more widespread use of environmental impact assessment under federal and provincial legislation as a planning tool before development or resource allocation decisions are made. It remains to be seen to what extent precaution will be widely applied with respect to the numerous fisheries of Canada, especially since resource management tends to be on a stock basis. Ecosystem science is still in its infancy and virtually every fishery is subject to intense political pressure.

Not surprisingly given the contextual complexity of ICOM in Canada, much of the *Oceans Strategy* and the accompanying *Policy and Operational Framework* focuses on integration and the integrated approach. Integration is defined as a "continuous process through which decision-making is made"[89] and "a commitment to planning and managing human activities in a comprehensive manner while considering all factors". It includes principles and concepts such as (1) holistic knowledge, information sharing, communication and education, (2) inclusive and collaborative structures and processes, (3) flexible and adaptive management as knowledge improves and in response to uncertainty, and (4) planning on the basis of a combined approach to natural and economic systems.[90] There are three major dimensions to its application, namely in relation to multiple ocean use planning, the management of the relationship between human uses and the environment (ecosystems), and the design and implementation of institutional responses. There seems to be a "conflict" bias in this approach to integration, in the sense that while conflict avoidance and management are writ large, there is no apparent emphasis on promotion of complementarities (e.g. complementary coastal and ocean uses).[91]

Legal factors

The *Oceans Strategy* is cautious in dealing with sensitive constitutional issues. Due respect is paid to the provinces' primary responsibility for provincial lands, shoreline and specific seabed areas; municipalities' responsibility for many land-based activities that have an impact on the marine environment; and aboriginal rights as recognized and protected by the *Constitution Act, 1982* and treaty rights.[92] The Strategy also recognizes that the various government bodies have to operate within their existing statutory mandates.

Canada is committed to playing an international role in ICOM. Canada, a main beneficiary of the LOS Convention, recognizes that the "maintenance and preservation of sovereignty over national ocean space is . . . a fundamental right in international law and is a priority for Canada."[93]

Institutional factors

The *Oceans Strategy* identifies collaboration between and within each level of government as a core commitment to ocean governance.[94] It also recognizes that almost every federal government department or agency (some 20 in total) is involved in ICOM.[95] Although many have a strong marine mandate (e.g. Department of Transport and Environment Canada), only two have a broad explicit or implicit ICOM mandate. DFO has the explicit lead role in integrated management in the *Oceans Act*. Environment Canada's jurisdiction is implicit in relation to all activities that have an adverse impact on the marine environment (e.g. pollution, wildlife protection). Other departments play a lead role for their particular sector or area of marine concern (e.g. Department of National Defence). In the marine transportation area, the Department of Transport is the maritime administration of Canada, but the Canadian Coast Guard is part of the DFO necessitating close cooperation between these two departments.

The integrated approach called for by the Strategy will also require federal collaboration and coordination with and among provincial and municipal governments. The Strategy proposes to use new and existing institutional mechanisms such as committees and boards for this purpose.[96] Given frequent departmental turfing at the federal level and occasional federal–provincial turfing, there is no indication of how the lack of cooperation and occasional competition will in fact be addressed. Moreover, line departments weigh their participation and expected benefits against actual costs. The Strategy is unclear as to whether and how line departments would be expected to shoulder the costs of an integrated management initiative under the lead of another department, when their primary concern will be with the performance of the "core" of their (sectoral) mandate.

Participatory processes

According to the Strategy, "[T]he governance model proposed for integrated management is one of collaboration."[97] In addition to the governmental cooperation discussed above, the Strategy foresees broader social participation in three ways. First, the process of integrated management will result in the establishment of advisory bodies. Although the Strategy is silent on their actual role and composition, it would seem that the collaborative approach advocated throughout the Strategy would lead to the inclusion of stakeholders. Second, where an integrated management body is created, composition would include governmental and non-governmental persons. Stakeholders will play more than an advisory role: "[P]articipants take an active part in designing, implementing and monitoring the effectiveness of coastal and ocean management plans, and partners enter into agreements on ocean management plans with specific responsibilities, powers and obligations."[98] Third, there will be specific situations where co-management can take place, although it is unnecessarily conceived only with reference to aboriginal communities and not to coastal communities generally.[99] Given the pervasiveness of co-management in many parts of the world and Canada, the Strategy does not pay much attention to this form of management, leaving it more as a prescription for "specific" cases, rather than promoting it as a more general practice.

An innovation is attached to stakeholder participation. Integrated management bodies (second and third situations above) will not only provide advice, but will "also assume responsibility for implementation of the approved management plan."[100] This is consistent with the "collaborative" governance approach advocated. How far this will be pursued remains to be seen. One experience with privatization of contingency planning and response led a federal commission to conclude that there are certain governmental responsibilities (i.e. contingency planning for environment protection) that should not be delegated.[101]

What the Strategy leaves uncertain in relation to these notions of active participation is who actually assumes responsibility when a problem arises as a result of the management action undertaken. While this could be a potential problem in terms of holding persons accountable for their actions, it could also encourage stakeholder participation. At its worst, participants could expose themselves to legal liabilities. In order to address this potential problem, it might be necessary (1) for the federal government to assume full responsibility, including for decisions that are not fully its own, (2) to provide participants with liability exemptions for the consequences of decisions of integrated management bodies, or (3) to incorporate integrated management bodies. The Strategy gives the impression that this potential problem has not been given sufficient attention.

The *Oceans Act* already went to great lengths in establishing a duty to consult on DFO. The *Oceans Strategy* goes much further in developing, in a Canadian administrative law context, a legitimate expectation for stakeholders to demand that the federal government live up to the stated policy for an inclusive decision-making process. Although the Strategy notes that consensus might not always be possible, there is no turning back to the *dirigisme* that was the case in the past. ICOM decisions might be judicially reviewable if procedural fairness is not observed.

Resources

The *Oceans Strategy* is silent on the cost of implementation. Currently, it has no separate budgetary allocation, and it is expected that elements of the Strategy would be pursued through current departmental programing and funding. The major challenge that this multi-sectoral strategy faces, and will continue to face, is how to receive a fairer share of a decreasing departmental budget that is oriented towards sectoral concerns. The Strategy gives examples of activities and initiatives that could be undertaken over a four-year period (with no indication as to start date). Some will require doing current business in new ways, suggesting that the activities could be pursued within current sectoral programing and budget. New multi-sectoral activities, however, will require new resources.[102]

Evaluation

The *Oceans Strategy* is conceived as an iterative or "rolling" strategy that will be updated regularly as a result of knowledge gained and lessons learned from adaptive management. The Strategy stresses the importance of measuring progress, relevance and effectiveness without suggesting how this might be done. Evaluation will have great value for what is in effect a management experiment. In order to do this it will be necessary to have a sophisticated evaluation process that currently does not exist. Such an evaluation process would need to integrate results-based management principles and approaches discussed earlier, and factor-in the introduction of change while the Strategy is still being assessed against stated performance indicators. Change that is introduced on an ongoing basis is likely to make it difficult to measure the Strategy's performance, especially if performance indicators are also changed.

Eastern Scotian Shelf Integrated Management (ESSIM) initiative

The *Oceans Act* mandates DFO to lead and facilitate the development and implementation of integrated management plans (IMPs).[103] To date, no IMP has been established although there are integrated management initiatives under way on the three oceans of Canada.[104] ESSIM is one such initiative under federal leadership and covers a large part of the marine area off the Atlantic coast of Nova Scotia. Since ESSIM is *in fieri in statu nascendi*, the analysis provided below must be considered provisional. This provisional analysis is useful because this initiative already provides valuable insights on the emerging practice of integrated management in Canada. There are similar DFO initiatives for the Gulf of St. Lawrence and Pacific Ocean. Another major initiative for integrated planning is in the Arctic (Beaufort Sea), but it antecedes the *Oceans Act*.[105] Of all these initiatives to date, ESSIM covers the most extensive management area, has the largest number of participants and provides "an important national policy roadmap for future oceans management."[106]

Trigger

Consequent to DFO's *Oceans Act* mandate concerning IMPs, ESSIM has at least a legislative trigger. However, this on its own does not fully explain why this area of the Atlantic region, among several candidates, was selected. Of particular relevance to the genesis of this initiative is public concern over perceived threats to the rich marine life in and over a submarine canyon on the Eastern Scotian Shelf known as the Gully. Much of the Scotian Shelf is covered by exploration licenses, mostly for natural gas. There are numerous other marine uses off Nova Scotia that are occasionally in conflict with one another, particularly fishing, the oldest and most widespread use of the area. These spatial interactions often result in uneasy relationships among coastal communities and between these communities and offshore developers.

Sable Island is located close to the Gully and is subject to conservation and management by the Sable Island Preservation Trust.[107] The Gully and its waters will be subjected to a separate regime of protection. Mounting conservation concerns had prodded the federal government to declare the Gully as an area of interest as a marine protected area and the adoption of the Sable Gully Conservation Strategy in 1997. This strategy recommended the initiation of integrated ocean management with an offshore focus in this area.[108] The *Oceans Act* also tasked DFO with the leadership and coordination of the development and implementation of a national system of marine protected areas for the purpose of integrated management planning in Canada.[109]

Problems and baselines

Although there are several actual or potential problems (e.g. conflicts between submarine cables and fishing in the same area; offshore oil and gas licensing and conservation concerns; etc.) the ESSIM initiative to date has left problem definition for later. The initiative will eventually address both long- and short-term objectives. It has been recognized that there should be prioritization of specific issues for immediate action. In effect, although the initiative has been triggered by perceived problems as well as a legislative mandate, the overall direction of the initiative is not problem-oriented. The key elements set out for integrated management plans rely heavily on the definition of an area and within that area the identification of actual issues to be the subject of integrated management planning.[110] Criteria for the inclusion of issues in the IMP have been anticipated, namely "(i) an issue that could involve multiple oceans use with social/economic impacts (i.e. intersectoral spatial and temporal conflicts); or (ii) activities that could result in ecosystem impacts."[111] An issue-based approach might preclude proactive planning, continuing the much-criticized issue-oriented response of the past. ESSIM will need to hold a steady course and operate at two levels, without one level derailing the other: first, defining medium- to long-term planning and management objectives, and second, responding to immediate problems, whether existing or as they arise.

Management area

It will be recalled that the *Oceans Strategy* provides for management at both large and small scales (LOMA, such as ESSIM, and coastal management areas yet to be initiated) that draw from a mix of ecological and administrative criteria.[112] The management area is perhaps one of the most unclear matters in the ESSIM initiative. At approximately 325,000 square kilometers, it covers only half of the Scotian Shelf. Although the *Oceans Strategy* asserts that IMPs will be driven by ecosystem-based approaches, definition of the ESSIM management area has very little ecosystemic basis and the marine space that will be encompassed is uncertain and confusing. The area currently covered by the initiative seems to have been inspired by various factors, mostly jurisdictional or administrative, and hardly any of which have any ecosystemic relevance.

First, if a large marine ecosystem were to be identified in the area, it would have to encompass the entire Scotian Shelf, whereas only the eastern part of the shelf is included in the initiative.[113] The DFO has identified natural divisions,[114] but these have been questioned.

Second, the actual area coincides with an administratively defined fishing zone: Northwest Atlantic Fisheries Organization (NAFO) division 4VW, bordered by 4WX to the west and 4V to the east.[115] This suggests that fishery interests dominated the first cut at the management area and raises the fundamental question of whether a sectoral boundary is useful for a multi-sectoral ecosystemic task.

Third, in the vicinity of the eastern boundary, there is another relevant boundary, i.e. the Newfoundland–Nova Scotia offshore boundary, pursuant to the federal offshore accord legislation with the two provinces. This boundary defined the limits of offshore licensing by the two federal–provincial offshore boards and thus provided some convenience.[116] An arbitration tribunal has reviewed this boundary, and there is a new boundary that is considerably closer to Sable Island (to Nova Scotia's disadvantage).[117] Thus the eastern limit and a significant portion of the ESSIM management area now falls within Newfoundland's offshore raising the issue of Newfoundland's participation in the ESSIM development process.

Fourth, to date, the ESSIM area has excluded the territorial sea (a 12 nautical mile belt along the coast), and there does not seem to be provision for inclusion of the important territorial sea around Sable Island.[118] Although the stated intention of ESSIM is to address the offshore, the exclusion of the key territorial sea area will exclude a range of activities that have a significant impact on the Scotian Shelf ecosystem and which in effect are also a springboard for offshore activities. There is very little human activity in the ESSIM area that does not emanate from the coastal zone. Divorcing the offshore from the inshore defeats the rationale and purposes of the integrated approach, which normally requires consideration of the full range of interacting activities and cumulative impact (cause–effect relationships) on the ecosystem. One reason put forward is that the impact of land-based activities on the marine environment tends to be up to the 12 nautical mile limit, but no supporting scientific evidence is provided.[119] Again, this suggests that jurisdictional concerns (federal–provincial constitutional limits) may have been one deciding factor in addition to inshore water use complexities, such as coastal fisheries. The consequence of this exclusion is that the ESSIM area as defined is not justifiable on either user or on ecosystemic grounds.

Fifth, the seaward limits of the ESSIM area are also uncertain. ESSIM is supposed to extend to the full extent of Canadian seaward jurisdiction, in this case the extended continental shelf in accordance with the *Oceans Act* and the LOS Convention. However, Canada has not defined its outer limits and what is legally certain at this time is the limit of the EEZ at 200 nautical miles. In the case of Newfoundland, offshore exploration and development licenses have been granted beyond 200 nautical miles despite the absence of a formally defined Canadian outer limit.[120]

Policy factors

The *Oceans Act* and the *Oceans Strategy* are the two major instruments that set out the policy framework for ICOM in Canada. In the case of the ESSIM initiative, however, the principal policy guidance has come not from the Strategy, but from the Act. Since 1999, the bulk of the initiative has run parallel to the development of the *Oceans Strategy*. By the time the *Oceans Strategy* was introduced in 2002, the ESSIM initiative was already well-defined and with a proposed structure. The *Oceans Act*, on the other hand, because it performs the function of legislating the policy framework and directions for ICOM, and despite its high level of generality, served to guide the development of ESSIM. It is conceivable that, because ESSIM had conceptually advanced ahead of the Strategy by the time of the 1st ESSIM Forum Workshop, the latter served to inform the former. There is little to suggest that the Strategy added much that was not already anticipated and factored into the ESSIM initiative. If this observation is correct, it suggests that insofar as this Atlantic region initiative is concerned, initiatives at the national level may lag behind regional initiatives and that leadership in ICOM is more likely to be exercised at the frontline than at headquarters. The "learning-by-doing" approach at the operational level is more likely to guide this initiative than a highly generic national ocean policy.

The complexity and novelty of the ESSIM initiative makes it difficult to judge its efficiency. It remains to be seen how the significant stakeholder input at the 1st ESSIM Forum Workshop will be used to strengthen the initiative.[121] However, as shown below, ESSIM has paid meaningful attention to inclusive participation.

Legal factors

ESSIM operates within the statutory framework of the *Oceans Act*, but should also be expected to be governed by other relevant federal statutes and regulations governing marine activities, such as the *Canada Shipping Act*,[122] *Fisheries Act*,[123] *Navigable Waters Protection Act*,[124] *Canadian Environmental Protection Act, 1999*,[125] and the *Canada-Nova Scotia Offshore Petroleum Resources Accord Implementation Act*.[126] This is a substantial legal framework. The actual operational relationships between the general (*Oceans Act*) and dedicated (other statutes) legislation, mandates they confer on government bodies, rights granted to ocean users, and standards set out for various activities remain to be tested by an experimental integrated approach.

The offshore focus of the ESSIM management area avoids potential difficulties of property and jurisdiction with the province of Nova Scotia. As noted earlier, Nova Scotia has historically maintained its position that it brought into Confederation maritime property, and much of this is arguably located within bays and inshore waters. In proximity to the ESSIM management area, Nova Scotia considers Sable Island part of the province and also levies a charge for the laying of transatlantic cables over submarine areas it considers to be part of the province. Property matters apart, the federal government retains jurisdiction for navigation, shipping and fisheries and this facilitates promotion of integrated management in the ESSIM area. There is no legal uncertainty in this regard.

Two constraining issues have already been alluded to. The first is the lack of definition of the extended continental shelf, meaning that the full formal seaward extent of jurisdictional authority in the ESSIM management area is not known and is thus subject to some uncertainty outside 200 nautical miles. The second is the implication of recognition of aboriginal resource constitutional rights in the ESSIM area. If the argument that aboriginal title is a form of encumbrance on Crown title is pushed to its logical conclusion, then Crown title over the resources of the EEZ and continental shelf could be encumbered.

There is also a substantive body of international law that applies to the ESSIM area, as implemented through statutes or simply applicable as a matter of treaty or customary law. This is relevant from two perspectives: first, Canadian subscription to international instruments and second, entitlements of the international community to use the ESSIM area. Examples of the latter include the right of unimpeded international navigation through the EEZ and the laying of submarine cables.[127] Canada should take into consideration whether the placement of offshore installations and structures or its conservation policies in the ESSIM area impede other legitimate uses and rights protected by customary law.[128]

Institutional factors

Four levels of government (federal, provincial, municipal and aboriginal) and over 20 federal and provincial bodies could potentially be engaged in the ESSIM institutional framework.[129] The ESSIM initiative proposes a planning and management structure that includes three major bodies, all of which are encompassed in the ESSIM Forum. The ESSIM Forum for this purpose should be distinguished from the 1st ESSIM Forum Workshop, which did not include the full range of senior level representation expected in the area management structure yet to be established. Participants at the 1st ESSIM Forum Workshop noted that the proposed structure was unclear as to where authoritative decision making lies and what the relationships among the major bodies would be.

The first major ESSIM body is a Federal-Provincial Working Group composed of senior government representatives at the federal and provincial levels. This is the first level of government engagement, and clearly this body will have executive decision-making authority.[130] Perhaps the real concern at the 1st ESSIM Forum Workshop may not have been the lack of clarity in the relationships between bodies as much as the actual *locus* of the decision making (i.e. a committee of senior government officials), which suggests that although the initiative is touted as collaborative and inclusive, in reality the ultimate authoritative decision remains a governmental one. A further issue is the extent of participation by aboriginal government. It is unclear whether the federal government will treat aboriginal governments as "governmental" partners or simply as stakeholders, although it is recognized that First Nation involvement would occur in all ESSIM bodies other than the Secretariat.[131] It has been submitted that aboriginal groups are more than stakeholders. In addition, although the initiative recognizes the importance of municipalities, municipal governments are not included in this governmental structure. Presumably, two justifications might be that municipalities are creations of provincial governments (and therefore a provincial government *de facto* represents municipalities) and that the inshore has so far been excluded in the management area. Whereas the exclusion of municipalities logically results from the exclusion of inshore waters, one can anticipate a constraint in embarking on effective efforts to address marine pollution from land-based activities, where municipal governments have a significant role to play.

This raises a question as to what is, or should be, the role of the second ESSIM body, the Oceans Management and Planning Group (OMPG). This body consists of a plenary (the OMPG proper, which includes stakeholders and government officials at the planning, but not senior decision-makers), the Plan Implementation Working Group (PIWG) (an OMPG sub-committee that works both within and outside the plenary and consists of government planning officials and reports to the Regional Committee on Governmental Affairs – RCGA), and issues-based working groups (consisting of stakeholders and government officials). Much attention has been placed on the need to improve intergovernmental and interdepartmental coordination in the framework. At the other end of the non-governmental spectrum, it is still moot as to how meaningful the structure and process in this second body really is (i.e. to what extent there is involvement in decision making). At the most, the OMPG can hope to influence decision making by providing "information and advice" to the RCGA.[132] Non-governmental participants at the 1st ESSIM Forum Workshop generally found this unsatisfactory.

The third body is the ESSIM Secretariat. Its role is facilitation and coordination throughout the structure. This body is generally viewed favorably.

ESSIM has not addressed sufficiently the sensitive subject of accountability. On the one hand transparency is evident throughout the whole process. The fact that the 1st ESSIM Forum Workshop has occurred and the manner in which it has occurred is a significant milestone in ICOM processes in Canada. Because of its frankness, the report of the workshop proceedings suggests that DFO (Maritimes) is committed to good process.

What remains to be addressed is the difficult question of accountability in the context of inclusive processes. As a public service provider, government is subject to an accountability system that elevates to a political level: "the buck stops here!" Clearly, any government department is responsible and accountable for its programs. However, in ICOM, DFO is a lead agency, and frequently its role will be coordination and facilitation, and not necessarily implementation. It will be interesting to see how collaborating governments and departments will share responsibility and accountability in this scenario, especially in a consensus-building process. If the intention behind the RCGA is to define a "decision-making moment," then participating institutions should be accountable for that decision.

Ironically, that "decision-making moment" can be obfuscated if decision making is further decentralized to enable more inclusive participation by stakeholders. Stakeholders are not necessarily public service providers and most likely represent special interests. The dominant view by participants at the 1st ESSIM Forum Workshop was that there should be wider stakeholder participation in the decision-making process. However, to what extent participating stakeholders could be held accountable together with government, and the form of this accountability, are moot. The potential problem arises where stakeholders participate in decisions that affect other interest groups (e.g. designation of MPAs in places and in a manner that cause loss to other users, e.g. offshore oil and gas industry).[133] The likelihood is that government, which cannot relinquish its public service responsibilities, will continue to be accountable for decisions, no matter how inclusive they are, especially those decisions that occur within its policy framework.

Participatory processes

Although the *Oceans Strategy* and ESSIM are initiatives of the same department, the participatory process devised for each is quite different. As seen earlier, after the initial launch of the discussion paper and subsequent cross-country consultations, development of the Strategy remained for the most part an in-house affair until the release of the final document. In contrast, with ESSIM's launch in 1999, various groups were consulted informally, two discussion documents were distributed in November 2001, an advisory committee of invited stakeholders was set up, and a major workshop convened in February 2002 with over 150 participants. The workshop was novel because stakeholders were able to react to components of ESSIM and to evaluate the workshop process. A survey was also conducted. At the end of the workshop, stakeholders were invited to participate in the follow-up. Irrespective of the substantive direction and content of ESSIM, the process to date is consistent with the collaborative approach advocated for ESSIM and suggests that stakeholder interest is likely to be maintained.

Still there are weaknesses in the ESSIM processes. First, ESSIM needs to re-examine its current characterization of First Nations as "stakeholders" considering the emergence of aboriginal government as an aboriginal right. This development is not restricted to ICOM and affects all federal and provincial initiatives.[134] Second, a significant gap in stakeholder participation to date is the absence of the transportation sector, which is responsible for Canada's seaborne trade. This sector includes regulators, port authorities, industry (shipowners and service providers), and the various professions servicing the sector. Its relevance is obvious: it affects the regulation of all navigation in the ESSIM area and consequently concerns all uses, and the designation of particularly sensitive sea areas affecting navigation needs the approval of the International Maritime Organization. Standards for safety and marine environment protection are set at the international level and implemented domestically (this will constrain the regulatory tools used in the ESSIM area). DFO has recognized this weakness and is committed to including this vital sector in the next ESSIM development phase.

Given the diversity of interests involved in the development and eventual implementation of ESSIM, it is unlikely that all differences will be immediately reconcilable. A genuinely inclusive participatory process should serve as a conflict avoidance tool. The ESSIM structure will need internal conflict management processes to address differences that mature into conflicts, e.g. between levels of governments or between departments, multiple competing users, federal or provincial government and First Nations, First Nations and other fishing communities, and regulatory bodies and specific users or special interest groups. The *Oceans Strategy* recognizes that consensus-based decision making may not always be possible, and tough decisions may need to be taken when consensus cannot be reached.[135] ESSIM does not have a similar proviso, but may well have to develop one since it operates within the framework of the Strategy. ESSIM could benefit from mediation and conciliation structures that operate independently of ESSIM and probably also of DFO.

Resources

Like the *Oceans Strategy*, ESSIM is not accompanied by separate financial appropriation, and no funding was announced at the 1st ESSIM Workshop.[136] DFO (Maritimes) has designated the Oceans and Coastal Management Division to service this initiative, and eventually to serve as its secretariat. Other federal departments and the provincial government have been involved and this suggests that there are institutional costs for these participants as well. Several stakeholders participating in the 1st ESSIM Forum Workshop expressed concern over the time demands of volunteer participation.[137] However, these resources will not be sufficient for plan implementation.[138]

Evaluation

At this early stage, no quality assurance system has been publicly discussed in anticipation of inclusion in the future IMP for the ESSIM region. A preliminary desk assessment of knowledge and institutional capabilities to address ecosystem objectives has been undertaken.[139] A draft IMP will be developed for submission to the 2nd ESSIM Forum Workshop scheduled for February 2003.[140] The drafting process is expected to include management objectives and indicators.[141]

The Atlantic Coastal Action Program

Not all ICOM initiatives in Canada derive from or necessarily operate under the auspices of the *Oceans Act*, *Oceans Strategy* or the leadership and facilitation of DFO. In fact, for the past several decades, several other federal departments and provincial agencies have developed and supported "unofficial" integrated coastal management initiatives that address and deliver on many of the principles and approaches espoused in current Canadian ICOM thinking.[142] These initiatives are due, in part, to a growing desire and capacity of coastal stakeholders for a more participatory form of democracy and a maturing attitude by governments toward sharing responsibility for planning and management in coastal areas.

One example of such a program is the Atlantic Coastal Action Program (ACAP), which was launched by the Atlantic Region office of the federal environment agency (Environment Canada) in 1991.[143] ACAP was established to build the capacity of ecosystem-based communities throughout the four Atlantic provinces so that they could assume the lead in determining their own long-term goals and environmental priorities, build multi-sectoral partnerships in their communities, and undertake direct action to address local issues that constrain the sustainability of their watersheds and adjacent coastal areas.

At the time of writing ACAP is a network of 14 community-driven, watershed-based ecosystem initiatives with five sites in each of Nova Scotia and New Brunswick and two each in Newfoundland and Labrador, and Prince Edward Island.[144] With over ten years of experience in ACAP, there are several lessons learned that derive from objective analysis and examination of day-to-day operations.[145]

Trigger

ACAP was established in response to both an increasing concern by the public about the environmental quality and sustainability of the Atlantic coastal zone and their growing demand to be more actively and meaningfully involved in decisions concerning their future.[146]

Problems and baselines

ACAP is an innovative attempt to overcome the litany of sectorally-oriented and government-controlled planning and management initiatives traditionally practiced in Atlantic Canada and elsewhere. In traditional public involvement processes, the public does not share in the responsibility or ownership of the proposed initiative and ongoing implementation is usually out of their hands. Before the establishment of ACAP, the most commonly held viewpoint within government was that problems, information needs and optimal solutions were "known" by government experts and the challenge was to convince others of what they already knew. Communities, for their part, had little incentive to develop creative and innovative solutions to local issues, and were often disappointed by government responses that did not appear to fit their circumstances. ACAP changed this mental model that both government and communities had of each other. Through ACAP, local citizens, Environment Canada staff, and other government and non-government stakeholders came together as peers to discuss concerns, exchange ideas and negotiate their own interests. Multi-stakeholder processes are by nature inclusive and recognize the rights of all interested parties to be at the decision-making table. Their decisions reflect a wide range of interests and ideas, and result in a better understanding of the constraints and opportunities facing each stakeholder. In ACAP, the local group became the proponent and champion of the project or initiative, leading to greater ownership and responsibility. Together ACAP participants have developed and implemented realistic solutions that meet communities' environmental concerns, as well as their economic and social goals. Many of the solutions go well beyond the immediate scope of any single department or level of government.

Management area

ACAP is designed spatially on coastal watershed-estuary management units[147] and functionally on community leadership. The 14 program areas range from medium-sized to large watershed-estuary complexes that contain several municipalities and vary from urban-industrial to rural-agricultural settings. Although the *Oceans Strategy* excludes rivers and lakes, and by implication, watersheds, the watershed-based model employed in ACAP should play a complementary role to integrated management initiatives developed by DFO.

The majority of the sites are small, and where the watershed is large, the participating communities tend to be well-defined. This combination enables participants to relate more easily to the local impact of their actions or inaction.

Knowledge base

The first five years of ACAP were focused on building the community organizations, their institutions, priorities and partners. These planning efforts culminated in the formulation of a Comprehensive Environmental Management Plan (CEMP), a long-term strategy for the local ecosystem. While there was no one prescribed methodology that all sites had to or chose to follow, six components generally describe the CEMP development process: (1) formation and incorporation of a multi-stakeholder organization that is representative of the community; (2) reaching consensus on an integrated community-based environmental, social and economic vision and well-defined use objectives for the future of the area; (3) developing a common set of goals and objectives for their ecosystem; (4) conducting an environmental quality assessment that includes gathering relevant data to determine baseline environmental conditions and the issues affecting environmental quality; (5) identification of remedial options to close the gap between existing and desired levels of environmental quality; and (6) reaching a broad consensus on an implementation schedule complete with timelines, a financial plan and the identification of those responsible for carrying out the necessary actions.

All sites have geographical information systems to integrate data from users, local knowledge and science.

Policy making

In ACAP, the traditional role of government is shared with the participating local organizations. The local communities assume the traditional government policy and priority-setting functions and the government agencies become partners in responding to the communities' identified needs.

There has not been any conscious and systematic treatment of principles of sustainable development (as espoused in the Rio Declaration), but ACAP partners have been guided by a generalized sense of sustainability of the local ecosystems. Interestingly, at the local level, the discourse is not dominated by polished concepts employed by scholarly and bureaucratic elites, rather by participants' basic perceptions of local problems and pragmatic responses.

Institutional factors

ACAP was founded on two basic premises. First, complex coastal issues cannot be resolved without a holistic, inclusive, participatory, ecosystem-based approach that can influence behaviors that impact negatively on environmental quality and community sustainability. Second, most solutions to environmental and natural resource management issues will not be effective unless the range of participants in coastal governance is expanded to include all those with a stake in decisions concerning coastal resources and uses. These stakeholders must be provided with the capacity and the opportunity to take ownership of issues and responsibility for their solution.

"Community" in the context of ACAP does not refer solely to traditional geographical or political conceptions. Community in this instance refers to the degree of "common interest and unity" amongst social, economic and environmental stakeholders. The institutional actors in the local ACAP organizations include: municipalities, businesses and industries, universities, federal and provincial government agencies, non-government organizations, First Nations and environmental groups. Citizens at large also participate. Thousands of volunteers and youth are engaged on local priorities. First Nations' involvement is also developing.

Perhaps the most important ingredient in keeping the ACAP organizations functioning is a capable and respected community coordinator who is hired, not by government, but by the local organizations. The coordinators and several project and administrative staff are the only paid individuals in the ACAP process at the community level; all other participants are volunteers. A concern, and occasional constraint, is "volunteer burnout." To address this problem, Environment Canada facilitates periodic volunteer training workshops and places a priority on recognizing and supporting the volunteers.

While the ACAP approach made intuitive sense to those at the community level, it was a bold step for the federal government in Canada. For government, the program has presented several challenges. These include: changing the corporate culture from hierarchical, linear delivery to one of horizontal, or team delivery; shifting from the command-and-control model to one of enabler and facilitator (Environment Canada sits on ACAP committees as a stakeholder, not as controller); adapting information and data to meet community needs; opening effective communication channels; redirecting current programs and resources to support community initiatives;

and recognizing management scenarios arrived at through community consensus.

To meet these challenges in their *modus operandi*, bureaucrats as individuals needed to develop new skills and perspectives. This shift occurred rapidly in those individuals who sit on the local ACAP committees directly (referred to as "windows").[148] These "windows" have repeatedly stated that they have found their work with communities to be one of the highlights of their careers in terms of what they have learned, and what they have accomplished. It took more time, however, for the concept to infiltrate into and up the bureaucratic system to the point where senior management and departmental scientists understood the needs and accepted this sharing of control.

ACAP organizations have had positive effects on their communities and individual and organizational behavior, and have become major contributors to local sustainable development. Strong partnerships, alliances and multi-stakeholder membership are key components to the success of this process. Persons interested in local sustainable development or who are impacting the local environment are encouraged to participate. This open and inclusive approach has provided a cooperative forum for persons with competing views or who do not normally come together to work for common ends.

Participatory processes

A basic premise of ACAP is that public participation in ICOM initiatives must go beyond the mere provision of information and consultation. Rather than being government-driven, the participatory process is led by locally-incorporated multi-stakeholder community organizations. The federal government provided seed funding and initial facilitation, but control rests with local participants.

Environment Canada, as program sponsor and partner in each ACAP initiative, participates in direction setting, issues identification and the selection of appropriate responses to issues and priorities on a par with other participants. Environment Canada contributes funds, information, expertise and services. Interestingly, by participating in this strategic manner, Environment Canada achieves departmental objectives and desired results, such as improvements in air and water quality, characterization and remediation of toxic contaminants, habitat protection and restoration, weather and environmental prediction, and understanding and preparing for the predicted impacts of climate change. Like other partners, Environment Canada participates in those projects that are consistent with their mandate and objectives.

Resources

The first phase of ACAP (1991–97) required a large amount of funding from Environment Canada for planning, institution building and direct action projects. As the program and the local institutions matured, departmental funding declined, partnerships multiplied and funding diversified. Today Environment Canada enters into annual contribution agreements with each ACAP organization and provides funds (CDN$80,000 per year) and technical, scientific and program support to undertake planning, management and action projects in pursuit of departmental and ACAP organization objectives. Environment Canada works with the ACAP organizations to bring in other federal and provincial departments as partners to local initiatives.[149]

In turn, local ACAP organizations have assumed responsibility for the implementation of plans of action. These organizations are expected to build local partnerships, to secure funding from other sources and to undertake work in the field. As a result, ACAP organizations have partnered with universities, foundations and industries. The diversification of partners and funding has increased ACAP sites' sustainability and independence.

Given recent federal government funding cuts, it is useful to consider whether ACAP sites are truly sustainable and could survive a hypothetical termination of Environment Canada funding. It is likely that most sites would survive, but that some might face a difficult adjustment process, if not struggle. However, it is useful to consider the federal government contribution and its impact in a larger context. Local resources appear to be accessed and used efficiently. For example, of the total cash funding provided to all ACAP groups only 32 percent has come from the federal ACAP program. ACAP groups have received large contributions of donated labor, services and materials. The modest CDN$12 million invested by Environment Canada since ACAP's inception was used to leverage almost CDN$100 million by the ACAP groups. In fact the argument in support of sustainability can be taken much farther. In terms of the impact of the local ACAP organizations on the Atlantic economy, a recent study of ACAP over the five-year period 1997–2001 calculated the total GDP impacts to be in excess of CDN$22 million, a taxation impact of CDN$8.03 million and the creation of 482 person-years of employment.[150] The same study estimated that if Environment Canada had attempted to undertake the same suite of projects and activities that the 14 ACAP organizations have completed over the same period, it would have cost the department over CDN$71 million; a significant increase over the CDN$6.1 million Environment Canada invested. In summary, it does not appear that the existence of most individual ACAP sites, as distinct from the federal program that supports them, would be threatened if federal ACAP funds were terminated.

Evaluation

Like the other federal initiatives discussed in this chapter, ACAP did not start with a well-defined set of baselines and performance indicators to enable continual performance measurement of the program as a whole.

There is no ongoing programmatic performance assessment but periodic external evaluations are conducted. However, separately from the programmatic level, each site has had a comprehensive environmental evaluation. Consequently baselines have tended to be defined on a site-by-site basis. Also each site hosts a major monitoring event, the annual general meeting, for both participants and the public. Performance objectives are set annually at this level, and progress on these is the subject of the annual report.

Phase II of ACAP (1998–2003) focused on implementing individual site strategies, expanding the ACAP network, and collaborating with others to understand science better and achieve measurable ecosystem goals. Phase III (2003–08) will continue with the existing model and community partners, but will add more sites, continue to work with multi-stakeholder coalitions at a larger regional ecosystem scale, support greater networking and knowledge sharing among the sites, plus take more cooperative, theme-based approaches (e.g. sewage).[151]

In terms of accomplishments to date, over 800 projects have been undertaken by the 14 ACAP organizations, involving hundreds of organizations and thousands of volunteers. Results have included pollution prevention programs for business and households, restored habitats, the establishment of new parks, the creation of artificial wetlands for enhanced sewage treatment, training and education workshops for youth and the unemployed, sustainable forestry management plans for industry and landowners, reforestation of riparian zones, the development of environmental farm plans, scientific research studies,[152] air and water quality programs, climate change projects and shellfish remediation activities. In some sites there has been a significant spill-over of activity outside the original environmental realm to include crime prevention, health education, youth training and employment.

Additionally, several ACAP organizations have established community resource centers providing the public, students, businesses and educators with sustainability information. Today, most ACAP organizations are considered reliable third parties in their communities, trusted by all stakeholders and depended on for reliable information. ACAP organizations are also tackling priority region-wide issues. For example, ACAP convened a Coastal Communities Sewage Workshop in Lunenburg in October, 1999, and is now developing a regional strategy among all ACAP organizations and others to address the pervasive regional issue of inadequate sewage treatment.

Assessment

The buzzword syndrome

"Sustainability," "governance," "integration," "process," "partnership," "precaution," "transparency," "responsibility," "accountability," "stewardship," "collaboration," "ecosystem approach," among others: the practice of ICOM in Canada is replete with buzzwords. It is unfortunate that concepts that should be useful for policy, planning and management are so frequently overused, misused or used loosely to the extent that their utility is severely diluted. Perhaps one of the major weaknesses of the *Oceans Strategy* is that it suffers from the buzzword syndrome: it places faith in repeated concepts without offering substantive and action-oriented content. The concepts obviously provide political kudos, but little management content. In fairness, these concepts do carry underlying values and all initiatives need to at least formally acknowledge them; however, beyond the ritual of respect lies the task of articulating the "who," "when," "how" and "at what cost," i.e. the *modus operandi*.

Comparing initiatives

The three initiatives considered here were not triggered by crises or any one individual event. At most, they were initiated as an exercise of a legal mandate and possibly under osmotic pressure from a changing operational environment and public expectations. All three had or have a significant gestation period. This suggests that integrated initiatives require trial-and-error and learning-by-doing accompanied by many inter/intra-institutional transactions. Compared with some past sectoral policies that had short gestation and specific triggers, such as a resource collapse or fiscal cuts, the three initiatives were not "jolted" into existence. Thus integration initiatives are less likely to be reflex reactions than crafted pioneering experiments.

With reference to a problem-oriented approach, ACAP is perhaps the clearest of the three initiatives in terms of what it hopes to achieve. A management area is a combination of community, watershed and estuary functioning as a system. This approach seems to have worked, but it should be remembered that the scale was local. ESSIM is also local, but on a larger scale. The definition of the management area is spatial, possibly use-biased, and not problem-oriented. The extent to which an ecosystem-based approach will work in a spatially defined management area remains to be seen, but the focus is offshore, not inshore, and thus fewer user conflicts can be expected. The *Oceans Strategy* is national in scope and provides a framework for LOMAs and coastal management areas. Again, it is more spatial in orientation than problem-oriented. The future relevance of the *Oceans Strategy* and ESSIM will depend on the extent to which they allow smaller scale local problem-oriented (proactive and reactive) approaches.

All three initiatives support an integrated approach to knowledge. The *Oceans Strategy* and ESSIM do not indicate how natural science will actually relate to social science, user and aboriginal knowledge. A challenge for DFO as a lead agency is to articulate an approach to knowledge building that is seen to be efficient and equitable, while not being unduly biased towards natural science, its traditional knowledge base. ACAP offers an example of how local knowledge of the area can be married to the scientific with technical support facilitated by a federal agency (in this case Environment Canada). All three cases are still experimenting with multidisciplinary inputs into decision making and interdisciplinary decision outputs.

The policy experience is different in the three initiatives. ACAP enjoys a simple policy environment, administered mostly by one federal department and applied at the regional level. Because Environment Canada provides seed money to generate local integrated planning and management initiatives, it is not *dirigiste*. The *Oceans Strategy*, in contrast, is the policy expression of a department statutorily mandated to play a national lead role. The dilemma for DFO is two-fold: (1) scope-versus-focus in the strategy and (2) not prejudicing other departments' mandates. The generality of the Strategy competes with the specificity of sectoral policies, while seeking to provide a coordinating framework that involves departments that do not have a counterpart statutory duty to follow DFO. As a result, the *Oceans Strategy* tries to be everything to everyone. ESSIM commenced before the policy environment of the *Oceans Strategy* was created and is unlikely to be affected by any perceptions of a weak national policy. On the contrary, ESSIM is more likely to be perceived as providing content to the Strategy at the regional level and perhaps further influence its development as a result of regional experience.

Legal factors have produced the same constitutional constraints for all three initiatives. Environment Canada had to consider the interests and views of Nova Scotia when introducing Sable Island as a new ACAP site. Environment Canada also needed to ensure support for initiatives in all four Atlantic provinces. The *Oceans Strategy* had to ensure intergovernmental partnerships and involvement of First Nations. This underlies Canadian federal and aboriginal constitutional realities and is further mandated by the *Oceans Act.* Coastal ICOM initiatives will not be possible without full provincial participation. ESSIM faces a further constitutional constraint: DFO has chosen to focus the initiative away from provincial shores and inshore waters in the offshore where there are potentially fewer constraints. Even in the offshore, however, ESSIM must contend with federal–provincial offshore development boards (themselves a politico-constitutional compromise) and claims of Aboriginal peoples. ESSIM also has to contend with rights of offshore actors (e.g. licensed offshore operators) acquired under other departmental sectoral legislation.

Together with the constitutional framework, institutional issues are a major constraint to these three ICOM initiatives. ACAP overcame potential institutional difficulties early by working out compromises with individual provincial governments. However, Environment Canada had less of a need to work closely with other federal departments than DFO in either the *Oceans Strategy* or ESSIM. As pointed out, although DFO has legal duty to lead, there is no counterpart duty for other departments in the *Oceans Act.* As a result, DFO's transaction efforts and costs can be expected to be significantly high, and with no new financial resources allocated to oceans, its ability to influence the behavior of other departments is necessarily constrained. The alternative for DFO is to avoid such jurisdictional conflicts and exercise less of a directing lead role in favor of a broader consensus-based approach with other federal departments to promote buy-in and cross-departmental commitment of resources.

There are significant novelties in the participatory processes promoted by ACAP and ESSIM. ACAP has an established local decision-making and implementation process led by coastal communities and facilitated, not directed, by the federal government. Although not all potentially important stakeholders are involved, the ESSIM development process likewise has a creative and successful approach to stakeholder involvement. The promotion of inclusive participation is no longer purely a matter for administrative discretion, but one of legal necessity under the *Oceans Act* and public expectation.

It remains to be seen whether ESSIM can maintain the process and pace it has set without new departmental financial resources. The danger is that demanding, but well-meaning stakeholders will expect follow-up at a time when resources are not available. If DFO were to slow down the process until funds are available, it could affect credibility and continued involvement of stakeholders. Thus, although the ESSIM participatory process is highly credible, the expectations and demands of participants may be difficult to meet. This is a major difficulty that could significantly constrain DFO in exercising its oceans mandate: resources for oceans will likely have to come from resources for fisheries. In addition to intra-departmental difficulties, DFO would also face the wrath of a historically influential fisheries constituency. ACAP, on the other, has no such difficulty. Environment Canada is a contributing partner and its monies are part of a diversified resource portfolio. As a result, ACAP has greater prospects of continuity than ESSIM at this time.

A difficulty shared by all three initiatives is the lack of baselines and specific performance indicators normally needed to monitor and measure progress. The *Oceans Strategy* speaks to indicators in a general sense, but omits reference to baselines altogether. ESSIM has not started to address this issue. Without such a framework in place, any assessment of progress is more likely to be open to subjective and political influences. It is suggested that Canada still needs to develop an ICOM evaluation approach.

Conclusion

A qualitative and comparative weighing of key factors suggests that the ICOM experience in Canada is mixed and significantly more complex than the exchange between the Parliamentary Committee on Fisheries and Oceans and the federal government presented at the beginning of this chapter suggests. There is much more to ICOM than the *Oceans Act*.

Although it is not possible to determine whether any of the three initiatives presented have improved the marine environment, this is a long-term evaluation issue, it is possible to assess the process of crafting, articulating and implementing an integrated approach. Clearly, the development of novel and creative participatory processes at the regional and local levels are major achievements. Inclusive participatory processes are consistent with the commitment to the integrated approach. In the case of ACAP, other significant achievements are consistency, resource diversification and continuity. Major under-achievements are the insufficient management of inter-institutional relations (departmental and constitutional), continued lack of federal resources for ICOM at the regional and national levels (which suggests a low profile for oceans on the national agenda) and absence of appropriate monitoring and evaluation frameworks.

Finally, would ICOM in Canada fare better if Canada were a unitary rather than a federal state? Probably yes. Federalism requires a continuous management of the relationships that keep the individual parts attached to the greater whole while allowing diversity to flourish. Most unitary states do not have to contend with this process. ICOM is complex enough in terms of the high degree of integration needed. As a result, ICOM in Canada competes with other national issues and is easily marginalized when resources at the federal level are scarce and issues are prioritized. Where should this leave the ICOM policy-maker, planner and manager? There is a lesson from ACAP that sticks out: local, small, community-centered, inclusive and low-resourced ICOM initiatives now have a proven track record in Canada.

Notes

* Written in December 2002, the analysis of the three case studies in this chapter reflects their status as of this date. The authors have updated references in the text and notes as required.

1 S.C. 1996, c. 31. Introduced as Bill C-98 in 1995, it was re-introduced as Bill C-26 and adopted in the 1995–96 legislative session. The *Oceans Act* was assented to on 18 December 1996 and came into force on 31 January 1997.

2 Section 52 of the Act (ibid) requires the review by the Standing Committee on Fisheries and Oceans and provides for the comprehensive review of the administration of the act and for recommendations for amendments or administration.

3 See the various review reports of the Standing Committee, in particular the fourth *Report on the Oceans Act*, Wayne Easter, M.P., Chair (House of Commons, Ottawa: October 2001) (hereafter *SCFO Fourth Report*).

4 *Government Response to the Fourth Report of the Standing Committee on Fisheries and Oceans, "Report on the Oceans Act"* (Department of Fisheries and Oceans, Ottawa: March 2002). Much of the federal government response emanates from the Department of Fisheries and Oceans, the principal institution addressed by the Standing Committee.

5 Responses to: recommendation one (enactment of regulations under the Act); recommendation two (references to fishermen and their organizations to be consulted under the Act); recommendation seven (converting the minister's discretionary duty to consult to an obligation in s. 33[3]); recommendation 12 (minister should play a proactive lead role), where the government avoids a direct response to the criticism implicit in this recommendation, and provides

information on various activities including the process for the long overdue national oceans management strategy; recommendation 16 (cost-effective manner of delivery of marine services). Ibid.

6 Responses to: recommendation four (establishment of an interdepartmental committee for stewardship and sustainable management); recommendation 11 (DFO's primary responsibility for ocean management in Canada); recommendation 13 (marine services or ice-breaking fees); recommendation 14 (application of marine service fees to ferries). Ibid.

7 Responses to: recommendation two (annual state of the oceans report), where the government response in essence agreed to produce such a report every three-to-five years; response to recommendation six (definition or clarification of terms in s. 35[1]), where government agrees with the need for definition of various terms but disagrees with a legislative intervention to do so. Ibid.

8 Responses to: recommendation five (publication of information on MPA sites in the Oceans Program Tracking System); recommendation eight (environmental assessment under federal legislation of offshore exploration in the Gulf of St. Lawrence); recommendation nine (offshore development guidelines to better inform developers of license limitations); recommendation ten (fisherfolk representation on the Canada-Nova Scotia Offshore Petroleum Board). Ibid.

9 Response to recommendation 15 concerning the provision of the results of the Treasury Board evaluation of marine services or ice-breaking fees to the Standing Committee. Ibid.

10 *Canada's Oceans Strategy: Our Oceans, Our Future* (Fisheries and Oceans Canada, Ottawa: 2002) at 2.

11 The National Marine Conservation Areas System Plan developed by the Parks Canada Agency lists 29 different marine regions in the Arctic Ocean, Atlantic Ocean, Great Lakes and Pacific Ocean, each with distinct physical and biological characteristics. Online at: www.parkscanada.pch.gc.ca/progs/amnc-nmca/plan/index_E.asp (accessed 25 February 2004). See also *Guiding Principles and Operational Policies* (Parks Canada, Minister of Supply and Services Canada, Ottawa: 1994).

12 *R.* v. *Marshall*, [1999] 3 S.C.R. 533, in a split 3-2 decision. Prior to *Marshall* the Supreme Court of Canada had already confirmed the existence of an aboriginal right to fish for food, social and ceremonial purposes. *R.* v. *Sparrow*, [1990] 1 S.C.R. 1075.

13 *Constitution Act, 1982*, s. 35(1).

14 For an overview of TAGS, see Newfoundland Fishery Education home page. Online at: www.stemnet.nf.ca/cod/tags1.htm (accessed 25 February 2004). See also, Standing Committee on Fisheries and Oceans, "The East Coast Report." Online at: www.parl.gc.ca/InfocomDoc/36/1/FISH/Studies/Reports/fishrp01/03presenta-e.htm (accessed 25 February 2004).

15 J.P. Ellsworth, L.P. Hildebrand and E.A. Glover "Canada's Atlantic Coastal Action Program: A community-based approach to collective governance" (1997) 36 *Ocean & Coastal Management Journal* 121–42.

16 "Integration" as a principle of environmental planning in Principle 13 of the *Declaration of the UN Conference on the Human Environment*. Online at: www.fletcher.tufts.edu/multi/texts/STOCKHOLM-DECL.txt (accessed 25 February 2004). On the integration in an oceans context, see Arild Underdal "Integrated marine policy: What? Why? How?" (1980) *Marine Policy* 159–69; Edward L. Miles "Future Challenges in Ocean Management: Towards Integrated National Ocean Policy" in Paolo Fabbri (ed.) *Ocean Management in Global Change* (Elsevier, London: 1992) 595–620; Jean-Pierre Levy "A national ocean policy: An elusive quest" (1993) *Marine Policy* 75–80.

17 Edward L. Miles "Concepts, approaches and applications in sea use planning

and management" (1989) 20 *Ocean Development and International Law Journal* 213–38.

18 Miles, note 16 at 599.

19 *The Law of the Sea: Official Text of the United Nations Convention on the Law of the Sea with Annexes and Index* (LOS Convention) (United Nations, New York: 1983) 1–157.

20 The *Oceans Act*, note 1 at ss. 5–6, provides for straight baseline delineation by regulation. See also, *Territorial Sea Geographical Coordinates Order*, C.R.C., c. 1550; *Territorial Sea Geographical Coordinates (Area 7) Order*, S.O.R./85-872.

21 Opinion expressed by the Legal Bureau, Department of External Affairs (at the time), reproduced in (1974) 12 *Canadian Yearbook of International Law* 277–79. In the past, the United States objected to Canada's internal water claims in the Bay of Fundy and Arctic archipelago.

22 *Oceans Act*, note 1 at ss. 4, 10, 13 and 17.

23 "Smuggling Chinese immigrants into Canada called big business" CNN.Com, 13 January 2000. Online at: www.cnn.com/2000/WORLD/americas/01/13/canada.smuggling/ (accessed 24 February 2004).

24 *Oceans Act*, note 1 at s. 21. The only such extension of provincial law under this act is the *Confederation Bridge Area Provincial (P.E.I.) Law Application Regulations*, S.O.R./97-375. See also *Canada-Newfoundland Atlantic Accord Implementation Act*, S.C. 1987, c. 3, s. 152 which extends the application of provincial health, safety and labor law to the offshore; *Canada-Nova Scotia Offshore Petroleum Resources Accord Implementation Act*, S.C. 1988, c. 28, s. 157 makes a similar extension.

25 LOS Convention, note 19 at article 86.

26 Ibid., at s. 77(4). Article 61 states the right and duty to conserve in the EEZ. Although there is not a similar stipulation in relation to the continental shelf, the exclusive right to explore and exploit sedentary species outside 200 nautical miles in s. 77(4) should be read against Article 193, which provides that state parties to the LOS Convention have a right to develop their natural resources pursuant to their environmental policies and in accordance with their duty to protect and preserve the marine environment.

27 *Oceans Act*, note 1 at ss. 7, 8, 15 and 19.

28 *Reference Re Ownership of Offshore Mineral Rights of British Columbia*, [1967] S.C.R. 792; *Reference Re Ownership of the Bed of the Strait of Georgia and Related Areas*, [1984] 1 S.C.R. 388.

29 *Reference Re Mineral and Other Natural Resources of the Continental Shelf* (1983), 145 D.L.R. (3d) 9 (Nfld. C.A.); *Reference Re Seabed and Subsoil of the Continental Shelf Offshore Newfoundland*, [1984] 1 S.C.R. 86.

30 For a discussion of potential entitlements of New Brunswick and Nova Scotia, see G.V. La Forest "Canadian inland waters of the Atlantic Provinces and the Bay of Fundy incident" (1963) 1 *Canadian Yearbook of International Law* 149–71; G.V. La Forest "The delimitation of national territory: Re Dominion Coal Company and County of Cape Breton" (1964) 2 *Canadian Yearbook of International Law* 233–44; E.C. Foley "Nova Scotia's case for coastal and offshore resources" (1982) 13 *Ottawa Law Review* 281–308.

31 Foley, ibid., at 308. For the purposes of offshore development "Nova Scotia lands" are "the land mass of Nova Scotia including Sable Island, and includes the seabed and subsoil off the shore of the land mass of Nova Scotia, the seabed and subsoil of the Continental shelf and slope and the seabed and subsoil seaward from the Continental shelf and slope to the limit of exploitability." *Petroleum Resources Act*, R.S., c. 342, s. 7.

32 Re Bay of Fundy, *R.* v. *Burt* (1832), 5 M.P.R. 112 (N.B.C.A); re Conception Bay and other bays in Newfoundland, *Direct United States Cable Co.* v. *Anglo-*

American Co. (1877), 2 App. Cas. 394 (P.C.); re Bay of Chaleurs, *Mowat* v. *McPhee* (1880), 5. S.C.R. 66; re Mahone Bay, N.S., *King* v. *Conrad* (1938), 12 M.P.R. 588 (N.S.C.A.). A case to the contrary in relation to Spanish Bay, Cape Breton, was *Re Dominion Coal Co. and County of Cape Breton* (1963), 40 D.L.R. (2d) 593 (N.S.C.A.).

33 (U.K.), 30 & 31 Vict., c. 3, ss. 91 (federal powers) and 92 (provincial powers).

34 Ibid., at s. 33.

35 *Reference re the Seabed and Subsoil of the Continental Shelf Offshore Newfoundland* (1984), 5 D.L.R. (4th) 385 (S.C.C.).

36 *Canada-Newfoundland Atlantic Accord Implementation (Newfoundland) Act*, R.S.N. 1990, c. C-2; *Canada-Nova Scotia Offshore Petroleum Resources Accord Implementation (Nova Scotia)* Act, S.N.S. 1987, c. 3.

37 See the Federal Aquaculture Development Strategy. Online at: www.ocad-bcda.gc.ca/eaqu_e.pdf (accessed 28 October 2002). For instance Nova Scotia and DFO have had a series of aquaculture development agreements since March 1986. The current agreement was renewed recently. See "Thibault and Fage Renew Canada/Nova Scotia Memorandum of Understanding on Aquaculture Development," DFO News Release, 18 June 2002, www.ncv.dfo.ca/media/newsreel/2002 (28 October 2002).

38 Bernd Christmas, Chief Executive Officer, Membertou, "Overview of the Legal Challenges Facing Land and Maritime Claims of the Mi'Kmaq" a public lecture delivered in the Marine and Environmental Law Programme Lecture Series, Dalhousie Law School, Halifax, Canada, 1 November 2001.

39 Signed on 24 October 2002, this agreement is the first such agreement in Canada. The Nunavik Inuit will be sharing royalties from any oil, gas and precious stone discoveries, as well as proceeds from fisheries development. The agreement also gives the Inuit 80 percent of the Nunavik islands and interconnecting waters in the Hudson Bay, Hudson Strait and Ungava Bay, a total of 250,000 square kilometers. "Inuit royalties deal a first for Ottawa: Quebec Natives to share proceeds of offshore resources" *National Post* 26 October 2002.

40 *Oceans Act*, note 1 at s. 28.

41 Ibid., at s. 31. Through text ("for greater certainty") and its location as the first section of Part II in s. 28, the legislator has chosen to remove any uncertainty regarding the exclusion of rivers and lakes.

42 For instance, specific matters such as water quality, vessel-source pollution and illegal fishing matters in inland waters could be addressed through the *Canada Water Act*, R.S.C. 1985, c. C-11; *Canadian Environment Protection Act* (CEPA), 1999, S.C. 1999, c. 33; *Fisheries Act*, R.S.C. 1985, c. F-14; and *Canada Shipping Act*, R.S.C. 1985, c. S-9.

43 Recommendation 1, *SCFO Fourth Report*, note 3.

44 DFO: *Oceans Act*, note 1 at s. 35; Canadian Wildlife Service: *Canada Wildlife Act*, R.S.C. 1985, c. W-9; Parks Canada Agency: *National Parks Act*, R.S.C. 1985, c. N-14. The three institutions are cooperating to harmonize approaches.

45 Prior to the *Oceans Act*, the Canadian Coast Guard was part of the Department of Transport. DFO is now responsible for the Coast Guard. *Oceans Act*, note 1 at s. 41.

46 Aldo Chircop, Hugh Kindred, Phillip Saunders and David VanderZwaag "Legislating for integrated marine management: Canada's proposed Oceans Act of 1996" (1995) 33 *Canadian Yearbook of International Law* 305–31 at 323.

47 For example, important provisions on limitation of liability for maritime claims and oil pollution compensation respectively do not apply to floating platforms for non-living resource exploration and development and vessels undertaking on-site exploration or exploitation of hydrocarbons. *Marine Liability Act*, S.C. 2001, c. 6, ss. 25 and 49.

48 See *Working Together for Marine Protected Areas: A National Approach* (Minister of Public Works and Government Services Canada, Ottawa: 1998). Online at: www.dfo-mpo.gc.ca/canwaters-eauxcan/infocentre/publications/brochures/wtogeth_e.asp (accessed 24 February 2004).
49 For example, CEPA, note 42, implements, among others, the Basel Convention on the Control of Transboundary Movements of Hazardous Wastes and their Disposal, 1989. Online at: www.fletcher.tufts.edu/multi/texts/BH937.txt (accessed 25 February 2004); and the Convention on the Prevention of Marine Pollution by Dumping of Wastes and Other Matter, 1972 and 1996 Protocol (London Convention). Online at: www.fletcher.tufts.edu/multi/texts/dumping.txt (accessed 25 February 2004).
50 The LOS Convention, note 19, is a case in point. The maritime zones in the LOS Convention have been implemented in the *Oceans Act*, note 1, even though Canada became a party to the Convention in November 2003.
51 For example, Canadian courts have applied the international customary right of ships in distress. "It is a well-recognized principle, supported by the jurisprudence as well as by the opinions of authors on international law, that a ship, compelled through stress of weather, duress or other unavoidable cause to put into a foreign port, is, on grounds of comity, exempt from liability to the penalties or forfeitures which, had she entered the port voluntarily, she would have incurred." *Cashin* v. *Canada*, [1935] Ex. C. R. 103.
52 This is the case in a wide variety of settings. See, for instance, *Suresh* v. *Canada*, [2002] S.C.C. 1, File No.: 27790; *Ordon Estate* v. *Grail*, [1998] 3 S.C.R. 437; *Baker* v. *Canada*, [1999] 2. S.C.R. 817.
53 In Canada, municipalities are creations of provincial governments. They have powers that are central to integrated coastal management, such as the power to zone, to set standards for construction, to tax real estate and to manage municipal waste.
54 On this subject, see "Chapter 3: Admiralty Jurisdiction," in Edgar Gold, Aldo Chircop and Hugh Kindred *Canadian Maritime Law* (Irwin Law, Toronto: 2003) 99–139.
55 See the commentary in Chircop *et al.*, note 46.
56 Fisheries and Oceans Canada *Toward Canada's Oceans Strategy: Discussion Paper* (Communications Directorate, Ottawa: 1997).
57 Edgar Gold (Chair), John Gratwick and Peter Yee *Canadian Oil Spill Response Capability: An Investigation of the Proposed Fee Regime: Final Report* (Fisheries & Oceans Canada, Ottawa: 1996).
58 *Oceans Act*, note 1 at s. 40.
59 Recommendation 11, *SCFO Fourth Report*, note 3.
60 *Oceans Act*, note 1 at s. 29.
61 Ibid., at s. 31.
62 Recommendation 12, *SCFO Fourth Report*, note 3.
63 *Oceans Act*, note 1 at s. 30.
64 "[S]ustainable development, that is, development that meets the needs of the present without compromising the ability of future generations to meet their own needs;" "the precautionary approach, that is, erring on the side of caution," ibid.
65 "[I]ntegrated management of activities in estuaries, coastal waters and marine waters that form part of Canada or in which Canada has sovereign rights under international law," ibid.
66 *Oceans Act*, note 1 at ss. 29 and 31.
67 See recommendation three, which advocates amendment of the *Oceans Act* "to include references to fishermen and fishermen's organizations in the sections of the act that require the Minister to consult." This recommendation seems to

have been arrived at after some pressure from Area 19 Snow Crab Fishermen's Association. *SCFO Fourth Report*, note 3.

68 For example, *Oceans Act*, note 1 at ss. 29 and 31 require the Minister to collaborate with governments, aboriginal communities, coastal communities and other persons. *Canadian Environmental Assessment Act*, 1992 S.C., c. 37, ss. 16(1)(c) re comments of the public are a factor to be considered, 18(3) re consideration of public comments, 19(2) re public notice among several other public notice requirements, 34 and 35(3) concerning the convening of public hearings by a review panel, 55 et seq. re access to information, and 58(1.1) re participant funding. Provincial statutes have similar provisions. For example, Nova Scotia's *Environment Act*, S.N.S., 1994–95, c. 1, s. 44 has a public consultation requirement.

69 See David J. Mullan *Administrative Law* (Irwin Law, Toronto: 2001), especially 147–75, on the reach of procedural fairness rights.

70 Ibid., at 177–86. See also J.M. Evans, N.H. Janisch and D.J. Mullan *Administrative Law: Cases, Text and Materials* 4th edition (Emond Montgomery, Toronto: 1995) 129–47.

71 For case law sources and analysis on procedural fairness, see ibid., at 45–206.

72 Miles includes equity as a criterion for evaluation of national ocean management regimes. Note 16 at 599.

73 Biliana Cicin-Sain and Robert W. Knecht *Integrated Coastal and Ocean Management: Concepts and Practices* (Island Press, Washington, DC: 1998) 241–48.

74 *Oceans Strategy,* note 10. The release of the *Oceans Strategy* was accompanied by a second document: *Policy and Operational Framework for Integrated Management of Estuarine, Coastal and Marine Environments in Canada* (Fisheries and Oceans Canada, Ottawa: 2002). Further information on the *Oceans Strategy* is available online at: www.cos-soc.gc.ca/.

75 Department of Fisheries and Oceans *Oceans Policy for Canada: A Strategy to Meet the Challenges and Opportunities on the Oceans Frontier* (Department of Fisheries and Oceans, Ottawa: 1987).

76 Marine Environmental Quality Working Group of the Interdepartmental Committee on Oceans, "Framework for the Management of Marine Environmental Quality within the Federal Government" (Ottawa, 1992).

77 Marine Environmental Quality Working Group of the Interdepartmental Committee on Oceans, "Federal Framework and Action Plan for Marine Environmental Quality" (Dartmouth and Ottawa, 1994).

78 "To enable Canada to effectively maintain and enhance marine environmental quality. This will in turn allow for the sustainable development of Canada's marine resources and for the enjoyment, use and good health of present and future generations and for the restoration, conservation, protection and enhancement of natural and cultural marine areas." Ibid.

79 "The quality of the marine environment is of local, regional, national and global importance and is essential for the sustainable development of marine resources; the federal government has a national leadership role for the overall conservation and protection of Canada's marine environment, as well as specific statutory responsibilities; the stewardship of the marine environment is the shared responsibility of the international community, federal and provincial governments, First Nations, and other stakeholders. Crucial to the achievement of the framework's objective are consultation and collaboration among all stakeholders; the maintenance of a healthy marine environment, the support of environmentally-sound economic activity, and the provision of environmental services are priorities of the federal government; scientific understanding of the marine environment is essential for sound management of marine environmental quality; and sound decisions by all levels of Canadian

society are based on the availability of timely information on issues relevant to the quality of the marine environment." Ibid.

80 "The federal government has identified nine major goals related to the management of marine environmental quality in Canada: honour Canada's international commitments and obligations for the management of oceans; maintain and enhance the marine environment through cooperation at national and regional levels with other levels of government, First Nations and others; fulfill the federal government's statutory and regulatory responsibilities in relation to the quality of the marine environment; integrate environmental, social and economic objectives in marine areas to meet sustainable development; promote safe and environmentally sound human activity in marine areas; establish and manage a comprehensive network of marine conservation areas, migratory bird sanctuaries and other protected areas; protect human health; improve scientific knowledge and understanding of the marine environment and maintain appropriate Canadian research, technical and managerial expertise to address marine issues; and communicate the benefits and importance of a healthy marine environment." Ibid.

81 *Oceans Act*, note 1 at s. 29.

82 For an overview of this plan see the discussion in Sakell in Chapter 3 in this volume.

83 *Oceans Strategy,* note 1 at 12–13.

84 Ibid., at 10.

85 Ibid., at 12–18.

86 Ibid., at 10–12.

87 For example, the role of coastal communities and aboriginal peoples and their knowledge, *Rio Declaration on Environment and Development*, 1992. Online at: www.fletcher.tufts.edu/multi/texts/RIO-DECL.txt (accessed 25 February 2004).

88 *Oceans Strategy,* note 10 at 11–12.

89 *Policy and Operational Framework*, note 74 at 36.

90 *Oceans Strategy*, note 1 at 11.

91 See, for example, the definition of integrated management (IM): "A continuous process through which decisions are made for the sustainable use, development, and protection of areas and resources. IM acknowledges the interrelationships that exist among different uses and the environments they potentially affect. It is designed to overcome the fragmentation inherent in a sectoral management approach, analyses the implications of development, conflicting uses and promotes linkages of development, conflicting uses and promotes linkages and harmonization among various activities." *Policy and Operational Framework*, note 74 at 36.

92 *Oceans Strategy*, note 1 at 7.

93 Ibid., at 17–18.

94 Ibid., at 18.

95 *Oceans Strategy*, note 10 at 6.

96 Ibid., at 19.

97 Ibid.

98 Ibid.

99 Ibid. Co-management is defined as "[A] management approach in which responsibility for resource management is shared between government and resource user groups." *Policy and Operational Framework*, note 74 at 36.

100 *Oceans Strategy*, note 10 at 20.

101 Gold, note 57.

102 *Oceans Strategy*, note 10 at 21–26, especially at 22.

103 *Oceans Act*, note 1 at s. 32.

104 Matthew King, Assistant Deputy Minister for Oceans in DFO, reported that since 1999 there have been as many as 18 integrated management pilot projects on the three coasts. See keynote address by King "Canadian Oceans Management" in S. Coffen-Smout, G. Herbert, R.J. Rutherford and B.L. Smith (eds) *Proceedings of the 1st Eastern Scotian Shelf Integrated Management (ESSIM) Forum Workshop*, Halifax, Nova Scotia, 20–21 February 2002, Canadian Manuscript Report of Fisheries and Aquatic Sciences 2604 (DFO, Halifax: 2002) at 57.

105 The Beaufort Sea Integrated Management Planning Initiative (BSIMPI) traces its origins back to the Inuvialuit Final Agreement in 1984 that set up a resource co-management arrangement.

106 King, note 104 at 62.

107 Online at: www.sabletrust.ns.ca/ (accessed 3 November 2002). Editor's Note: On 1 April 2005, the federal government resumed responsibility for staffing and operating the Sable Island station.

108 *Sable Gully Conservation Strategy* (Environment Canada, Canadian Wildlife Service, Atlantic Region, Halifax: 1998). Online at: www.ns.ec.gc.ca/reports/pdf/sable.pdf (accessed 3 November 2002). On 27 September 2002 DFO released a discussion paper entitled "The Gully Marine Protected Area" with a proposal for formal zoning. In May 2004, regulations were enacted to formally designate the Gully MPA. Online at: www.mar.dfo-mpo.gc.ca/oceans/e/essim/gully/gully-regs-e.html (accessed 26 April 2005).

109 *Oceans Act*, note 1 at s. 35(2).

110 Coffen-Smout *et al.*, note 104 at 29.

111 Ibid., at 28.

112 *Policy and Operational Framework*, note 74 at 15–16.

113 This was pointed out by participants to the *First ESSIM Workshop*, Coffen-Smout *et al.*, note 104 at 20. See also R. O'Boyle (ed.) *Proceedings of a Workshop on the Ecosystem Considerations for the Eastern Scotian Shelf Integrated Management (ESSIM) Area* (Proceedings Series 2000/014, Canadian Stock Assessment Secretariat, Ottawa: 2000). Online at: www.mar.dfo-mpo.gc.ca/science/rap/internet/pro2000-14.pdf (accessed 25 February 2004); Large Marine Ecosystems of the World home page, LME#8 Scotian Shelf. Online at: www.edc.uri.edu/lme/Text/scotian-shelf.htm (accessed 25 February 2004).

114 "In terms of marine ecosystem considerations, the Oceans Management Area corresponds to an offshore ecozone based on oceanographic and bathymetric features. The Area is comprised of four distinct physiographic zones: (i) the inner shelf bordering the Nova Scotia coastline; (ii) the middle shelf consisting of several banks and basins; (iii) the outer shelf area with wide banks and the shelf break; and (iv) the continental slope and oceanic waters. Based on oceanographic conditions, such as currents, salinity and temperature regimes, a natural division is recognized between the eastern/central and western part of the Scotian Shelf (i.e. west of Halifax)." *ESSIM Initiative: Development of Collaborative and Management Process* (DFO/Oceans and Coastal Management Division, Dartmouth: November 2001) at 10. In response to this, several participants (including DFO) at the 1st ESSIM workshop expressed the view that the proposed ESSIM boundaries carried an element of superficiality and did not reflect ecological concerns. See Coffen-Smout, *et al.*, note 104 at 5–6, 14, 20, 31, 45.

115 See *Development of Collaborative and Management Process*, ibid., at 9.

116 Schedule 1, *Canada-Nova Scotia Accord Act*, note 24 at Schedule 1. The Newfoundland counterpart does not contain such a precise and detailed definition of the offshore area.

117 Arbitration Between Newfoundland and Labrador and Nova Scotia Concerning Portions of the Limits of Their Offshore Areas as Defined in the *Canada-*

Nova Scotia Offshore Petroleum Resources Accord Implementation Act and the *Canada-Newfoundland Atlantic Accord Implementation Act*, Award of the Tribunal in the Second Phase, Ottawa, March 26, 2002, at paras. 4.32–4.36, 5.13–5.15. Online at: www.boundary-dispute.ca> (accessed 25 February 2004). This award will probably necessitate an amendment to the *Canada-Nova Scotia Accord*, note 25, Schedule 1 definition of the boundary between this province and Newfoundland.

118 *Development of Collaborative and Management Process*, note 114 at 9–10.

119 Ibid., at 10.

120 The formal definition of the outer limit would be by the Governor-in-Council acting upon the recommendation of the Minister of Foreign Affairs. *Oceans Act*, note 1 at s. 25(a)(iv).

121 Editors' note: The 2nd ESSIM Forum Workshop was held as scheduled in February 2003 to review a draft Strategic Planning Framework for the development and implementation of the Eastern Scotian Shelf Ocean Management Plan (see R. J. Rutherford, S. Coffen-Smout, G. Herbert, and B. L. Smith (eds.) *Proceedings of the 2nd Eastern Scotian Shelf Integrated Management (ESSIM) Forum Workshop, Halifax, Nova Scotia, 18–19 February 2003*. Canadian Manuscript Report of Fisheries and Aquatic Sciences 2637. Ottawa: Department of Fisheries and Oceans, 2003). The 3rd ESSIM Forum Workshop was held on 22–23 February 2005 to review a draft Integrated Ocean Management Plan and to assist in the development of an Action Plan that will set out more immediate actions to be undertaken through the ESSIM process. Details of these initiatives are available on the ESSIM website, at <http://www.mar.dfo-mpo.gc.ca/oceans/e/essim/essim-intro-e.html>.

122 R.S.C. 1985, c. S-9. A new *Canada Shipping Act, 2001*, S.C. 2001, c. 26, has been legislated but is not yet in force.

123 R.S.C. 1985, c. F-14.

124 R.S.C. 1985, c. N-22.

125 S.C. 1999, c. 33.

126 An overview of federal, provincial and international regulatory frameworks applicable to the ESSIM region is currently being prepared by DFO. Coffen-Smout *et al.*, note 104 at 47–8.

127 LOS Convention, note 19 at article 58.

128 Ibid., at article 60. Article 60(1) provides that artificial islands, installations and structures and their safety zones "may not be established where interference may be caused to the use of recognised sea lanes essential to international navigation." The right of innocent passage and the duty of a coastal state to warn of dangers to international passage is protected by international law. *Corfu Channel Case* (United Kingdom v. Albania), *ICJ Reports*, Judgment, 9 April 1949.

129 See the introductory comments to the 1st ESSIM Forum Workshop of Neil Bellefontaine, Regional Director-General, DFO Maritimes Region, Coffen-Smout *et al.*, note 104 at 54.

130 *Development of Collaborative and Management Process*, note 114 at 19. The precursor of the Plan Implementation Working Group is the federal–provincial government working group established in January 2001. It consists of representatives of over 20 federal and provincial government bodies. Bellefontaine, note 129 at 53.

131 *Development of Collaborative and Management Process*, note 114 at 19. See also "Participatory processes" discussion below.

132 Ibid., at 18.

133 For example, MPA designation in areas already under offshore oil and gas licenses. See Aldo Chircop and Bruce Marchand, "Oceans Act: Uncharted seas

for offshore development in Atlantic Canada?" (2001) 24 *Dalhousie Law Journal* 23–50 at 45–8.

134 This difficulty has been recognized. See Coffen-Smout *et al.*, note 104 at 12, 38.

135 With reference to integrated management bodies, "Even without the full endorsement or participation of some interests, some management actions will still proceed to meet existing jurisdictional responsibilities. For example, actions necessary for conservation can proceed under the authority of the Minister of Fisheries and Oceans." *Policy and Operational Framework*, note 74 at 20.

136 Assistant Deputy Minister King's keynote address is conspicuously silent on this point. King, note 104 at 56–62.

137 Coffen-Smout *et al.*, note 104 at 13, 21–3, and 32.

138 One discussion table at the 1st ESSIM Forum Workshop had the following to say: "The table considered the plan elements as being too broad and unattainable. It stressed the need for prioritization with timelines for the short- and long-term, with ongoing monitoring. It also expressed the need to define the area and process for developing the plan. Support is needed – who will pay for the plan? It was cautioned that one industry could not fund all the elements." Ibid., at 32.

139 S. Coffen-Smout, R. G. Halliday, G. Herbert, T. Potter and N. Witherspoon *Ocean Activities and Ecosystem Issues on the Eastern Scotian Shelf: An Assessment of Current Capabilities to Address Ecosystem Objectives* (Research Document 2001/095, Canadian Stock Assessment Secretariat, Ottawa: 2001). Online at: www.dfo-mpo.gc.ca/csas/Csas/English/Research_Years/2001/2001_095e.htm (accessed 25 February 2004).

140 Coffen-Smout *et al.*, note 104 at 47.

141 Ibid., at vii. See also note 121.

142 For a discussion on past initiatives and historical development of ICOM in Canada, see L.P. Hildebrand *Canada's Experience with Coastal Zone Management* (Oceans Institute of Canada, Halifax: 1989).

143 For a more in-depth description and analysis of ACAP, see Ellsworth *et al.*, note 15.

144 The sites/organizations are: Nova Scotia: Clean Annapolis, Pictou Harbour, Bluenose ACAP, ACAP Cape Breton and Sable Island; New Brunswick: Eastern Charlotte Waterways Inc., St. Croix, ACAP Saint John, Miramichi River, Madawaska; Newfoundland: St. John's Harbour, Humber Arm; Prince Edward Island: Bedeque Bay, Southeast Environmental. For details on and contact information for each ACAP organization see the ACAP website. Online at: www.atl.ec.gc.ca/community/acap/ (accessed 25 February 2004).

145 See "Lessons Learned: Atlantic Coastal Action Program (ACAP)." Online at: atlantic-web1.ns.ec.gc.ca/community/acap/C2EBBA88-F416-4E4B-A13B-F6BFE511B033/lessonslearned_e.pdf (accessed 25 February 2004). For the full report see S. Moir, *ACAP: Lessons Learned*, a Moir Consulting report prepared for Environment Canada (Moir Consulting, Dartmouth: 1997).

146 Various studies have documented environmental and resource degradation, increasing use conflicts and consequent socio-economic hardship. See for example P. B. Eaton, L. P. Hildebrand and A. A. D'Entremont, *Environmental Quality in the Atlantic Region 1985* (Environment Canada, Environmental Protection Service, Atlantic Region, Dartmouth: 1986).

147 Each community initiative sets its boundaries pragmatically based upon the interests and the issues at hand; in each case, the watershed serves as the organizing spatial framework for their identified communities.

148 An ACAP "window" is an Environment Canada scientist, engineer, economist, program manager or technical expert assigned (normally on a voluntary basis)

to work directly with one of the ACAP organizations over the long term. "Windows" function as a two-way liaison between the department and the community organization's board of directors, sitting as an *ex officio* member.

149 Environment Canada has developed memoranda of understanding with provincial agencies in New Brunswick and Prince Edward Island that facilitate coordination of and cooperation with community-based initiatives in those provinces. In Nova Scotia, 35 federal and provincial agencies are collaborating in support of government harmonization and streamlined service delivery to communities through an effort known as the Sustainable Communities Initiative. Online at: nsaccess.ns.ca/sci/ (accessed 25 February 2004).

150 Gardner Pinfold Consulting Economists Ltd. "An Evaluation of the Atlantic Coastal Action Program (ACAP): Economic Impact and Return on Investment" (Report prepared for Environment Canada, Halifax, 2002).

151 The Bay of Fundy Ecosystem Partnership (BoFEP) and the Southern Gulf of St. Lawrence Coalition on Sustainability are two larger regional ecosystem-based initiatives in Atlantic Canada that receive support and partnership participation from Environment Canada and many other government agencies. Online at: www.bofep.org/ (accessed 25 February 2004) and www.coalition-sgsl.ca/ (accessed 25 February 2004).

152 The ACAP "Science Linkages" component makes available CDN$250,000 per year for science and research projects that are jointly proposed, developed and implemented by one or more ACAP organizations in full cooperation with one or more Environment Canada scientists. S. Dech, "ACAP's Science Linkages Initiative: A Sound Investment in Science and Community" (Internal report prepared for Environment Canada–Atlantic Region, Dartmouth: 2002).

3 Operationalizing integrated coastal and oceans management in Australia

The challenges

Veronica Sakell

Introduction

Australia's coasts are currently subject to many pressures including declining water quality, record population growth, coastal development, increased tourism, loss of habitat, weeds and introduced pests, lack of information, capacity, and compliance and enforcement. These pressures need to be managed for future economic growth and environmental health. Such management presents Australia with a continuous challenge. The challenge is enhanced by the competing interests of coastal uses, the complex jurisdictional arrangements that exist for coastal zone management, and cumulative impacts of decision making in the coastal zone – the "tyranny of small decisions." To address these challenges, the Coasts and Clean Seas Initiative was launched in 1996, the Australian Coastal Atlas project in 1997 and numerous marine protected areas (MPAs) have been announced since 1996. Additionally, a new National Coastal Policy to facilitate further integrated coastal management is currently being investigated by the Commonwealth Government.

Since *Australia's Oceans Policy* was launched in December 1998,[1] the National Oceans Office has been tasked with its implementation. The cornerstone of implementing the Policy is regional marine planning. The first large marine ecosystem to be investigated in this context is the South-east Marine Region. An extensive scoping and assessment phase has been undertaken as part of the planning process, including wide-ranging stakeholder consultation. As part of the assessment phase, new information has been gathered in relation to biological and physical characteristics, impacts, uses, community and cultural values, indigenous uses and values, and management and institutional arrangements. The National Oceans Office has developed an implementation framework for regional marine plans. Implementing regional marine plans presents many and varied challenges, including how to move from theory to practice and how to provide for cross-jurisdictional coordination, cross-sectoral coordination, ecosystem-based management, performance assessment, outcomes-based management and participation, and transparency in decision making.

This chapter will identify and examine the challenges that face coastal and oceans management in Australia with specific reference to operationalizing *Australia's Oceans Policy*.

Operationalizing integrated coastal management

Marine governance in Australia is shared between the Commonwealth, state, Northern Territory and local governments. While the state and Northern Territory governments provide the legislative basis for coastal zone planning and management, local governments are responsible for day-to-day management. Offshore area responsibility is split between the Commonwealth and the state and Northern Territory governments. Generally speaking, the states and Northern Territory are responsible for offshore management from the coast to three nautical miles and the Commonwealth from three to 200 nautical miles.

Under the Australian constitution, the Commonwealth has power to legislate with respect to a number of different subject matters in relation to oceans and coasts, including interstate and overseas trade and commerce, fisheries in Australian waters beyond territorial limits, taxation, defence, lighthouses, quarantine, corporations, petroleum and minerals beyond three nautical miles, Aboriginal and Torres Strait Islander affairs, and territories and external affairs (in relation to matters physically external to Australia and in relation to giving effect to Australia's international obligations). The Commonwealth relies on these powers to legislate for oceans and coastal management.[2]

In the 1970s and 1980s, the Commonwealth influenced coastal and oceans policy by developing environmental protection and management legislation.[3] A number of institutions and agreements between the Commonwealth and the states and Northern Territory were set up as a result, including the Offshore Constitutional Settlement (OCS) and the *Intergovernmental Agreement on the Environment* (IGAE).

Constitutional and intergovernmental agreements

The OCS was negotiated between the Commonwealth, state and Northern Territory governments in 1979 to resolve increasingly complex issues of sovereignty over offshore areas. Under the OCS, the Commonwealth granted title and legislative power over marine and seabed resources from the low water mark to three nautical miles to the states and the Northern Territory. The Commonwealth retained jurisdiction over the marine environment from three to 200 nautical miles.[4]

The Commonwealth and the states and territories adopted the IGAE in 1992. It provides that Commonwealth responsibilities and interests in national environmental matters include ensuring that the policies or practices of a state or territory do not result in significant external adverse effects on maritime areas within Australia's jurisdiction, subject to existing Commonwealth legislative arrangements relating to maritime areas. It also provides that these Commonwealth interests and responsibilities cover implementation of environmental treaties.

The 1997 Council of Australian Governments (COAG) Heads of Agreement on Roles and Responsibilities for the Environment states that the Commonwealth has an interest in cooperation with the states to develop strategic approaches to ensure the management and protection of Australia's marine and coastal environment. As a result of this set of complex arrangements, one of the major challenges in relation to integrated coastal and oceans management is cross-jurisdictional coordination.

Cross-jurisdictional coordination

All of the states and the Northern Territory have coastal policies and state-level coastal advisory boards. The common thread to these policies is the overall objective of achieving ecologically sustainable development of coasts and coastal resources. Using this common objective across each level of government, Australia is advancing integrated coastal management through a joint Commonwealth/state and territory ministerial council – the Natural Resource Management Council (NRMC). NRMC has a separate marine and coastal committee to provide a forum specifically for oceans and coastal management discussions across and between spheres of government. NRMC plays an important role in cross-jurisdictional and cross-sectoral communication on marine and coastal issues. It also provides a forum for the linkages between coastal and oceans policy development. Another body, the Intergovernmental Coastal Reference Group, consisting of state coastal management agencies, local governments and state coastal council chairs, provides further cross-jurisdictional integration on coastal management issues.

A National Coastal Policy

To further integrate management of coasts and coastal resources in Australia, the Commonwealth government is developing a *National Coastal Policy* with state and the Northern Territory governments. The focus of the new national policy will be:

- improving water quality in coastal and marine areas;
- conserving and restoring important coastal and estuarine habitats and biodiversity; and
- protecting the economic base of coastal areas, particularly fishing and tourism.

Further, the Policy will be integrated with the regional catchment management approach in the *National Action Plan on Salinity and Water Quality*

(NAP) and the extension of the Natural Heritage Trust, and with regional marine planning under *Australia's Oceans Policy*, to complete a comprehensive natural resource management approach.

Marine protected areas

All levels of government in Australia are committed to developing a National Representative System of Marine Protected Areas (NRSMPA). Development of an NRSMPA fulfills Australia's international responsibilities and obligations under several international instruments including the 1992 Convention on Biological Diversity[5] and the 1979 Convention on the Conservation of Migratory Species of Wild Animals (Bonn Convention).[6] State, territory and Commonwealth governments initiated the NRSMPA through the Australia and New Zealand Environment and Conservation Council (ANZECC) in the early 1990s. According to ANZECC, MPAs contribute to the long-term sustainablility and ecological viability of marine and estuarine systems and protect Australia's biodiversity.[7] MPAs can be declared under Commonwealth, state or Northern Territory legislation within their own jurisdictional waters.

Due to the varying legislation within each jurisdiction, Australian MPAs use a range of naming conventions, subtypes and zones, including marine parks, marine national parks, marine and intertidal habitat areas, coastal reserves, marine management areas, fish habitat protection areas, aquatic reserves, seaward extensions of national parks, marine nature reserves and marine reserves. However, they share the common intent of protecting the marine and estuarine environment, particularly habitats such as reefs, seagrass beds, tidal lagoons, mangroves, rock platforms, coastal, deep water and underwater seabeds areas, and any marine cultural heritage. In excess of 2,000 MPAs have been established in Australia, including the Heard Islands and McDonald Islands (HIMI) Marine Reserve covering 65,000 km^2.[8]

Integrated coastal management programs

One of the predominant ways in which the Commonwealth can influence coastal management in Australia is through financial assistance grants to the state, Northern Territory or local governments. Programs such as the Coasts and Clean Seas Initiative, Coastcare and the Australian Coastal Atlas have been instrumental in driving a national approach to integrated coastal management in Australia.

The Coasts and Clean Seas Initiative, launched in 1996, was established to combat pollution problems and threats to water quality and marine life in the coastal zone. Several complementary programs are supported under the initiative at community, regional, state/territory and national levels, including Coastcare and the Australian Coastal Atlas.

Coastcare promotes and encourages community involvement in the protection, management and rehabilitation of Australia's coastal and marine environments. The program encourages linkages between communities and local management. Coastcare is a cross-jurisdictional program that is supported by every level of government in Australia. The Commonwealth, state and Northern Territory governments provide matching funding for Coastcare community grants while local governments provide in-kind support. Over 1,700 projects have been funded around Australia with the focus on assisting on-ground work such as:

- protecting or rehabilitating dunes, estuaries and wetlands;
- monitoring beach conditions, and coastal flora and fauna;
- helping to develop and implement local management plans; and
- providing education and training activities that raise community awareness, knowledge or skills on coastal and marine conservation issues.

Since the commencement of the Australian Coastal Atlas project in 1997, in consultation with a wide forum of users, a network of nodes have been established delivering almost 900 data layers over the Internet.[9] The purpose of the Atlas is to increase knowledge further about Australia's coastal zone, providing an accessible information base to support decision making for coastal zone management. A diverse range of data is available at a variety of scales to be mapped and queried, including administrative boundaries, indigenous sites, species distribution, aquaculture, bathymetry, marine habitats, cyclone tracks, beach safety and coastal reorganization.

Operationalizing integrated oceans management – regional marine planning

Australia's Oceans Policy provides the framework for integrated ecosystem-based planning and management for all of Australia's marine jurisdictions. The cornerstone of the Policy is regional marine planning and the key plank of regional marine planning is the ecologically sustainable development of Australia's oceans resources.

In developing regional marine plans (RMPs), Australia aims to integrate sectoral interests and conservation requirements and as far as possible integrate planning and management across Commonwealth, state and territory waters. In accordance with *Australia's Oceans Policy*, each regional marine plan will:

- identify ocean resources and economic and other opportunities;
- identify current and emerging threats to ecosystem health and determine planning and management responses to those threats;
- within the region, set out what is known of ecosystem characteristics and a broad set of objectives for those systems;
- identify the requirements and priorities for environmental baselines and basic biological inventory and other surveys in the development of regional marine plans;

- identify priorities and put measures in place to meet conservation requirements and determine those areas that should be assessed for marine protected area declaration;
- identify community and sectoral interests, including the interests of Aboriginal and Torres Strait Islander communities;
- identify priorities for industry and economic development of the region;
- put in place a planning regime to prevent conflict between different sectors over resources access and allocation;
- provide a framework within which there is increased certainty and long-term security for marine-based industries; and,
- establish indicators of sustainability and requirements of monitoring, reporting and performance assessment.[10]

Regional marine planning uses large marine domains (ecosystems) as one of the starting points for the planning process by creating regional boundaries that are based on ecosystem characteristics, representing a major step towards ecosystem-based management. Large marine domains are extensive areas of ocean that have relatively uniform broad-scale internal structures such as topography, ocean currents and fish species groupings. There are 13 large marine domains for which regional marine plans will be developed (see Figure 3.1).

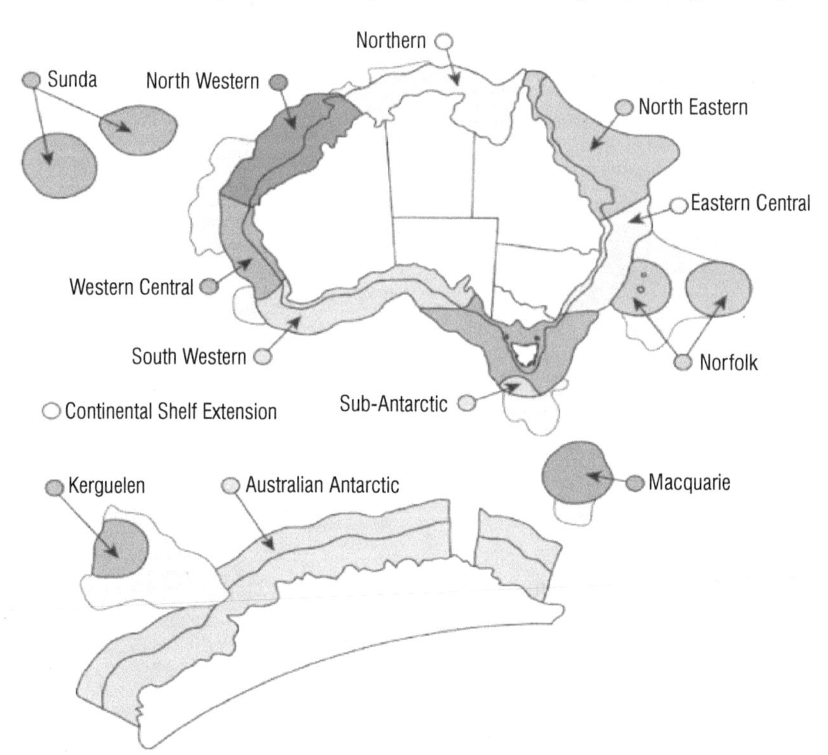

Figure 3.1 Australia's large marine domains.

South-east Regional Marine Plan

The South-east Marine Region is the first region in Australia to be investigated using this planning process. It includes waters off Victoria, Tasmania, Macquarie Island, southern New South Wales and eastern South Australia – capturing three of the large marine domains including the South-eastern, Sub-Antarctic and Macquarie (see Figure 3.2).

There are four objectives for the South-east Regional Marine Plan, which draw on the goals of *Australia's Oceans Policy*. The objectives are to:

1 Develop a shared understanding and appreciation of the characteristics of the South-east Region;
2 Design a RMP that is a decision making and planning framework for management across sectors that:
    - identifies shared values of the Region, including environmental, economic, social and cultural values;
    - identifies new information needed;
    - integrates resource management on an ecosystem basis;
    - identifies the methods for assessing performance;
    - is adaptive to changing conditions and improved knowledge; and
    - adds value to existing management arrangements;

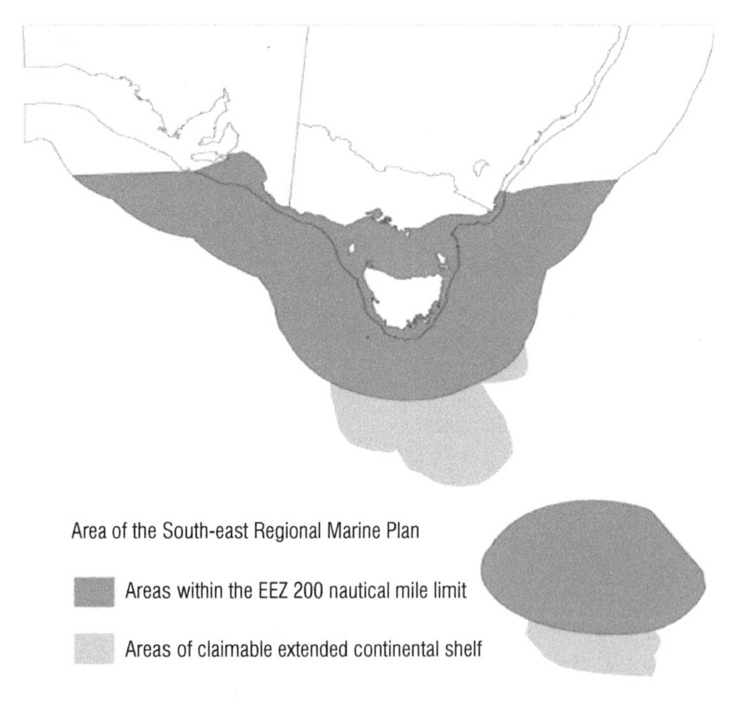

Figure 3.2 The South-east Marine Region.

3 Use the shared values of the Region to guide development of economic, social and conservation opportunities; and,
4 Accommodate community needs and aspirations by encouraging involvement and being inclusive, fair and transparent at all stages of the plan.

The South-east regional marine planning process began in April 2000 with an Oceans Forum. On 9 May 2002, seven assessment reports were launched, providing new information across six broad themes relating to ecosystems and human activities over state and Commonwealth waters. Those six assessment streams are:

- Biological and physical characteristics – identifying the key ecological characteristics in the Region, their linkages and interaction;
- Uses within the Region – describing our knowledge of the nature and dimension of human uses and their relationship with each other;
- Impacts on the ecosystem – providing an objective analysis of how activities can affect the Region's natural system;
- Community and cultural values – ensuring community wishes and aspirations are reflected in the planning process;
- Indigenous uses and values – gaining an understanding of and support for indigenous interests in the Region; and
- Management and institutional arrangements – analyzing current legislative and institutional frameworks to determine the best mechanisms for implementing regional marine plans.

As part of the process, scientific data collection, stakeholder workshops, community surveys and consultations provided unprecedented information in relation to Australia's deep ocean's ecosystems and marine resource use. Specialist working groups of stakeholders and experts were also established to provide direction and input to the process.

Biological and physical characteristics

The assessment of the South-east Marine Region's biological and physical characteristics provided an overview of the structure and function of its ecosystems and produced two key inputs for developing an ecosystem-based regional marine plan:

- An Interim Bioregionalization which identifies bioregions based on ecological attributes (geology, ocean currents, biota) between the continental shelf-break and the limits of Australia's exclusive economic zone (EEZ) – the bioregions provide an ecosystem-basis for developing planning units for the Region; and
- Ecosystem conceptual models that illustrate how the ecosystems of the

> Region function. These conceptual models provide a starting point for developing more formal models for specific management issues that are addressed by regional marine planning. They also assist in the development of ecosystem objectives and indicators, which are key elements in evaluating the success of the plan so that management can change depending on the outcomes of the evaluation (see Figure 3.3).

This biological and physical assessment of the South-east Marine Region has significantly improved Australia's knowledge of the deep-water ecosystems of the region and its physical and biological characteristics, including the characteristics of the marine ecosystems, seascapes, oceanographic characteristics and diversity of marine life.[11]

Uses

The uses assessment provided an understanding of the current uses and pressures in the South-east Marine Region, along with future uses and opportunities, and the value of the marine resources. Two assessment reports were produced, *Resources – Using the Oceans* and *Resources – Macquarie Island's Picture* documenting information about activities carried out in the South-east Marine Region. The users of Australia's oceans and its resources face many challenges. These challenges were identified by the two reports as:

- Economic and market-based – those affecting users including changes in demand for products and costs of inputs;
- Lifestyle – those brought about through changes in peoples' preferences or attitudes;
- Resource use – impacts of resource use on the environment and the provision of environmental services;
- Institutional – those arising from legal, regulatory or other institutional requirements, including resource management arrangements; and
- Cross-cutting – those that arise when one use affects a number of others or where a particular issue has potential implications across a range of uses.

The report *Resources – Using the Ocean* also investigated how uses might develop within the Region in the future, for example petroleum and minerals discovered in new areas of known economic potential which, depending on market forces, may come under production within the next 25 years.

Figure 3.3 Large-scale process conceptual model.

Large scale process Conceptual Model – primary productivity. Satellite measurements of sea surface productivity. Dark and mid-gray colours indicate low levels of productivity; the East Australian Current imports a low-nutrient, low-productivity water mass into the northeastern part of the Region. Light gray colours indicates higher productivity; evidence of local plankton blooms can be seen in East Australian Current eddy east of New South Wales, along the Bonney Coast of southeastern South Australia/western Victoria, and in the Subtropical front around Tasmania. Reproduced with permission from David Griffin, CSIRO Marine Research.

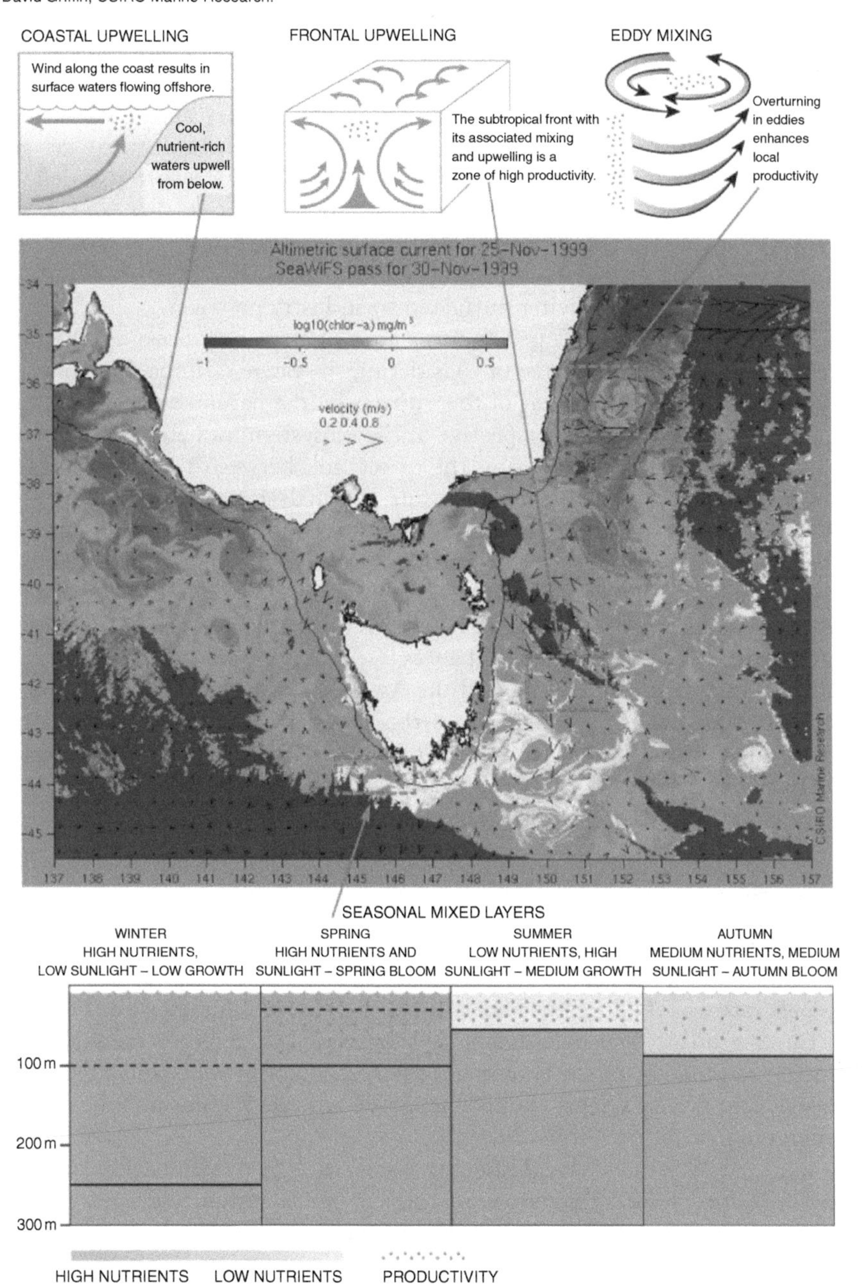

Impacts on the ecosystem

The impacts assessment categorized human activities and actions and their effects on the marine environment in the South-east Marine Region. Impacts were defined by the assessment phase as any human activity, action or process that has an effect on the ecosystems in the Region. The purpose of the impacts assessment stream was not to duplicate existing work on specific impacts, but to consider the range of impacts across the whole South-east Marine Region. *Impacts – Identifying Disturbances* defines and describes 12 categories of disturbance and provides an overview of where disturbances are known to occur in the environment (see Table 3.1). The report also defines 13 categories of activities and describes those activities that are either "known", or considered "possible" to cause the disturbance (see Table 3.2). It includes impacts that may be negligible, temporary and/or localized, as well as impacts that are being mitigated by industry practices.

The two matrices developed as part of the assessment process are the first step in meeting the challenge of considering the range of impacts across the whole region. The assessment of the impacts on the natural system analyzed the information from the perspective of the ecosystem, rather than the more traditional approach of exploring the direct link between the activity itself and any potential disturbance. As a result, the analysis described which parts of the ecosystem are affected by each disturbance category. The outcome of this analysis is illustrated in the matrix "Ocean environs and disturbances" (see Figure 3.4). Direct links between each activity and the type of disturbance it causes are also explored. The outcome of this analysis is illustrated in the matrix "Activity and disturbances" (see Figure 3.5).

The assessment process followed the *Australian and New Zealand Standard for Risk Management* as a general methodology for analyzing information about impacts on the ecosystem (see Figure 3.6). The matrices represent the initial stage of "identifying the risks" and aim to identify the broad range of impacts that affect the marine ecosystem.

Community and cultural values

The assessment of community and cultural links with the marine environment provided a snapshot of the community's values and aspirations for the deeper waters of the Region. People living within 50 kilometers of the coast of the South-east Marine Region, key national and regional conservation groups, and marine-focused community interest groups were consulted. The coastal community's level of knowledge about the Region and its broad demographic data were collected.

The coastal area of the Region is home to approximately 1.4 million people. A survey and workshop of marine-focused community groups identified their vision for the Region, which includes:

Table 3.1 Disturbance categories used to define impacts to the South-east Marine Region

Disturbance category	*Description*
Chemical change	Changing the concentration or properties of compounds naturally occurring in the ocean, such as changes to salinity, nutrients and dissolved oxygen
Contaminants	Introducing substances that are not normally found in the marine environment, such as heavy metals, PCBs and litter
Temperature change	Changing the marine environment's natural temperature range
Mechanical change	Removing or changing structural (biological and physical) components of the ecosystem, such as building dams
Nuclear radiation	Introducing radioactive isotopes into the marine environment
Electromagnetic radiation	Introducing radiation that consists of electromagnetic waves
Noise	Increasing the level or amount of sound in the marine environment beyond its natural range
Biological interaction	Removing or damaging organisms such as discarding by-catch
Introduced pathogens	Introducing disease-producing organisms to the marine environment, either from terrestrial or marine sources
Introduced marine species	Introducing species from outside their natural or historical ranges
Turbidity/light	Changing the extent to which light penetrates the water column
Artificial light	Introducing a source of light that would not naturally occur in the marine environment

- better management of the marine environment through use of management tools, including marine protected areas;
- protection of endangered species;
- a reduction in pollution;
- resource and environmental sustainability; and
- increased education to promote a greater sense of community stewardship.

Indigenous uses and values

In assessing the indigenous uses and values of the Region, the underlying message received was that indigenous people do not distinguish between land and sea. Instead, land and sea exist irrespective of the boundaries put in place over the last 200 years. Together land and sea form "country" – a

Table 3.2 Sources of disturbance in the South-east Marine Region

Source of disturbance	*Description*
Aquaculture	Activities associated with cultivating the food resources of the sea or inland waters, e.g. feeding, waste disposal and physical location
Defence	Activities specific to defence activities in the marine environment, e.g. sonar, live firing exercises and underwater explosions
Emerging	Activities which are new or recent to the marine environment, such as biotechnology
Harvesting	Activities that relate to fishing activities, including discarding of fish, diving and fishing gear disturbance (shipping-related activities are included under "Shipping")
Human changes coastal zone	Activities by humans that cause changes to the coastal zone such as coastal construction and dredging
Indigenous customary use	Activities associated with indigenous customary use, including customary harvest and ceremonial activities
Land-based	Activities that are distinguished from human changes by the types of input that they have to the environment, including industrial discharge, sewage and urban discharge
Ocean dumping	Activities that are associated with the disposal of waste and other products (e.g. ammunition) at sea
Petroleum	Activities that are associated with petroleum exploration and production in the marine environment, e.g. seismic survey, rig establishment and produced formation water disposal (ship-related activities are included under "Shipping")
Recreational activities	Recreational activities that do not fit into the tourism category, including collecting species and diving
Shipping	All shipping-related activities, including those from harvesting, petroleum and defence, including hull-fouling, ballast water discharge and shipping maintenance
Submarine cables	Activities associated with submarine cables including the physical presence of cables
Tourism	Activities associated with tourism (not including shipping, these are covered under "Shipping") including interactions with wildlife, and the development of tourism sites

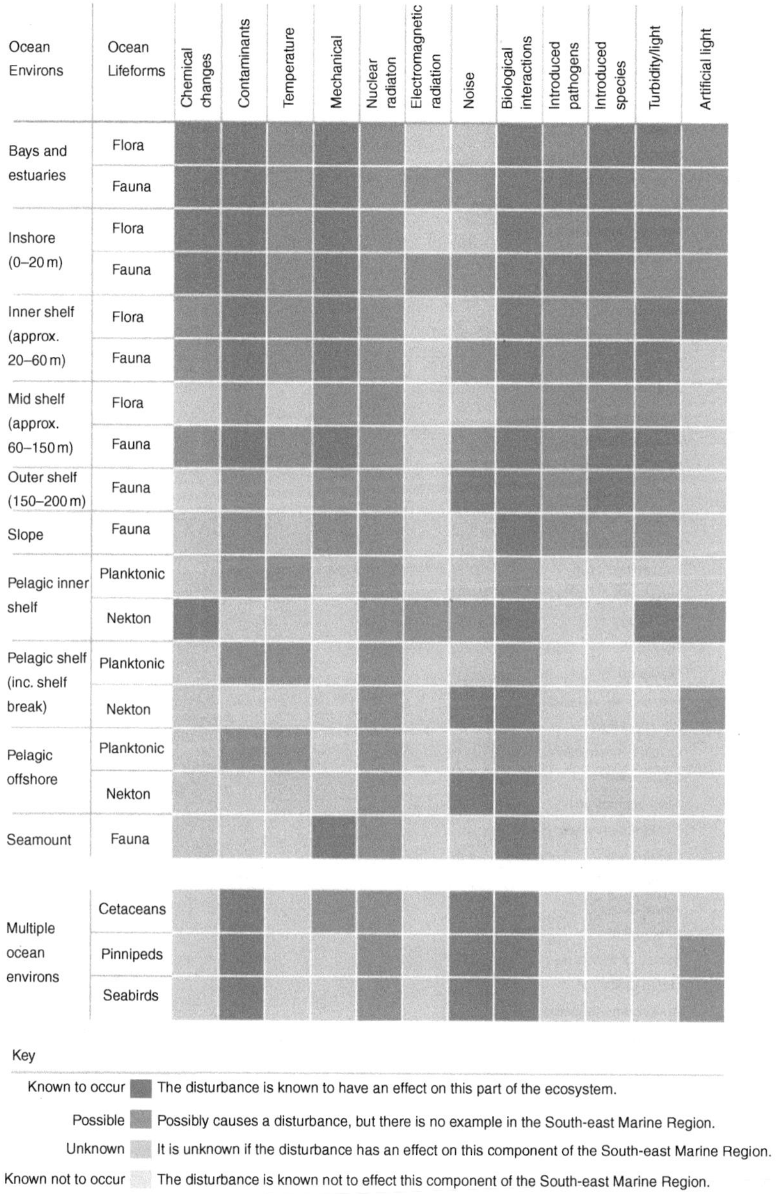

Ocean Environs	Ocean Lifeforms	Chemical changes	Contaminants	Temperature	Mechanical	Nuclear radiaton	Electromagnetic radiation	Noise	Biological interactions	Introduced pathogens	Introduced species	Turbidity/light	Artificial light
Bays and estuaries	Flora												
	Fauna												
Inshore (0–20 m)	Flora												
	Fauna												
Inner shelf (approx. 20–60 m)	Flora												
	Fauna												
Mid shelf (approx. 60–150 m)	Flora												
	Fauna												
Outer shelf (150–200 m)	Fauna												
Slope	Fauna												
Pelagic inner shelf	Planktonic												
	Nekton												
Pelagic shelf (inc. shelf break)	Planktonic												
	Nekton												
Pelagic offshore	Planktonic												
	Nekton												
Seamount	Fauna												
Multiple ocean environs	Cetaceans												
	Pinnipeds												
	Seabirds												

Key

Known to occur – The disturbance is known to have an effect on this part of the ecosystem.

Possible – Possibly causes a disturbance, but there is no example in the South-east Marine Region.

Unknown – It is unknown if the disturbance has an effect on this component of the South-east Marine Region.

Known not to occur – The disturbance is known not to effect this component of the South-east Marine Region.

Figure 3.4 Ocean environs and disturbances.

Note

This report lists the range of impacts on the natural system that occur in the South-east Marine Region, but does not make any judgments about the relative importance or consequence of those impacts or the activities that cause those impacts, nor does it explain the many mitigation mechanisms in place.

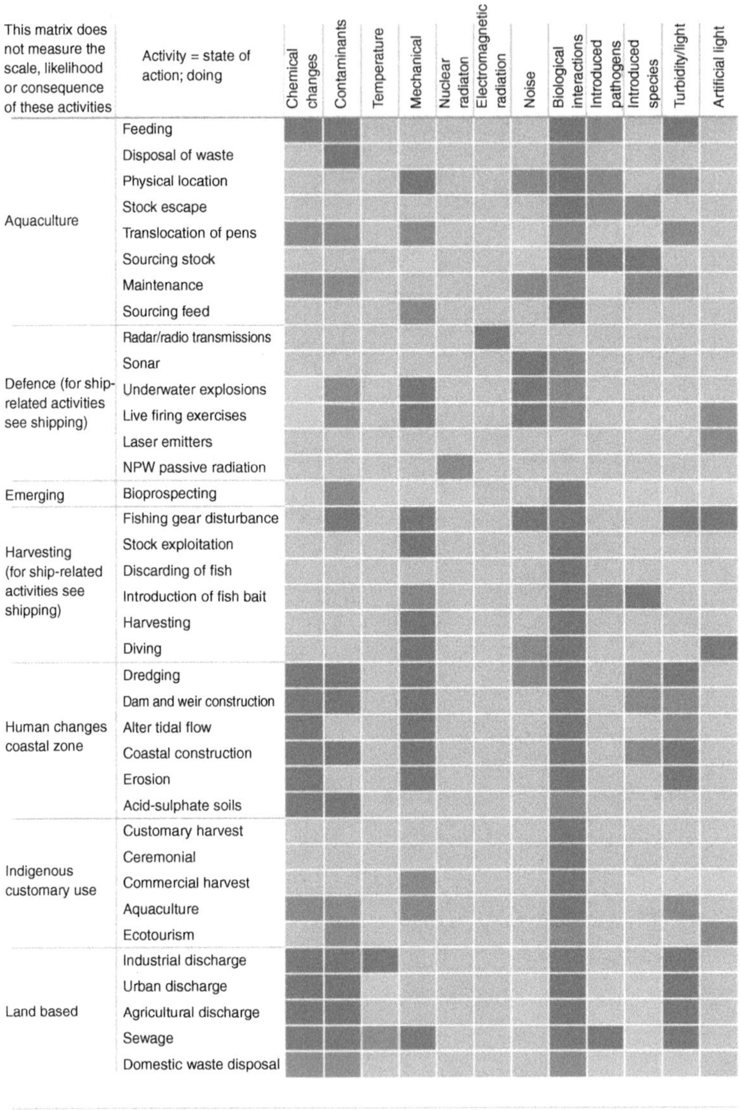

This matrix does not measure the scale, likelihood or consequence of these activities	Activity = state of action; doing	Chemical changes	Contaminants	Temperature	Mechanical	Nuclear radiaton	Electromagnetic radiation	Noise	Biological interactions	Introduced pathogens	Introduced species	Turbidity/light	Artificial light
Aquaculture	Feeding												
	Disposal of waste												
	Physical location												
	Stock escape												
	Translocation of pens												
	Sourcing stock												
	Maintenance												
	Sourcing feed												
Defence (for ship-related activities see shipping)	Radar/radio transmissions												
	Sonar												
	Underwater explosions												
	Live firing exercises												
	Laser emitters												
	NPW passive radiation												
Emerging	Bioprospecting												
Harvesting (for ship-related activities see shipping)	Fishing gear disturbance												
	Stock exploitation												
	Discarding of fish												
	Introduction of fish bait												
	Harvesting												
	Diving												
Human changes coastal zone	Dredging												
	Dam and weir construction												
	Alter tidal flow												
	Coastal construction												
	Erosion												
	Acid-sulphate soils												
Indigenous customary use	Customary harvest												
	Ceremonial												
	Commercial harvest												
	Aquaculture												
	Ecotourism												
Land based	Industrial discharge												
	Urban discharge												
	Agricultural discharge												
	Sewage												
	Domestic waste disposal												

Figure 3.5 Activity and disturbances.

Note

This report lists the range of impacts on the natural system that occur in the South-east Marine Region, but does not make any judgments about the relative importance of consequence of those impacts or the activities that cause those impacts, nor does it explain the many mitigation mechanisms in place.

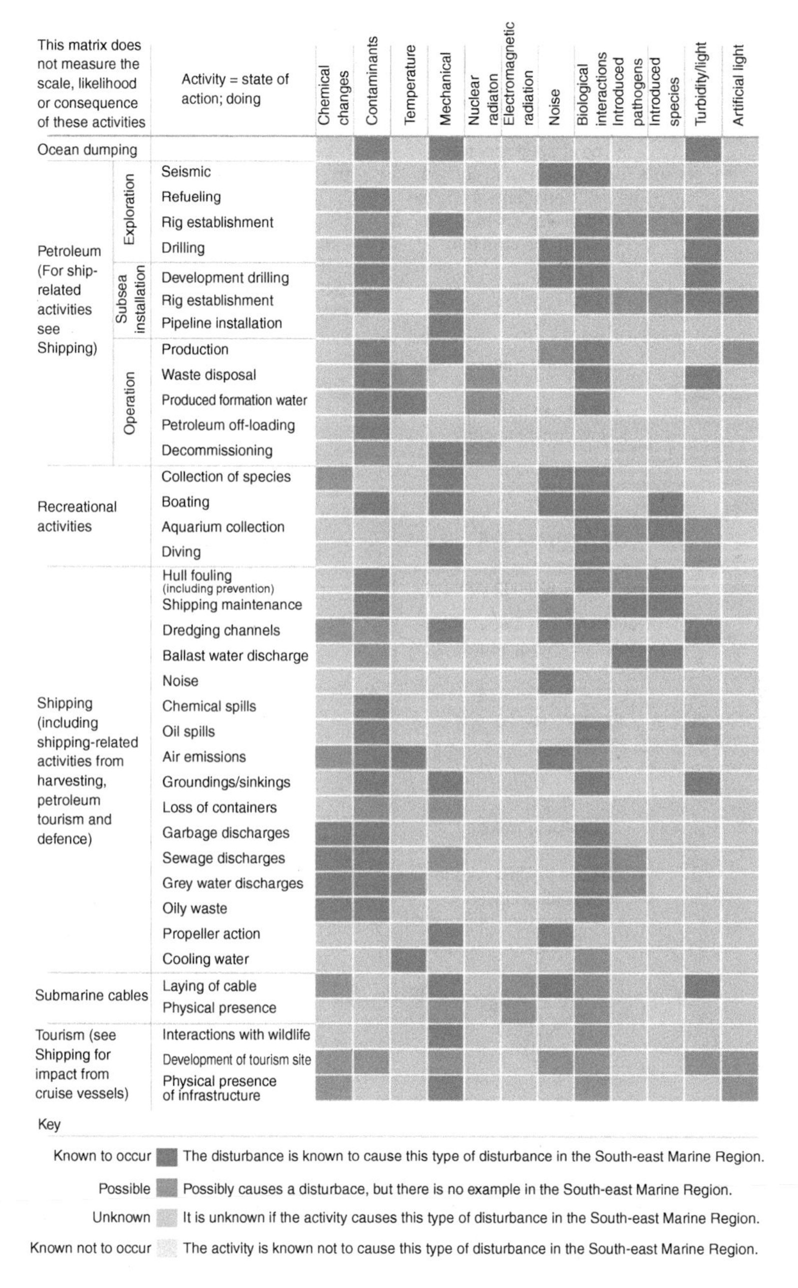

Figure 3.5 continued

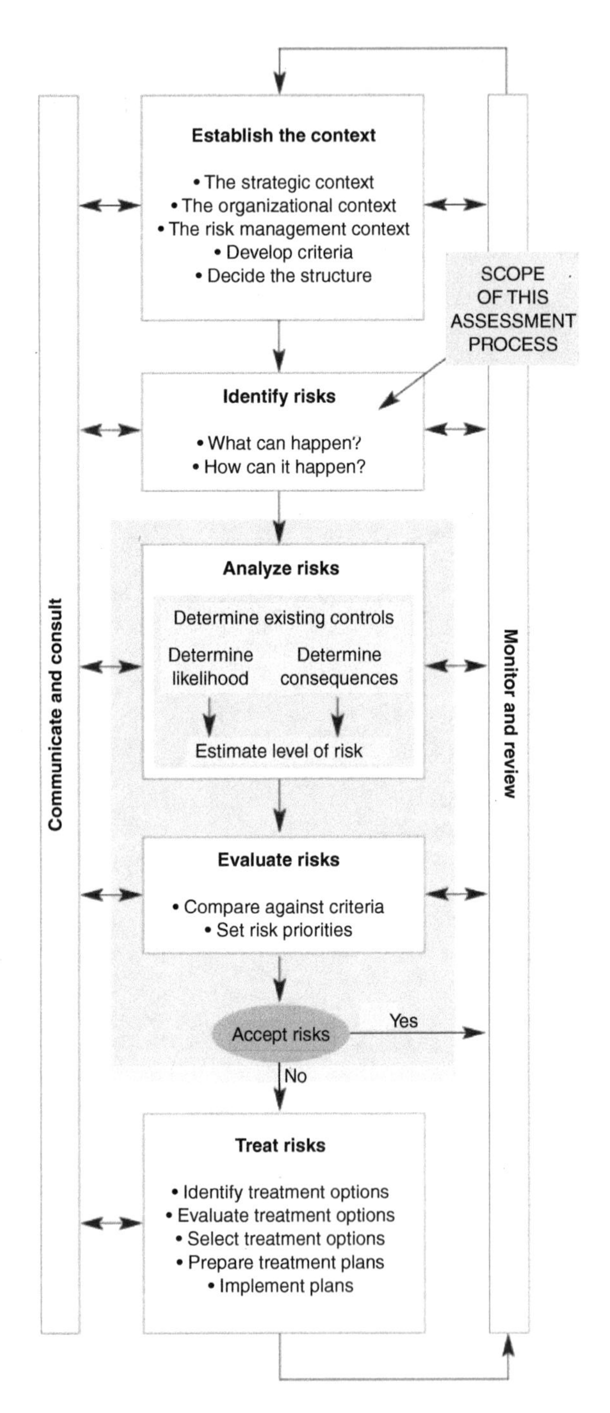

Figure 3.6 The risk assessment process.

country of significant cultural sites and "dreaming tracks" of the creation ancestors. During the consultation, Indigenous people expressed their desire to improve their health and well being, particularly through:

- recognition and respect for culture;
- co-management and resource sharing;
- culturally appropriate education and training; and
- economically, environmentally and culturally sustainable employment opportunities.

Management and institutional arrangements

The management and institutional arrangements assessment broadly described the Commonwealth legislation affecting how Australia uses and protects its oceans in the South-east Marine Region. The management arrangements for oceans use in Australia arise from historical management based on individual sectors in various state and Commonwealth waters, and these are characterized by a complex system of legislation.

The overarching framework of marine regulation in Australia is dominated by international law, principally the law of the sea, and Australia's constitutional structure. International law, as contained in the 1982 United Nations Convention on the Law of the Sea (LOS Convention),[12] sets out the basic rules for the exercise of jurisdiction by a coastal state. This varies according to the zone of jurisdiction, but usually extends to the limit of the 200 nautical mile exclusive economic zone.

The sovereignty and sovereign rights of Australia under the LOS Convention are given force in Australia by the *Seas and Submerged Lands Act 1973* (Cth), which vests such rights in the Commonwealth for all waters except those within the limits of state or territory waters. This legislative framework is then linked to the OCS in terms of management of offshore areas, and is given effect under the *Coastal Waters (State Powers) Act 1980* (Cth) and the *Coastal Waters (State Title) Act 1980* (Cth).[13]

The assessment of Commonwealth legislation presents issues regarding the current framework and suggestions for implementing regional marine plans as raised by stakeholders. Issues include the gaps and duplications in existing management arrangements of each sector.

Northern Planning Area

In 2001 it was announced that the second regional marine planning area would be in northern Australia encompassing the eastern part of the Arafura Sea, the Gulf of Carpentaria and the Torres Strait. The initial stages of this planning process focused on establishing contact and developing administrative and consultative arrangements. This time the process of identifying strategic information and planning priorities included a dedicated effort to bridge the cross-jurisdictional divide between the Queensland, Northern Territory and Commonwealth governments. The positive response from all levels of government to the initial scoping phase of the Northern Planning Area means that the context of Australia's second regional marine plan may be markedly different from the first. Several issues and challenges have already been identified for the process, including:

- Indigenous populations in the Northern Planning Area have significant ownership and involvement in management of coastal resources. The remoteness of communities combined with access issues are likely to require the development of specific stakeholder liaison mechanisms which may augment existing government or non-government arrangements.
- The Torres Strait also has remote communities that may be difficult to access. Furthermore, the Torres Strait has independent institutional arrangements from its neighboring state government.
- The waters of the Northern Planning Area abut waters of Papua New Guinea and, to a lesser extent, Indonesia providing an international dimension to regional marine planning not encountered before.[14]

Operationalizing integrated oceans management: challenges ahead for regional marine planning

Eleven planning principles, drawn from the vision, goals and directions provided by *Australia's Oceans Policy*, inform the regional marine planning process. They embody recent developments in natural resource management and direct the way that options for management will be developed and negotiated. These principles highlight both the challenges and direction of regional marine planning. Regional marine plans will:

- implement ecosystem-based management as the basis for decision making and management, and embed the principles of ecologically sustainable development, including precaution into all decision-making processes;
- promote ecologically sustainable marine-based industries that contribute to regional development;
- develop integrated management of sectoral activities and achieve strong efficient cross-sector linkages;
- work towards consistency in management across jurisdictional boundaries when impacting upon the same oceans resource or sector;
- lead to clearly defined and agreed regional marine plan outcomes that are integrated across all sectors;
- lead to fair decision making and conflict resolution regarding access to oceans resources within and between generations;
- increase involvement of resource users and the community at large in planning and decision making;

- engender long-term responsible use of oceans resources, i.e. stewardship;
- provide flexible management arrangements that focus on measurable outcomes coordinated across sectors;
- contribute to adaptive management based on monitoring and evaluation of outcomes of management against expected performance including providing for auditing and review processes; and
- establish clear and agreed definitions of issues and terminology.

Operationalizing the South-east Regional Marine Plan – moving from theory to practice

The major challenges facing integrated oceans management in Australia are moving from theory to practice; identifying shared values and objectives; meeting objectives and defining planning issues and providing for cross-sectoral coordination; ecosystem-based management; cross-jurisdictional coordination; outcomes-based management; and participation and transparency in decision making. Figure 3.7 indicates how the South-east Regional Marine Plan moved from the theoretical to the practical level. In 2003 the "Draft South-east Regional Marine Plan"[15] was publicly released and public responses were sought from Stakeholders.[16] The final Plan was released on 21 May 2004 (see Postscript). However, the Plan is intended to be a dynamic management document that will be progressively modified and improved as new information becomes available and new mechanisms for oceans management are established.

Identifying shared values and objectives

The specific objectives of each regional marine plan will vary depending on the characteristics of each Region. Shared values are derived through community, industry and government consultation; they focus on those attributes of the region that Australians wish to maintain. Using these shared values it is possible to define what Australians want to achieve through the plan and this will be clearly stated in objectives. The following shared values were distilled from the South-east regional marine planning process:

- healthy thriving ecosystems;
- strong regional communities;
- partnerships with Indigenous peoples;
- competitive, diverse and innovative industries;
- responsive and responsible governance; and
- fair, efficient and participatory decision making.

Meeting objectives and defining planning issues

A further challenge of moving from theory to practice is determining if the

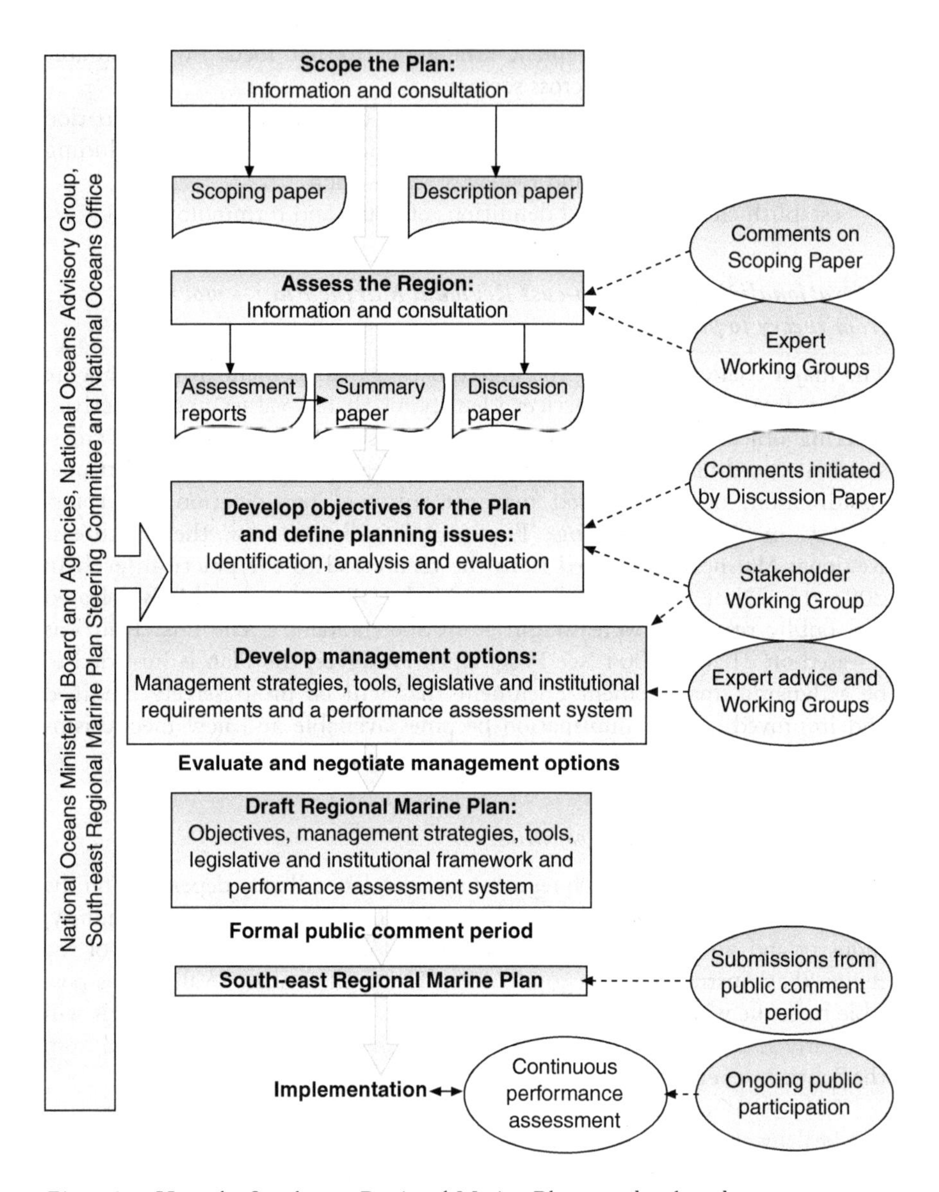

Figure 3.7 How the South-east Regional Marine Plan was developed.

objectives of the RMP are being met. If an objective is not being met, finding out why will assist in defining planning issues. Guidance provided by *Australia's Oceans Policy* will be used to determine whether an issue can be dealt with by the RMP. Also, each planning issue will be analyzed so that the right management options can be developed. It may be that some planning issues require an immediate and specific response, for example, a list of potential marine protected areas. Other issues might require activity-based restrictions, for example, fishing gear modifications, or planning responses in the case of the development of a threat abatement plan. Through a framework of institutional arrangements, planning and management processes and a built-in performance assessment system, planning issues will be managed with a view to achieving the objectives of each plan.

Managing planning issues – options

Management options consist of different combinations of management tools (management strategies) and implementation arrangements, including a system to assess the performance of each strategy. Management tools are measures used to control or direct particular aspects of an activity, methods for reaching management decisions, or ways of influencing patterns of use. Management tools range from very specific tools such as minimum mesh size on fishing gear, to environmental impact statements for individual activities, to large-scale spatial zoning of activities and/or market-based incentives. The management options for regional marine planning would focus on integrating existing management tools (through the responsible agency) and addressing gaps and developing specific additional tools where necessary. Examples of existing management tools that could form a management strategy include fishing quota systems, marine protected areas and codes of conduct.

Management options may include several management strategies to resolve regional marine planning issues. These strategies will focus on setting management outcomes and may range from policy initiatives to more specific measures such as legislation. It is likely that management options will become apparent when the planning issues are being analyzed, but they will need to be compared with each other in order to be specific about the details of each option.

The management options will then need to be evaluated considering the costs (economic, social and environmental) of implementation against the benefits derived from the option. The practicality of implementation requirements for each option will also need to be considered. Advice from stakeholders and experts in their fields will be used in negotiating which management options to include in the South-east Regional Marine Plan.

Cross-sectoral coordination

Australia's Oceans Policy acknowledges that progress towards improved marine management has been made within separate sectors, but that management and decision making have not been sufficiently integrated across the various sectoral interests. It also recognizes that:

> management of our oceans purely on an industry-by-industry basis will not be sustainable in the long run. Activities such as fishing, tourism, shipping, aquaculture, coastal development and petroleum production must be collectively managed to be compatible with each other and with the ecological health of the oceans.[17]

A Stakeholder Working Group for the South-east Marine Region have assisted in finalizing the regional marine planning objectives and analyzing planning issues identified during the assessment phase for the South-east Marine Region. Further, the South-east Regional Marine Plan Steering Committee formed in the early stages of the regional marine planning process will continue to guide the development of the Plan, with clear links to the Stakeholder Working Group. To achieve this, the Chair of the Steering Committee chairs the Stakeholder Working Group.

Ecosystem-based management

To provide an ecosystem basis for regional marine planning, the South-east Marine Region boundary has been designed to recognize ecological characteristics by using large marine domains (see Figure 3.1). In addition, a National Work Program for the continued collection and monitoring of scientific marine data has been implemented.

Under the LOS Convention, Australia has one of the largest exclusive economic zones (EEZs) in the world, covering an area of 11.1 million km^2, which equates to about one and a half times Australia's land area.[18] Unlike the land and even the atmosphere, very little is known or understood about Australia's marine environment.[19] This lack of knowledge and understanding reduces Australia's capacity to manage its oceans resources. Further, it is not always possible to transfer management approaches based on terrestrial environments to marine environments.[20]

Although our understanding of the marine environment has increased as exploitation of resources and other uses has increased, our knowledge rapidly decreases from the shoreline outwards. Gathering data for the whole of Australia's marine environment is a formidable task, both logistically and financially. To address this lack of knowledge and understanding, a National Work Program (NWP) has been implemented to build on the existing knowledge base, to map the seabed and resource potential of Australia's marine jurisdiction, to measure performance of regional marine plans, to establish access to the existing marine data and to facilitate international cooperation via joint projects and arrangements. Key inputs to the development of regional marine plans have been identified by four major projects under the NWP, including:

- National bioregionalization – to provide fundamental inputs for the development of ecosystem-based regional marine plans and planning units.

- National socio-economic survey and assessment – to address the need to give appropriate consideration to potential socio-economic implications of ocean use and management arrangements.
- NORFANZ survey – a proposed multilateral international survey of the Norfolk Island Ridge and the Lord Howe Rise to increase knowledge of the biodiversity and resource potential of this biologically and geologically unique region.
- Relationship between hydrocarbon fluxes and halimeda mounds (algal beds) – to map and sample deep-water halimeda beds in northwestern Australia to investigate the relationship between deep algal beds and natural hydrocarbon seeps.

A number of expert bodies will be formed to assist with the NWP process.

Cross-jurisdictional coordination

The OCS, IGAE and COAG Heads of Agreement on Roles and Responsibilities for the Environment create a complex jurisdictional division of oceans governance in Australia. This poses substantial challenges to regional marine planning that is inherently ecosystem-based rather than organized by jurisdictionally defined "lines on maps." The Natural Resource Management Council is a joint Commonwealth, state and territory forum with the aim of facilitating cross-jurisdictional coordination on natural resource issues. The Council's Marine and Coastal Committee is seeking agreement on integrated oceans management principles between Commonwealth, state and territory governments. This is another important step in providing a platform to integrate further oceans management in Australia.

Outcomes-based management

To be outcomes based, a RMP would need to include a system to monitor whether it is working – a performance assessment system. To assess performance, the objectives for the RMP should be further defined into measurable operational objectives that relate directly to the management strategies. By regularly monitoring these operational objectives, evaluation of whether the management strategies are achieving the objectives of the RMP can be conducted.

Participation and transparency

Australia's Oceans Policy clearly points out the need and value of community and industry participation in the regional marine planning process. The South-east regional marine planning process provided a series of opportunities for public participation. The release in 2003 of the Draft Plan was followed by further public consultation and an opportunity for public comment. However, public participation in oceans planning and management does not stop with the release of a regional marine plan. Rather, public participation will play an important role in the periodic assessment and review of RMPs.

Conclusion

Australia is facing many challenges in operationalizing integrated coastal and oceans management. One of the main challenges is balancing the increasing pressure on coastal and marine ecosystems while promoting sustainable development for the benefit of all Australians now and in the future. Arising from this challenge, the need for an integrated approach to coastal and marine management is clear. Working in the context of complex jurisdictional arrangements, Australia is beginning truly to operationalize integrated coastal and oceans management. An essential link between coastal and oceans governance is provided by the Natural Resource Management Council through high level representation from Commonwealth, state and territory governments. Through the National Representative System of Marine Protected Areas, the Coasts and Clean Seas Initiative, and the development of a *National Coastal Policy*, Australia is beginning to address the increasing pressures on water quality in coastal and marine areas, coastal and estuarine habitats and biodiversity, and the economic base of coastal areas.

In the marine environment, Australia is applying the principles of cross-jurisdictional coordination, cross-sectoral coordination, ecosystem-based management, outcomes-based management, and participation and transparency in decision making through regional marine planning under *Australia's Oceans Policy*. Integrated coastal and oceans management in Australia is now moving from theory to practice – guided by input from all levels of government, experts in their fields, stakeholders and the community.

Postscript

The first regional marine plan under *Australia's Ocean Policy* was completed in May 2004 with the launch of the South-east Regional Marine Plan.[21] The Plan covers a marine region of more than two million square kilometers surrounding the states of Victoria and Tasmania, and parts of South Australia and New South Wales. It also extends to parts of the Southern Ocean around the sub-Antarctic territory of Macquarie Island. The Plan contains more than 90 designated actions, including the development of a system of marine protected areas. However, the Plan has no legislative underpinning, which is consistent with the approach Australia has taken to date in implementing the Oceans Policy. Rather, the Plan provides an operational blueprint for the federal, state and local government authorities responsible for its implementation. The Plan has an important focus on the role of marine industries, and seeks to create mechanisms to encourage and support marine industries within the South-east region. In addition, the Plan supports greater engagement by Indigenous communities in the management of the region's marine resources. A Supplement to the Plan is anticipated in 2005 that will contain the complete system of marine protected areas within the region.

Design and preparation has commenced for the second regional marine plan that covers the "Northern Planning Area," which extends from the Torres Strait across northern Australia to include the waters of the Gulf of Carpentaria and the eastern Arafura Sea.[22] This Planning Area will encompass up to 700,000 km^2 and includes waters over which indigenous peoples have long exercised various access rights. In recognition of the unique characteristics of the Torres Strait, including the presence of the Australia/Papua New Guinea maritime boundary, a separate planning process is proposed for those waters. The Scoping Phase of the Northern Planning Areas was completed between 2002 and 2004 and it is currently anticipated that the Plan should be finalized by December 2005. The next proposed regional marine plan is for the southwest, which encompasses waters offshore South Australia and southwest Western Australia.

Notes

1 Commonwealth of Australia *Australia's Ocean Policy* Volumes I and II (Environment Australia, Canberra: 1998). Online at: www.oceans.gov.au/the_oceans_policy_overview.jsp7 (accessed 24 April 2005).

2 See the discussion in Donald R. Rothwell and Marcus Haward "Federal and international perspectives on Australia's maritime claims" (1996) 20 *Marine Policy* 29–46.

3 M. Haward and B. Davis "Current developments in Australian coastal zone management" in P.G. Wells and P.J. Ricketts (eds) *Coastal Zone Canada '94 – Cooperation in the Coastal Zone: Conference Proceedings* Volume 1 (Coastal Zone Canada Association, Bedford Institute of Oceanography, Dartmouth: 1994) 19–29.

4 For a further discussion, see Rothwell and Haward, note 2.

5 (1992) 31 *International Legal Materials* 818.

6 1651 UNTS 333.

7 Australian and New Zealand Environment and Conservation Council *Strategic Plan of Action for the National Representative System of Marine Protected Areas: A Guide for Action by Australian Governments* (Environment Australia, Canberra: 1999). For further discussion, see Chapter 7 by Kriwoken, Fallon and Rothwell in this volume.

8 For details see Environment Australia "Coasts and Oceans: Marine Protected Areas." Online at: www.deh.gov.au/coasts/mpa/index.html (accessed 17 December 2003). In March 2005, the Australian Antarctic Division released a draft Management Plan for the HIMI Marine Reserve for public comment that expands the marine protected area component of the territory. Online at: www.heardisland.aq/protection/management_plan/public_comment.html (25 April 2005).

9 The Australian Coastal Atlas. Online at: www.deh.gov.au/coasts/atlas/index.html (14 January 2004).

10 Commonwealth of Australia *Australia's Ocean Policy* Volume 1 (Environment Australia, Canberra: 1998) at 13.

11 For a detailed description of the region based on this assessment process see National Oceans Office Draft South-east Regional Marine Plan (National Oceans Office, Canberra: 2003) 4–15, and Appendix 3.
12 1833 UNTS 396.
13 See discussion in Rothwell and Haward, note 2.
14 For a discussion of Australia's maritime boundaries in this area, see Stuart Kaye "Australian Ocean Boundaries: An Overview" in Lorne K. Kriwoken *et al.* (eds) *Oceans Law and Policy in the Post-UNCED Era: Australian and Canadian Perspectives* (Kluwer Law International, London: 1996) 97 at 102–5.
15 Note 11.
16 David Kemp (Minister for the Environment and Heritage) "Howard Government Achieves World First in Sustainable Oceans Management" Media Release (18 July 2003).
17 Commonwealth of Australia, note 10 at 11.
18 Department of Environment, Science and Technology *Our Sea, Our Future: Major Findings of the State of the Marine Environment Report for Australia* (Great Barrier Reef Marine Park Authority, Townsville: 1995).
19 B. Bowen *et al.* "Estuaries and the Sea" in *Australia: State of the Environment 1996* (CSIRO Publishing, Collingwood: 1996) Ch. 8.
20 B. Jones *Australia's Exclusive Economic Zone: Achieving Sustainable Multiple-use* (National Centre for Development Studies, Australian National University, Canberra: 1996).
21 David Kemp, Federal Minister for the Environment "Marine Plan leads worlds in sustainable oceans management" Media Release K87 (21 May 2004). Online at: www.deh.gov.au/minister/env/2004/mr21may04.html (accessed 27 April 2005). Details of the Plan can be found at www.oceans.gov.au/se_implementation_plan.jsp (accessed 28 April 2005).
22 See National Oceans Office *Scoping Report for the Northern Planning Area* (National Oceans Office, Hobart: 2004). Details of the planning process for the Northern Planning Area can be found at www.oceans.gov.au/North-rmp.jsp (accessed 28 April 2005).

4 The application of compliance and enforcement strategies on Canada's Pacific coast

François N. Bailet, Janna Cumming and Ted L. McDorman

Introduction

Integrated maritime compliance and enforcement strategies must be geographically specific since from one region to another, ocean users differ, legislative and policy goals vary and, as a consequence, capacity requirements will be different. The one commonality is the goal of integrated maritime compliance and enforcement – the efficient and effective use of traditional and non-traditional enforcement methods and resources to ensure a high level of ocean-user compliance with the applicable laws and appropriate practices.

This chapter will look at the application of integrated maritime compliance and enforcement strategies on Canada's Pacific coast by first outlining a structured approach to integrated maritime compliance and enforcement, then providing a survey of the ocean uses that create compliance and enforcement issues on Canada's Pacific coast before moving to a review of the way in which these issues are being met. The primary focus will be on the mandates and capacities of Canadian federal departments and agencies in inducing compliance through enforcement of the laws that are applicable to particular activities. The concluding section sets out a series of future challenges for compliance and enforcement strategies.

A structure for integrated maritime compliance and enforcement

Introduction

It is customary to look at maritime compliance and enforcement in terms of attaining or attempting to attain maritime security which is defined as the protection of maritime national interests from threats in, on or concerning the sea. Our understanding of maritime security has significantly expanded in recent years beyond the traditional concepts that dealt primarily with threats of a naval or military nature. The newer understanding of maritime security can include:

> [T]hreats and acts of violence to coerce, extort or accomplish a political goal; direct challenges to national sovereignty; disregard of national and international law; illegal resource exploitation; the illegal transportation of goods and people; and the deliberate or unintentional creation of an environmental hazard.[1]

To address effectively this broad range of threats and challenges to maritime security, flexible and cooperative approaches to surveillance, monitoring and control,[2] including traditional strategies of naval sea control and presence, are required. Essential to this is the ability to gather, manage and understand a comprehensive database on activities in ocean areas. Only by knowing what is happening and where, can a state respond to and formulate strategies to address maritime security issues.

Related to the ability to respond to maritime security threats relying on surveillance, monitoring and control techniques is the goal of the reduction of maritime security threats, in short, achieving compliance with the laws of the nation state. The relationship between reducing maritime security threats and responding to maritime security threats is a mutually reinforcing one. There are several strategies that can be utilized in reducing maritime security threats by increasing compliance through education, training, codes of conduct and self-regulation.[3] Another important strategy is the involvement of the relevant ocean-user communities, which would also include the private sector entities to be regulated, in the creation of laws and, where appropriate, in the implementation of that law. It is through the effective involvement of constituents in the formulation and implementation of law and policy and the adoption of other compliance inducement strategies that enhanced compliance can be achieved and, as a result, decrease the need for expensive state-centric enforcement operations.

The emphasis on the structure of integrated maritime compliance and enforcement being developed herein is less on the "softer" compliance strategies and directed more to the enforcement strategies that play a role in inducing or re-enforcing compliance. Fundamental to these enforcement strategies is the ability of a state to utilize surveillance, monitoring and control techniques.

Taking a broad view of maritime security results in the identification of five areas of marine activity for which a state may face a series of responsibilities, challenges and threats that may require enforcement: management of marine resources,[4] protection and preservation of the marine environment,[5] maintenance of maritime sovereignty, prevention of illegal activities, and marine safety.[6]

Responses to marine challenges

In order for coastal and island states to ensure a certain level of maritime security in the five areas of marine activity listed above, states may deploy various types of enforcement responses. These may include a combination of the use of operational, legal, political and non-state means. Within each of these response categories, there exist a variety of surveillance, monitoring and control options that the states may exercise. The combination or mix of enforcement responses utilized by a state is determined by its particular enforcement requirements and priorities, as well as the actual enforcement resources and response capabilities available to it.

Operational responses

Resources for the control of marine activities can encompass a wide range of technical platforms, equipment and personnel, as well as the communications equipment and infrastructure to support and maintain them. The resources will include: surface platforms and vessels; fixed- and rotary-wing aircraft, and other aerial platforms; underwater platforms such as submarines, submersibles and fixed sensors; space-based satellites with sensors and radar; as well as land-based platforms, such as coastal radar systems. Information systems are critical to the effective use of operational resources.

Legal responses

For maritime security, a coastal state must have a system of laws, regulations, standards and procedures applicable to its ocean jurisdiction, which will form the legal basis for maritime enforcement strategies and general control activities. National laws regarding marine activities are inevitably reflective of international ocean law, conventions and agreements, including the 1982 United Nations Convention on the Law of the Sea (LOS Convention).[7]

Political responses

Political and bureaucratic arrangements to achieve maritime security may occur bilaterally or multilaterally with other states or actors, as well as internally within the state. Internationally, states can pool their respective resources to meet common maritime security needs. The regional approach to oceans development, management and enforcement offers economy and efficiency in addressing maritime security requirements of member states and in enhancing levels of maritime security in shared ocean areas. The national, or internal, components relate to the interaction among levels of government, departments and agencies responsible for oceans management and enforcement. An integrated approach to maritime compliance and enforcement requires careful management of relations among all the national and regional actors involved.

The macro-legal and political context of Canada's Pacific coast

The political geography of Canada's Pacific coast waters

Canada's Pacific coastline stretches approximately 27,200 kilometers, just over 11 percent of Canada's total coastline. Canada exercises jurisdiction over internal waters, the territorial sea, the contiguous zone and the exclusive economic zone (EEZ) on the West Coast. Canada's Pacific waters, however, constitute mere percentage points of Canada's overall offshore area. The EEZ has been legislatively defined,[8] but the location is contested by the United States both in the north and south.[9] The territorial sea location has been defined,[10] but there are northern and southern gaps involving the United States.

The northern ocean boundary problem between Canada and the United States concerns the Dixon Entrance and the waters seaward of Dixon Entrance. Fishers from both countries are occasionally arrested in the Dixon Entrance area, although there appears to be a general forbearance by both countries to take direct enforcement action except in the most obvious of cases. The southern overlapping offshore claim problem is a small sliver of ocean seaward of the Strait of Juan de Fuca.

In the most heavily utilized west coast waterway, the Strait of Juan de Fuca between the southern part of Vancouver Island and the state of Washington, an ocean boundary has existed since 1846.[11] There is no dispute over the location of the international boundary in these waters.[12] There has been some uncertainty as to the precise international legal status of the waters of the Strait of Juan de Fuca and the international legal capacity of the United States to impose certain regulations on foreign vessels using the Strait on their way to Canada.[13] This uncertainty has not been an acute problem between the two states.

The reality of the United States on Canada's West Coast

Unlike on the east or Arctic coasts of Canada, the United States has an imposing presence on the Pacific coast. Canadian waters being shouldered by Alaska to the north and the state of Washington to the south have both positive and negative effects. Positively, most of the ocean traffic in Canada's Pacific waters is either American, going between the lower states and Alaska, or, if vessels are not calling on Canada, they are headed for the United States. There is little vessel traffic in Canadian west coast waters that is not ultimately subject to either U.S. or Canadian regulation and on issues of marine pollution and vessel safety, U.S. standards are broadly comparable with Canadian standards. Negatively, the United States' sensitivity to freedom of navigation rights, particularly in waters separating the American states, makes it difficult for Canada to undertake unilateral initiatives.[14]

Moreover, the United States can and does impose vessel regulations that may be beyond what Canada deems reasonable or in Canada's best interest.[15]

It also must be noted that many of the commercial vessels and virtually all the visiting pleasure craft that frequent Canadian Pacific waters are American. Thus, foreign vessels that breach Canadian laws on the West Coast are almost always American vessels. While seizures and arrests often take place, there can be political and neighborly undercurrents to Canadian seizure and arrests of U.S. vessels that are different than when Canada seizes a non-U.S. vessel.

The sharing of the high traffic and population areas of the Georgia-Puget Basin and the Strait of Juan de Fuca has implications respecting fisheries enforcement, vessel safety, smuggling, and pollution prevention and response in these areas. There is an obvious need for inter-state cooperation (Canada and the United States), cooperation amongst regional governments (British Columbia and Washington), and cooperation amongst ocean users in these areas for a wide variety of ocean management issues and for efficient and effective maritime compliance and enforcement. Several joint initiatives/arrangements exist to meet these cooperative needs.[16]

The federal–provincial–Aboriginal constitutional and political context

Constitutionally-recognized Aboriginal rights over land, ocean space and particularly regarding fisheries, the full extent of which are still subject to negotiation and litigation, are a critical component of the Canadian west coast picture respecting legal and political authority over and responsibility for ocean use management.[17] Access to fisheries and the integration of Aboriginal claimants into the management structures of the fisheries has been an important part of most land claim settlements in British Columbia.[18]

Unlike Canada's other coasts, on the west coast there is a single political sub-unit, the province of British Columbia, which shares with the federal government of Canada, subject to the above-noted Aboriginal claims and rights, constitutional authority regarding ocean space and activities. Respecting federal–provincial constitutional jurisdiction over offshore areas,[19] Supreme Court of Canada cases have determined that the federal government has, vis-à-vis British Columbia, exclusive legislative authority regarding all waters and seabed areas west of Vancouver Island and the Queen Charlotte Islands.[20] The waters and ocean floor between Vancouver Island and the mainland have been held by the Supreme Court of Canada to be within provincial jurisdiction.[21] There is legal uncertainty regarding federal–provincial jurisdiction over the waters and sea floor between Vancouver Island and the Queen Charlotte Islands (Queen Charlotte Sound) and the waters and sea floor landward of the Queen Charlotte Islands (Hecate Strait).[22]

West coast ocean uses, activities and issues

Marine living resources

In British Columbia, commercial fishing involves the harvesting of more than 80 species of aquatic plants and animals from oceans and rivers. However, the primary focus is on wild salmon and shellfish. There are five major species of Pacific salmon: coho, chinook, sockeye, pink and chum. Since 1996, the Pacific salmon fleet has undergone a significant reduction as a result of the federal Department of Fisheries and Oceans (DFO) policy aimed at resource conservation. Through a process of buying back fishing licenses, DFO has reduced the Pacific salmon fleet by nearly 50 percent.[23]

Sport fishing occurs in both tidal and fresh waters throughout British Columbia. More than 600,000 annual recreational angler licenses are now issued in the Pacific region. Given the crossover with the tourism industry, the sport fishing industry includes operators in the transportation, accommodation, food and beverage services, boat and sporting goods retailing, marinas and other recreation industries.

Aquaculture, particularly salmon farming, has emerged as a sensitive issue on the West Coast. Until recently, there has been a moratorium on the expansion of the finfish aquaculture industry in British Columbia. Under the moratorium, pre-existing operations continued with the ban extending only to new operations. Aquaculture has raised serious environmental concerns, particularly regarding its impact on wild salmon stocks and marine habitat.[24] There have also been resource-use conflicts amongst industry, First Nations and other communities.[25]

First Nations and the Aboriginal Fishery Strategy

The *Aboriginal Fishery Strategy* (AFS),[26] in place since 1992, is a nationwide federal government initiative that seeks to provide a framework for the management of the aboriginal fishery. The objectives of the AFS include providing opportunities for aboriginal people to fish for food, social and ceremonial purposes and providing a role for the aboriginal community in the management of aboriginal fisheries through the development of arrangements such as pilot projects for the sale of fish, cooperative fisheries management, and the creation of economic development opportunities for First Nations. Also, the AFS seeks to facilitate the transfer of commercial licenses from non-native fishers to native communities.

Several B.C. First Nations have been working on agreements designed to build on and create new DFO/First Nations partnerships and co-management schemes within the policy framework of the AFS. One such group is the Confederated Aboriginal Resource Management Agencies (CARMA). This group is led by the Skeena Fisheries Commission, which is made up of a conglomeration of B.C. First Nations, including the Tsimpshien, Gitxsan, Haisla, Carrier-Sekani, Shuswap and Okanagan Nations. According to one member, CARMA's approach is geared toward "creeping co-management" encouraging a situation whereby more First Nations will be working with and eventually within DFO.[27]

The policies of the AFS have generated some criticism and protest action from some members of the B.C. commercial fishery.[28] According to one DFO fisheries officer working in Richmond, at various times there have been significant numbers of commercial vessels fishing illegally as a protest against commercial selling of fish harvested pursuant to the AFS.[29]

Commercial shipping

Vancouver is Canada's busiest port with an annual throughput of about 70 million metric tonnes per year, of which some five million are containerized.[30] It is a major gateway to North America's Pacific Northwest for wheat and grain exports and has a major coal terminal. Vancouver is also a major port of call for the West Coast–Alaska cruise industry. There is little intra-Canadian vessel traffic except for the B.C. ferry system from Vancouver Island to the mainland. Virtually all the Vancouver vessel traffic is international and uses Haro Strait and the jurisdictionally shared Strait of Juan de Fuca.

One consequence of the high level of vessel traffic within B.C. waters and Canada's west coast waters is vessel-source pollution. Public and environmental concerns over the high volumes of tanker traffic and transported oil were at their height in the 1970s when the tanker routing was first being discussed. Awareness was revived in 1988 with the *Nestucca* spill where 875,000 liters of Bunker C oil was released off the coast of Washington.[31] When this crude oil washed from U.S. waters to the Pacific Rim National Park, it became British Columbia's worst oil spill incident.

Recreational activities

The principal focus of recreational activity – pleasure boating, kayaking, recreational fishing – is within the Puget Sound–Strait of Juan de Fuca–Haro Strait area since this is where most of the Canadian west coast population is concentrated. This creates a potentially serious maritime safety issue as pleasure craft share the same ocean space with larger commercial vessels utilizing the ports of Vancouver and Seattle. The presence of numerous whales in Canada's west coast waters has created a significant new ocean industry directed primarily at tourists – whale watching. Although largely confined to near-shore waters, whale watching can take place on the open ocean. The industry has increased the vessel traffic in already congested waters and has become a concern for the protection of the whales themselves.[32]

Drug and people smuggling

In 1999, several vessels attempted to deposit Chinese migrants[33] on British Columbia's shores.[34] These vessels were in various states of disrepair, ranging from complete "rust buckets" to only somewhat less dangerous. Other Chinese migrants have arrived in Canada in container vessels, usually in soft-top containers. The migrants are reported to come from Fujian Province in China and are part of an international persons trafficking system controlled by crime syndicates.

The 1999 arrivals resulted in a significant public interest and outcry. However, Canada receives many migrants every year by air, sea and land. Such arrivals to B.C.'s shores are nothing new, they have been occurring for the last 25 years.[35] A number of factors contributed to the public furor around the migrant arrivals: the large numbers, the decrepit vessels "dumping" their human cargoes on Canadian shores, the horrible living conditions that clearly existed on the vessels, and the subsequent intense media attention presenting the image that Canada cannot control its own borders.

Drug smuggling is another maritime issue on British Columbia's coast, which is firmly rooted in the region due to, *inter alia*, the high level of traffic in the Port of Vancouver (particularly containerized cargo), the region's proximity to the United States and the prevalence of Asian crime syndicates in the Vancouver area who are actively involved in drug trafficking. The number of enforcement agencies involved in counter-drug activities underscores the prevalence of drug trafficking on the West Coast. The Royal Canadian Mounted Police (RCMP) is the lead department in terms of drug enforcement, however, the Port of Vancouver also has a waterfront drug unit made up of different agencies. Municipal police departments like the Vancouver Police Department – Marine Unit are also involved.

Marine protected areas

There is considerable interest in the development and designation of marine protected areas (MPAs) on the West Coast.[36] The federal government has created four pilot west coast MPAs: Race Rocks, at the eastern end of the Strait of Juan de Fuca; Gabriola Passage in the Gulf Islands, offshore from Nanaimo; the Bowie Seamount Area off the coast of the Queen Charlotte Islands; and the Endeavour Hot Vents Area, 250 km southwest of Vancouver Island and part of the Juan de Fuca Ridge system. These pilot MPAs may become part of a system of MPAs under the mandate of Parks Canada, Environment Canada, and Fisheries and Oceans Canada. It is also worth noting that British Columbia has moved to protect approximately 15 percent of the province's coastline, and certain limited marine areas have been included in these coastal protection plans.

Defence

Canada's naval presence on the West Coast is headquartered in Victoria at Canadian Forces Base Esquimalt. While the Canadian Navy has an important maritime defence role, the Canadian Navy's role has had an increasing focus on the protection of Canadian sovereignty in terms of "peacetime surveillance and control, countering illegal activities, fisheries protection, search and rescue, and environmental surveillance in Canada's territory, airspace and maritime areas of jurisdiction."[37] Thus, in recent years, "the navy's strategy, doctrine and current fleet operations (have been) realigning in support of . . . other government departments in achieving various national goals in such areas as fisheries protection, drug interdiction and environmental protection."[38]

Enforcement responsibilities, capacities and operations on the West Coast

Marine policy objectives

For the purposes herein it is necessary to identify what can be described as the principal marine policy objectives for the purposes of compliance and enforcement on the West Coast. The following four appear to be most prominent. First is the sustainable management of marine living resources. Second is cooperation and coordination of the many ocean users in the Georgia Basin, the principal geographic area of concentration, particularly as regards maritime traffic. A third objective is marine pollution prevention and protection of the marine environment from continued degradation, again specifically in the southern British Columbia geographic area of population concentration. The fourth marine policy objective is an offshore "presence" (naval, coast guard, fisheries patrols, police) for purposes of projecting and exercising control (sovereignty) respecting suspect foreign activities, enhancing compliance with laws and standards, and responding to maritime emergencies.

Management of marine living resources

Department of Fisheries and Oceans

It is the Conservation and Protection Division within the federal Department of Fisheries and Oceans (DFO) that has the responsibility for the enforcement of fisheries regulations and management plans. In the Pacific region, 175 fishery officers carry out this mandate.[39] This involves maintaining an enforcement presence at fishery openings; checking vessels to ensure compliance; issuing warnings, tickets and appearance notices to violators; collecting evidence with respect to charges; monitoring and assessing enforcement activities; training fishery officers; and coordinating with other enforcement agencies.

Prior to 1995, the Conservation and Protection Division used DFO patrol vessels to carry out the fisheries enforcement mandate. The Canadian Coast Guard (CCG) was under the authority of Transport Canada and had its own separate fleet and mandate. In the mid-1990s, CCG was merged into DFO[40] with the goal being the creation of one civilian "multi-tasked" federal fleet to maximize resources and reduce duplication. With the merger, two different approaches/philosophies to enforcement/compliance have had to meet and find a way to cooperate in the context of the shared fleet and a general atmosphere of scarce resources. The changes brought about by the merger and the scarcity of resources have made for a rocky transition and at times a strained relationship between the two DFO bodies.[41] Effective 12 December 2003, responsibility for the control and supervision of those portions of the Directorate-General of Marine Programs responsible for policy related to pleasure craft safety, marine navigation services, pollution prevention and response, and navigable waters were transferred from DFO to the Department of Transport.[42] The operational arm of the CCG will be reconstituted as a "separate operating agency" of DFO.

Coordination and cooperation

In general, coordination between the Conservation and Protection Division and other government departments is common. Examples include the Department of National Defence/DFO high seas driftnet enforcement operations off the coast of northern British Columbia, coordination with the RCMP in support of the drug enforcement units and, where necessary, coordination with local police forces at the regional and field levels.

The Conservation and Protection Division coordinates with a number of U.S. bodies in carrying out its enforcement mandate. For example, Conservation and Protection meets once a year with the U.S. Coast Guard, the National Marine Fisheries Services, and Washington and Alaska state enforcement bodies. This forum provides an opportunity to exchange planned patrol activities and helps to avoid duplication and waste of resources.[43] Also, Canada and the U.S. have a *Reciprocal Enforcement Agreement for Fisheries* that allows both countries to rely on each other's evidence regarding suspect vessels.[44] For example, if Canadian authorities in British Columbia have evidence of suspected violations by a U.S. vessel, but the vessel moves into U.S. waters before being apprehended, Canadian authorities can give their evidence to U.S. authorities that may use it to prosecute in the United States.

Management of vessel traffic

The Marine Safety directorate, a component of Transport Canada, has authority for the regulation of commercial shipping in Canadian waters. On the West Coast, Pacific Marine Safety is responsible for the delivery of programs to control foreign and domestic commercial shipping, to prevent ship-source pollution, to ensure safe carriage of regulated cargoes, to oversee certification and training of marine crew, and for vessel certification, licensing and registration.[45]

In terms of regulation of domestic commercial vessels, Marine Safety ship inspectors are empowered under the *Canada Shipping Act* to board any vessel and to detain it if it lacks proper certification.[46] If an alleged criminal offence is involved, then the RCMP takes over and may prosecute. However, it is worth noting that vessel detention can be a fairly significant penalty, given the costs that ensue when a vessel's schedule is delayed.

In terms of the regulation of foreign commercial vessels, port state control inspections occur in the form of spot checks. In 2002, 360 vessels were inspected in Vancouver under the port state control program.[47] Several vessels were detained for as little as a couple of days to as long as a few months. Detention creates enormous costs for vessels with schedules to maintain. Over the last couple of years, Marine Safety has seen improvements in the quality of ships coming into the Port of Vancouver. Apparently, the message seems to be out that if your vessel does not meet Canadian requirements, there will be costs.

While CCG is now under the authority of DFO, it continues to carry out its more traditional transport-related functions and, as such, has enforcement responsibilities respecting commercial vessel safety and regulation.[48] CCG Pacific Region, is responsible for ensuring marine safety throughout Canada's west coast waters. There are 11 major vessels, 20 smaller vessels and two air cushion vehicles in the Coast Guard Pacific fleet. Some of the vessels are multi-tasked, carrying out combinations of search and rescue, marine navigation services, conservation and protection, fisheries management and ocean science responsibilities. Of the major vessels, five are capable of taking on any role within CCG's mandate.

Search and rescue

The Joint Rescue Coordination Centre Victoria (JRCC Victoria), located at CFB Esquimalt, is one of three JRCCs in Canada. The RCC is responsible for coordinating responses to aeronautical and maritime search and rescue (SAR) incidents within Victoria's Search and Rescue Region. The Victoria region consists of approximately 920,000 km^2 of mainly mountainous terrain of Yukon and British Columbia and 560,000 km^2 of the Pacific Ocean, extending to approximately 600 nautical miles offshore including over 27,000 km of British Columbia coastline.[49] CCG Pacific Region responds to over 2,000 emergency incidents annually.[50]

International coordination

The Marine Safety Directorate of Transport Canada has a high degree of interaction and cooperation with U.S. authorities, mainly because of the jurisdictionally shared Juan de Fuca Strait, within which a shared traffic management system operates. For example, there are three Canadian and American radio stations that are in constant communication regarding vessel traffic. Cooperative Vessel Traffic Services (CVTS) is a joint U.S.–Canada body that meets twice a year. CVTS has a framework/strategy for handling vessel problems in the shared zone that provides a decision-making process that allows Canadian or American authorities to contact the proper Canadian and/or American authority in the event of a problem in either Canadian or American waters.

Marine pollution prevention

For fisheries and vessel traffic management, including vessel-source marine pollution, it is relatively easy to point to a lead federal agency with the primary responsibility for dealing with enforcement issues. Such is not the case with marine pollution prevention that does not involve vessels. For land-based sources of marine pollution the enforcement agency is most often under the jurisdiction of British Columbia or a municipality within British Columbia. What is generally recognized as the most effective federal authority regarding marine pollution has been DFO and its responsibility to enforce the provisions of the *Fisheries Act* that deal with marine pollution and waters frequented by fish.[51]

Drugs and immigration

The RCMP has primary authority respecting the enforcement of Canada's laws on illegal drugs and immigration. The RCMP's west coast resources are limited to four high-speed three-person catamarans and some air support. The RCMP and the Department of National Defence (DND) have a working memorandum of understanding (MOU) regarding joint counter-narcotic operations that can result in DND providing operational equipment and personnel and access to defence capabilities and expertise.[52]

The Canada Customs and Revenue Agency (CCRA) plays a supporting role to the RCMP regarding drug smuggling at west coast ports. Related to this is the innovative arrangement between CCRA and the Vancouver Port Authority through which the two parties seek to work together to suppress contraband smuggling through the Port of Vancouver.[53] This successful cooperative relation led to the acquisition by the Vancouver Port Authority of gamma ray scanning equipment. This technology, designed to complement manual inspections, allows the Port inspectors to scan containers safely, non-intrusively and quickly (at the rate of one per minute) for hidden compartments associated with the transport of contraband. Local CCRA inspection officers will be able to view these radiographic images as cargo is loaded and unloaded from vessels with minimum interruption to the rapid flow of cargo.[54] Citizenship and Immigration Canada plays a similar supporting role to the RCMP regarding immigration issues.

Coastal Watch, an eight-year-old RCMP community-based program that relies on the public to call in regarding suspicious activities on the water, also has a role. The program is based on the premise that individuals living and working on the coast or at sea possess intimate knowledge of activities of their surroundings and are thus the best placed to notice abnormal activities. The Coastal Watch program enhances participants' ability to identify abnormal activities through public education that presents profiles, possible indicators of criminal actions and contact information for local authorities.[55]

Surveillance and control of the offshore area: the Canadian Navy

In addition to the capabilities and responsibilities of the Coast Guard on the West Coast regarding surveillance, control and enforcement and the more limited capabilities of the RCMP, there is the role and responsibilities of the Canadian Navy, specifically Maritime Forces Pacific (MARPAC). The Canadian Navy has numerous maritime assets, which are generally equally distributed between MARLANT (Atlantic) and MARPAC, and are tasked approximately 30 ship-days on a yearly basis to support RCMP counter-narcotics. On the West Coast, the naval fleet is generally composed of five patrol frigates, one destroyer, one operational support vessel, six maritime coastal defence vessels (MCDV), one long-range patrol submarine and several smaller auxiliary vessels. The Navy also has access to marine air support resources including five Aurora patrol planes, six Buffalo planes, six Sea King helicopters and four Labrador helicopters.

It is important to note that the Navy's role in enforcement is distinct from that of civilian agencies. Unlike, for example, CCG or CCRA, who possess a mandate to enforce the laws of Canada, Department of National Defence operations are limited by the principle of *posse comitatus*. As such, DND operations can only be aimed at Canadian citizens in order to enforce the laws of Canada under certain circumstances such as a civil emergency, or under the *Fisheries Act* (where Canadian Forces personnel have fishery officer status).[56] Notwithstanding these exceptions, MARPAC seldom utilizes such powers and generally limits its involvement in counter-narcotic operations to one of operational support as exemplified by the DND-Solicitor General MOU.[57]

The important role of MARPAC is to provide support to other government departments involved in enforcement. MARPAC has increasingly been the focal point for interdepartmental communication/operations related to the broad concept of sovereignty protection. Some of the major areas of concern include illegal migrant and drug smuggling, environmental pollution surveillance, search and rescue, and emergency response. Interdepartmental communication between the Navy and other governmental departments (primarily the RCMP, DFO and CCG) occurs on an ongoing basis through informal and formal mechanisms. In addition, MARPAC has representatives on numerous formal interdepartmental committees, such as the Interdepartmental Maritime Security Working Group (IMSWG). IMSWG, established after 11 September 2001, is the coordinating body for the federal response to marine security matters. IMSWG is chaired by Transport Canada and 17 federal departments and agencies are members, including all federal departments and agencies with fleet assets. IMSWG replaced the Interdepartmental Program Coordination and Review Committee (IPCRC),[58] which had been in existence in the 1990s to improve interdepartmental coordination to maximize unused available ship capacity of Canada's two main federal fleets – the navy, and since 1995, the consolidated Coast Guard fleet – but had fallen into disuse.

The Canadian Maritime Network (CANMARNET) is an unclassified, near real-time geographic, information system that receives surveillance and information inputs from a number of ocean-related government departments, agencies and the private sector. It comprises databases that are networked between departments and the key product is a composite and shared visual database: the Recognized Maritime Picture (RMP). CANMARNET is primarily controlled and utilized on both the east and west coasts by Canada's Maritime Forces, with the participation and support from Fisheries and Oceans/Coast Guard, Transport Canada, Environment Canada, Customs and Excise, Citizenship and Immigration, and the RCMP. It is utilized for information collection, management and dissemination among enforcement agencies, increasing the efficiency of Canada's maritime enforcement regime and improving oceans management.[59]

Conclusion and challenges

This snapshot of Canada's Pacific coast ocean activities indicates the diversity of ocean uses being undertaken in and on west coast waters. The geography of ocean activities on the West Coast, largely centered in the Georgia Basin, is more confined than on the East Coast. It is also worth noting that the number of coastal communities in British Columbia is minute compared with the number on the Atlantic coast. The ocean activities themselves are somewhat different. For example, there is no offshore hydrocarbon development as of yet, the dominant fisheries (salmon and shellfish) are conducted near-shore, and there is little non-Canadian/American vessel traffic or fishing activity in or near Canadian west coast waters.

What has been provided herein is a snapshot of the enforcement responsibilities, capacities and operations of the principal federal agencies that deal with the various ocean activities on the West Coast. Government enforcement capacity and operations is only one component of an integrated maritime compliance and enforcement structure, yet it is an important one since enforcement capacity provides the monitoring, surveillance and sanctioning that can encourage and reinforce voluntary compliance. The "enforcement" snapshot indicates that for different marine policy objectives there are different lead agencies that rely on inter-agency cooperation, on federal–provincial cooperation and, frequently, on bi-national (Canada–United States) cooperation. Reconfiguring enforcement resources to enhance use efficiency and to meet compliance objectives is an ongoing quest.

Canada's constitutional structure, the necessarily differing mandates of government departments and the equally valid but numerous marine policy objective priorities work to inhibit the operationalization of a heroic or macro-model approach to integrated maritime compliance and enforcement. However, for specific marine policy objectives, for example, sustainable management of fisheries, or on a micro-level, it may be possible to overcome some of the inhibitions that exist on the broader scale. To some extent this has been attained or is being striven for on Canada's Pacific coast.

The challenges that exist respecting integrated maritime compliance and enforcement in Canada are not Pacific coast specific but are common on all three Canadian coasts. The challenges can be grouped into three interlinked categories: terrorism, resources for implementing enforcement strategies and balance of responsibilities.

Concerns about terrorism create challenges for integrated maritime compliance and enforcement as well as the possibility of enhanced capabilities. It can be expected that the attention of enforcement strategies will shift respecting immigration (movement of terrorists), customs (movement of hardware that terrorists may use) and protection of offshore "assets" that may be targets of terrorism. One of the challenges will be to monitor whether this shift in attention is at the expense of enforcement strategies designed to enhance marine resource sustainability and responsible use of the oceans. It cannot be assumed that attention on terrorism necessarily means less attention on other compliance and enforcement priorities since improving capacity to implement enforcement strategies directed at terrorism may have the collateral benefit of augmenting enforcement capacities for more traditional marine purposes.

Irrespective of the terrorism challenge, integrated maritime compliance and enforcement still faces the challenge of determining the appropriate amount and kind of resources that should be dedicated to enforcement strategies. At-sea enforcement operations are very expensive and the information requirements of surveillance, monitoring and control techniques are costly. The challenge will be to make better use of existing technologies and to develop lower cost technologies. One of the unspoken challenges to maritime enforcement strategies is that of enforcement and compliance costs versus the benefits. This arises most prominently in the fisheries context: is the cost of enforcement strategies commensurate with the benefits to Canada and its citizens?

Another future issue in the deployment of at-sea enforcement resources is the existing divide between the role of military vessels and non-military vessels. Traditionally, gray-hulled vessels defend the state from "military" risks, while non-gray-hulled vessels take on the role of policing and enforcing national laws.[60] In light of terrorism concerns and the costs–benefits of enforcement strategies, is this a divide that continues to make sense? While there is a comfort level with a "yes" answer, this will be a future challenge.

The balance of responsibilities and challenges for integrated maritime compliance and enforcement highlights the relationship between enforcement strategies and "softer" approaches to attaining compliance. Having communities and user-groups "buy in" to policy and legislation through involvement in the process of development of policy and legislation is often, though not always, an effective way to enhance compliance while keeping more costly enforcement strategies to a minimum. However, there can be a public perception of coziness and favoritism amongst the governing and governed in such arrangements. The challenge is to make these "softer" strategies work better and to be perceived by the public as being effective.

The role of non-governmental organizations (NGOs) in enhancing compliance and enforcement is often touted as a low-cost, "softer" component of integrated maritime compliance and enforcement. However, NGOs frequently have their own agendas, suffer from non-transparency, have selective and shifting memberships, and have little responsibility or accountability for their actions or statements, which can make them uncertain and unreliable players in compliance and enforcement strategies. The challenge is to make room for those NGOs that can add value to compliance and enforcement strategies, while being aware of their limitations.

The final challenge regarding integrated maritime compliance and enforcement is an over-arching one embedded in the word "integrated": finding new and better ways to enhance cooperation and coordination of all the "players" (government agencies, user-groups, communities, the public, NGOs) to keep enforcement strategy costs reasonable while inducing higher degrees of compliance.

Acknowledgments

This contribution draws from François Bailet, Fred Crickard and Glen Herbert *A Handbook on Integrated Maritime Enforcement* (Dalhousie University, Halifax: 2000) (unpublished); Janna Cumming "The Pacific Coast Component of the Integrated Maritime Enforcement Project" (November 2000) (unpublished); and Ted L. McDorman "The Pacific Component in a Canadian Integrated Management Strategy" in Glen J. Herbert and Fred W. Crickard (eds) *Canada's Three Oceans: Strategies for Maritime Enforcement* (Workshop Proceedings) (Centre for Foreign Policy Studies, Dalhousie University, Halifax: 1998) 115–34.

Notes

1 Fred W. Crickard and Peter T. Haydon *Why Canada Needs Maritime Forces* (Naval Officers' Association of Canada, Nepean, Ontario: 1994).

2 Surveillance, monitoring and control are the terms most often associated with maritime enforcement. The following is offered as definitions for the three terms:

> *Surveillance* involves detection of activities or events of interest within the offshore area, and is the basic building block of both the exercise of sovereignty and enforcement.
>
> *Monitoring* involves the systematic observation of specific conditions, activities or events of interest within a coastal state's maritime jurisdiction. It entails locating, identifying, tracking and inspecting objects, events or activities. These may include illegal vessel or aircraft activity, meteorological phenomena (such as hurricanes and icebergs), inspection of fish catches (at sea or alongside), and accidents, such as oil spills or vessels in distress.
>
> *Control* includes the execution and the rendering effective of international and national rules and regulations. In this sense, it may entail sanctioning and implementation. In the first case, action is designed to ensure compliance with government regulation and has as its object a potential or real offender. In the second case, action is designed to ensure availability of a type of control and has as its object a particular service or function. For example, the enforcement of fishing quotas through arrest and fines would be sanction, while the enforcement of safe navigation through the provision of current charts, tide tables, marks and buoys would be implementation.

3 See more generally Francois Bailet "Achieving Compliance in the Maritime Sectors" paper presented to the Maritime Institute of Malaysia (MIMA), Kuala Lumpur, October 2000 (on file).

4 1982 United Nations Convention on the Law of the Sea (LOS Convention) 1833 UNTS 396; Arts. 61 and 62; see generally William T. Burke *The New International Law of Fisheries* (Clarendon Press, Oxford: 1994) 43–65.

5 LOS Convention, ibid., Part XII, "Protection and Preservation of the Marine Environment" in particular Arts. 192–5; see generally Patricia W. Birnie and Alan E. Boyle *International Law and the Environment* (Clarendon Press, Oxford: 1992) 254–7.

6 See LOS Convention, note 4, Arts. 60(3), 24(2), 43. Nevertheless, this responsibility on coastal states regarding navigational aids is controversial; see generally Philippe Boisson *Safety at Sea: Policies, Regulations and International Law* (translated by D. Mahaffey) (Bureau Veritas, Paris: 1999) 324–6.

7 See note 4.

8 *Oceans Act*, S.C. 1996, c. 31 (Canada) Section 13: Zone 3, Fishing Zones of Canada (Zones 1, 2 and 3) Order, C.R.C., c. 1547; Schedule III and IV, *Fishing Zones of Canada (Zones 4 and 5) Order*, C.R.C., c. 1548; and *Territorial Sea Geographical Coordinates Order*, C.R.C., c. 1550.

9 See discussion in Ted L. McDorman "Canada's Ocean Limits and Boundaries: An Overview" in Lorne K. Kriwoken, Marcus Haward, David VanderZwaag and Bruce Davis (eds) *Oceans Law and Policy in the Post-UNCED Era: Australian and Canadian Perspectives* (Kluwer Law International Law, London: 1996) 113 at 124–5.

10 *Oceans Act*, ss. 4 and 5, and C.R.C., c. 1550, *Territorial Sea Geographical Coordinates Order* 1996 (Canada).

11 1946 Treaty Establishing the Boundary in the Territory on the Northwest Coast of America Lying Westward of the Rocky Mountains (The Oregon Treaty), 100 Consolidated TS 39, art. 1.

12 Canada and the United States have a vessel traffic management agreement over these waters; see 1979 Exchange of Notes Between the Government of Canada and the Government of the United States of America Constituting an Agreement on Vessel Traffic Management for the Juan De Fuca Region [1979] CTS No. 28.
13 See Craig Allen "The Pacific Northwest boundary straits: High seas or internal waters" (February 1995) 13(2) *Pacific Maritime Magazine* 20–4.
14 For example, see the fishing vessel licensing issue that arose in 1994 noted in Ted L. McDorman "The West Coast salmon dispute: A Canadian view of the breakdown of the 1985 Treaty and the transit license measure" (1995) 17 *Loyola of Los Angeles International and Comparative Law Journal* 477 at 477–9 and 497–503.
15 See Allen, note 13.
16 The most recent high-profile cooperative arrangement between Canada and the United States is the 1999 Pacific Salmon Agreement. Online at: www-comm.pac.dfo-mpo.gc.ca/pages/release/p-releas/HQ/hq9929_e.htm and www.oceanlaw.net/texts/psc.htm (accessed 1 January 2004).
17 See discussion by Jones in Chapter 12 in this volume.
18 See *Aboriginal Fisheries Strategy – Pacific Coast* and discussion further below. Online at: www.pac.dfo-mpo.gc.ca/ops/fm/AFS/Treaties.htm (accessed 11 June 2002).
19 For complete detail on the federal–provincial powers over the offshore, see Chircop and Hildebrand in Chapter 2 in this volume.
20 *Reference Re Ownership of Off-Shore Mineral Rights* (1967) 65 D.L.R. (2d) 353.
21 *Re Ownership of the Bed of the Strait of Georgia* [1984] 1 S.C.R. 388.
22 See generally Peter Finkle and Alastair Lucas "The concept of the British Columbia inland marine zone" (1990) 24 *University of British Columbia Law Review* 37–52.
23 For more about the 1996 *Pacific Salmon Revitalization Plan* and the 1998–99 *Canadian Fisheries Adjustment and Restructuring Program*, see "Fleet reduction objectives achieved with the third round of voluntary commercial salmon license retirement" DFO *News Release* NR-PR-00-10E (26 January 2000). Online at: www-comm.pac.dfo-mpo.gc.ca/pages/release/p-releas/2000/nr0010_e.htm (accessed 1 December 2003).
24 "Report on Aquaculture and the Protection of Wild Salmon" in *Speaking for the Salmon: Aquaculture and the Protection of Wild Salmon*: *Aquaculture and the Protecting of Wild Salmon* (Conference Documents, Continuing Studies, Simon Fraser University, 2–3 March 2000).
25 See the discussion by Jones in Chapter 12 in this volume.
26 For more information on the AFS, see www.dfo-mpo.gc.ca/COMMUNIC/FISH_MAN/AFS_e.htm (Fisheries and Oceans Canada Backgrounder) (accessed 1 December 2003).
27 Mark Bowler, member of the Haisla Fisheries Commission (Personal Communication, 20 July 2000).
28 "Native fishery upheld by Appeal Court" *Toronto Globe and Mail* 7 July 2000 at A6.
29 Herb Redekopp, DFO Fishery Officer, Richmond, B.C. (Personal Communication, June 2000).
30 Port Vancouver. Online at: www.portvancouver.com> (accessed 1 December 2003).
31 David VanderZwaag *Canada and Marine Environmental Protection* (Kluwer Law, The Hague: 1995) 87.
32 Kim Lunman "Ottawa aims to save whales from those who love them" *Toronto Globe and Mail* 24 June 2000 at A1, A4.

33 The term "migrants" was used most often in the press to describe the visitors. Several of the "migrants" were granted refugee status within Canada. Hence, the term is not accurate.
34 John Geddes "Children at sea in a new land" *Maclean's* 30 August 1999 at 22; Tom Fennell "Terror and hardship" *Maclean's* 22 November 1999 at 24–8.
35 Colonel (Air Force) T.B. Rogers (Personal Communication, 28 June 2000).
36 *Oceans Act*, note 8, s. 35 allows for the creation of marine protected areas (MPAs).
37 Fred W. Crickard "Oceans Policy and Naval Policy 1970 to 1997: Divergent Courses or Making the Rendezvous? The Canadian Experience" No. 7/8 *Maritime Security Working Papers* (Centre for Foreign Policy Studies, Dalhousie University, Halifax: 1997) at 19.
38 Ibid., at 20.
39 Fisheries and Oceans Canada, Pacific Region. Online at: www-hr.pac.dfo-mpo.gc.ca/pages/fo-recruit_e.htm (accessed 1 December 2003).
40 *Oceans Act*, note 8, s. 41.
41 This was commented upon by Steve Cochrane, Coast Guard Officer (Personal Communication, 20 June 2000); Robert Martinolich, Assistant-Director, Conservation and Protection, DFO, Vancouver (Personal Communication, 23 June 2000); Larry Paike, Detachment Supervisor, Conservation and Protection, DFO, Victoria (Personal Communication, 20 June 2000); and Herb Redekopp, DFO Fishery Officer, Richmond, B.C. (Personal Communication, June 2000). The Osbaldeston Report had predicted there would be problems to work out and a period of transition during a merger of this scale. See Government of Canada (Gordon F. Osbaldeston, chair, Study Team on Canadian Government Fleet Utilization), *All the Ships That Sail: A Study of Canada's Fleets* (Report of the Study on the Utilization of the Federal Government's Marine Fleets) (Treasury Board, Ottawa: 15 October 1990), known as the "Osbaldeston Report."
42 Privy Council Office, Government of Canada, *Public Service Rearrangement and Transfer of Duties Act*, PC Number 2003-2090, 12 December 2003.
43 Robert Martinolich, Assistant-Director, Conservation and Protection, DFO, Vancouver (Personal Communication, 23 June 2000).
44 1990 Agreement between the Government of Canada and the Government of the United States of America on Fisheries Enforcement [1991] CTS No. 36.
45 Marine Safety, Transport Canada. Online at: www.tc.gc.ca/pacific/marine/marine.safety/menu.htm (accessed 1 December 2003).
46 *Canada Shipping Act 1985* (Canada) R.S.C. 1985, c. S-9, ss. 301 and 310 as amended.
47 Transport Canada, *Marine Safety Port State Control Annual Report 2002* (Marine Safety Directorate, Transport Canada, Ottawa: 2003). Online at: www.tc.gc.ca/MarineSafety/TP/TP13595/2002/menu.htm (accessed 1 December 2003).
48 Captain Brian Kennifick, Marine Safety, Transport Canada (Personal Communication, 23 June 2000).
49 Joint Rescue Coordination Centre, Victoria. Online at: www.pacific.ccg-gcc.gc.ca/sar/jrcc/index_e.htm (accessed 1 December 2003).
50 CCG-Pacific Region Overview. Online at: www.pacific.ccg-gcc.gc.ca/summary-sommaire/index_e.htm (accessed 1 December 2003).
51 *Fisheries Act* 1985 (Canada) R.S.C., 1985, c. F-14, ss. 35–38 and 40–42 as amended.
52 1994 Memorandum of Understanding between the Minister of National Defence and the Solicitor General of Canada on the Provision of Assistance by the Canadian Armed Forces in Support of the RCMP in its Drug Law

Enforcement Role. For a comprehensive study of this MOU, see François Bailet "Le protocole d'entente entre le Solliciteur Général (G.R.C.) et le Département de la Défense" in *L'Approche Juridique du Système International D'Interdiction du Narco-Trafic Maritime dans la Région Caraïbe/Atlantique Nord-Ouest* (University of Nice, Nice: 2001) 371–76 (Ph.D. dissertation).

53 1998 Memorandum of Understanding on the Suppression of Contraband Smuggling between the Government of Canada represented by the Minister of National Revenue and the Vancouver Port Authority as represented by the President Chief Executive Officer on Behalf of the Vancouver Port Authority.

54 Port Vancouver, Press Release, 25 January 2002: www.portvancouver.com/media/news_2002_1_25.html> (accessed 1 January 2004).

55 Gary Frail, RCMP Constable (Personal Communication, 23 July 2001). For more information on the Coastal Watch Program, see www.rcmp.ca/html/cwatch_e.htm (accessed 1 December 2003).

56 *Fisheries Act 1985* (Canada) R.S.C. 1985, c. F-14, S.C. 1991, c.1, s. 5(1). Note that this section allows the minister to designate any person as a fisheries officer and may limit the powers that person may exercise. There is no provision that explicitly provides the Canadian Forces personnel with fisheries officer status.

57 Greg Stead, LCDR., RCN, TRINITY (Personal Communication, 6 and 9 July 2001).

58 IPCRC was established based on a recommendation from the 1990 Osbaldeston Report, note 41.

59 See generally Standing Senate Committee on National Security and Defence *Canada's Coastlines: The Longest Under-Defended Borders in the World* Volume 1 (Senate of Canada, Ottawa: October 2003) at 110–13.

60 Vice-Admiral Gary L. Garnett, Commander Maritime Command, explained the "tradition" of the distinction between "military and civilian enforcement roles" as necessary so as to avoid "confusing . . . Canadians and . . . our military allies;" see Gary L. Garnett "The Navy's role in the protection of national sovereignty" in Peter T. Haydon and Gregory L. Witol (eds) *An Oceans Management Strategy for the Northwest Atlantic in the 21st Century* NIOBE Papers No. 9 (Naval Officers Association of Canada, Halifax: 1998) at 11. More generally and for a slightly different perspective, see Ken Booth *Law, Force and Diplomacy at Sea* (George, Allen and Unwin, Boston: 1985) 192–205.

5 Integrated maritime enforcement and compliance in Australia

Sam Bateman, Anthony Bergin, Martin Tsamenyi and Derek Woolner

Introduction

Australia's current arrangements for the enforcement of national legislation and the protection of national interests at sea are far from integrated. They owe much to their historical antecedents and the recommendations of successive reviews of the maritime surveillance and enforcement function.[1] Changes made to arrangements over the years have generally been ad hoc and in response to a specific crisis. Divisions of responsibility are spread widely between agencies of both the Commonwealth and the states. To the extent that maritime enforcement and compliance is a system, it is one of "distributed responsibility," often characterized by less than optimum coordination and cooperation between the agencies involved. The implementation of *Australia's Oceans Policy* through the process of regional marine planning (RMP) will add, however, new dimensions to maritime enforcement and compliance in Australia.

It has frequently been noted that implementing oceans policy in Australia requires better coordination between the federal, state and territory governments in integrating planning and management to ensure that jurisdictional boundaries do not hinder effective management.[2] This applies as much to maritime enforcement and compliance as it does to any other sectoral area of activity.[3] This chapter discusses the structure of the current regime for maritime enforcement and compliance in Australia and the prospects of achieving a higher degree of integration and coordination than exists at present.

Geographical and jurisdictional dispersal

Challenges of geography

Australia's geography and the significant distances involved in reaching both offshore territories and remote areas of the mainland pose great challenges for maritime enforcement and compliance. An unavoidable consequence is that Australia's enforcement of legislation in maritime zones under national jurisdiction cannot be a simple extension of terrestrial jurisdiction.

The coastline of mainland Australia is nearly 36,000 km in length.[4] The Australian Fishing Zone (AFZ) and exclusive economic zone (EEZ) around Australia and the offshore territories measure 8.15 million square kilometers,[5] nearly 20 percent larger than the Australian mainland. If the waters contiguous to the Australian Antarctic Territory (AAT) are added, the EEZ is approximately twice as large as the continental land mass. The distances to some of Australia's island territories are considerable. The Heard and McDonald Islands are over 2,400 nautical miles (nm) southwest of the mainland. Christmas Island is 1,500 nm west of Darwin but less than 200 nm south of Indonesia, as are the Ashmore and Cartier Reefs, 500 nm west of Darwin.

Most attention regarding maritime surveillance and enforcement in Australia has been given to the northern approaches to the mainland and the offshore island territories. Due to the geographical proximity of northern Australian waters to the archipelagos to the north, these areas are where the levels of threat are deemed to be higher. By far the largest amount of surveillance and patrol effort is expended in northern areas, with only intermittent fisheries protection patrols and occasional surveillance flights in southern waters.[6] The one exception is the area around the sub-Antarctic islands (primarily Heard and MacDonald Islands) where considerable illegal fishing has been evident in recent years.

Jurisdictional responsibilities

The *Seas and Submerged Lands Act 1973* (Cth) and the 1979 Offshore Constitutional Settlement (OCS) sought to reflect developments in international law and re-organize the structure of Australian offshore legislation. The *Coastal Waters (State Powers) Act 1980* (Cth) is the key legislation that establishes the legislative framework for maritime enforcement and compliance. This Act vests within each state/territory the power to make laws governing waters out to three nautical miles. Under the broad framework of the OCS, some state/territory agencies may have authority for particular areas of activity beyond the limit of their waters, such as port facilities and works, and management of specified fisheries.[7]

In reality, however, most maritime enforcement falls to agencies of the Commonwealth, sometimes in conjunction with their state/territory counterparts. Regardless of legislative authority, it is most often only the Commonwealth that has resources of sufficient distribution and capacity to determine what is happening in Australia's maritime zones and to do something about it. Thus an important characteristic of Australian maritime enforcement and compliance is the requirement for cooperation between agencies. There are numerous practical examples of cooperation, such as Western Australian fisheries officers accompanying Royal Australian Naval

(RAN) vessels on missions to the Heard and McDonald Islands, while Australian customs officers are authorized to act as fisheries officers under Western Australian legislation.

Roles of relevant Commonwealth agencies

There are at least 12 Commonwealth agencies with an ongoing role in maritime enforcement and compliance, whilst others may request assistance on an ad hoc basis. State or territory agencies can request assistance through their Commonwealth counterparts. Coastwatch is a branch of the Australian Customs Service (ACS) with a RAN two-star officer seconded as Director-General Coastwatch. It coordinates the aerial surveillance program and the surface response operations when required by "client" agencies, manages the aerial surveillance contractors and develops intelligence systems for maritime surveillance and enforcement. ACS is responsible for controlling the importation of illicit drugs and illegal goods and border protection generally. It administers the National Illicit Drugs Enforcement Strategy (NIDS) and is the parent organization for Coastwatch. It also controls the National Maritime Unit (NMU) of Australian Customs Vessels (ACVs).

The Australian Federal Police (AFP) is responsible for Commonwealth law enforcement, often in conjunction with state police forces. AFP may be involved in the prosecution of offences against Commonwealth law in virtually all areas of maritime jurisdiction such as fisheries, navigation, marine environmental protection and illegal importation. The Australian Fisheries Management Authority (AFMA) manages Australian and licensed foreign fishing within the AFZ. By agreement, it manages some fisheries that include state waters. It takes enforcement action against illegal fishing both by foreigners and nationals, but has limited operational and investigative resources of its own, relying instead on other agencies for enforcement and compliance and on the AFP to conduct criminal investigations on its behalf. The Department of Environment and Heritage is responsible itself and with associated portfolio agencies such as the Great Barrier Reef Marine Park Authority for preserving ecosystems in Australian waters, including the establishment of marine parks and marine protected areas. The Department monitors pollution, sea dumping, marine poaching and vandalism, and counters illegal taking of flora and fauna.

The Office of Transport Security (OTS) is a new key player in maritime enforcement and compliance in Australia. Established in the Department of Transport and Regional Services (DOTARS) in Canberra in 2004, it is the principal security regulator for maritime industry, including the implementation of the International Ship and Port Facility Security (ISPS) Code by Australian ports and shipping. OTS includes a Transport Security Operations Centre operating 24 hours a day and seven days a week, and staff outposted in major ports and overseas.

The responsibilities of the Australian Quarantine and Inspection Service (AQIS) extend to the prevention of the spread of exotic diseases through importation of infected insect, animal or vegetable material. These include national arrangements for the management of ballast water and introduced marine pests. The Department of Immigration, Multicultural and Indigenous Affairs (DIMIA) manages entry programs and the entry of individuals into Australia. It takes enforcement action against alleged illegal immigrants, including their removal from unauthorized transport to appropriate accommodation.

The Australian Maritime Safety Authority (AMSA) is responsible for shipping safety and the prevention of ship-sourced pollution in Australian waters. This includes implementation of port state control (PSC) measures in Australian ports including a network of regional offices around the country. AMSA provides maritime safety services in Australia and Australia's allocated area of search and rescue (SAR) responsibility (approximately 10 percent of the Earth's surface). This includes SAR operations for vessels in distress and for aircraft at sea through Australian Search and Rescue (AusSAR), which is part of AMSA. AMSA is also responsible for the management and development of the national plan for response to pollution by marine oil spills and other chemicals and for the maintenance and operation of the Commonwealth's pollution response equipment.

The Department of Defence has no legislative enforcement responsibilities (other than in respect of "naval waters") but in certain circumstances, the Australian Defence Force (ADF) may be used for the enforcement of civil law and Defence officers may be granted powers under particular pieces of Commonwealth legislation.[8] Defence is the major supplier of Commonwealth resources for maritime enforcement and compliance. In normal circumstances, Defence bears 70–80 percent of the total cost of the current system. The Department of Foreign Affairs and Trade (DFAT) is responsible for treaties with other countries, including maritime boundary agreements. This includes the Torres Strait Treaty with Papua New Guinea.[9] DFAT has a Torres Strait Treaty Liaison Officer based in Thursday Island who manages the treaty arrangements on a day-to-day basis.

Pieces of Commonwealth legislation that give offshore enforcement responsibilities to ADF or other Commonwealth agencies include the *Migration Act 1958* (Cth), the *Fisheries Management Act 1991* (Cth), and the *Customs Act 1901* (Cth), which also contains powers available to ADF personnel to deal with offences under the *Quarantine Act 1908* (Cth).

In addition to the Commonwealth agencies mentioned above, several others potentially have some involvement in the process of maritime enforcement and compliance.[10] These include the Marine Investigation Unit of the Australian Transport Safety Bureau (ATSB), which is responsible for the investigation of marine accidents, and the Australian Security and Intelligence Organization (ASIO), which has a major role in ensuring Australia's security against all forms of terrorism. The Commonwealth's Director of Public Prosecution (DPP) plays a major role in prosecutions for breaches of maritime legislation.[11]

Managing the Commonwealth's interests

Operational roles

Although a large number of agencies have responsibility for some element of control in Australia's maritime zones, a much smaller group provides most of the means for this to happen. Coastwatch coordinates requests from "client" agencies for surveillance in support of their legislative responsibilities, manages the contracts of commercial operators that provide aerial surveillance and develops the intelligence base necessary to maximize use of its comparatively limited resources across the extent of Australia's maritime zones. ACS also runs the NMU with eight sea-going patrol vessels, the *Bay* class, which can be deployed in response to possible breaches of law. However, RAN still supplies most of the response "muscle," primarily through its fleet of 15 patrol boats.[12] The Royal Australian Air Force (RAAF) provides approximately 250 hours of aerial patrol with P3C aircraft, whilst the Army's Regional Surveillance Units have some role in providing intelligence, particularly 53 Far North Queensland Regiment, which conducts open boat patrols in the Torres Strait area. Some agencies have a limited capacity to respond to certain aspects of their responsibilities. AFMA chartered a vessel for patrols of fishing areas in the Southern Ocean from 2000 until this responsibility was transferred to NMU of ACS in 2003. AusSAR has a network of commercial operators of light aircraft, suitably equipped for the role, with which it is able to respond to search and rescue incidents.

In December 2004, the Prime Minister announced that the Commonwealth government would assume direct responsibility for counter-terrorism prevention, interdiction and response in all offshore areas of Australia.[13] ADF will take responsibility for offshore counter-terrorism prevention, interdiction and response capabilities, including the protection of offshore oil and gas facilities. Coastwatch will retain responsibility for civil maritime surveillance. Central to these new arrangements is a new Joint Offshore Protection Command to be established by March 2005 and headed by the Director-General of Coastwatch. It will have a joint accountability structure being responsible to the Chief of the Defence Force (CDF) for its military functions and to the Chief Executive Officer of the ACS for its civil functions. The new command will manage the Maritime Identification Zone extending up to 1,000 nautical miles from Australia's coastline. On entering this Zone, vessels proposing to enter Australian ports will be required to provide comprehensive information such as ship identity, crew, cargo, location, course, speed and intended port of arrival.

All states and territories maintain water police elements although the capabilities vary from state to state. The water police deal with SAR incidents and criminal activities in state waters. Their work may extend beyond the three nautical mile limit, particularly for SAR incidents. In some circumstances, state or territory police officers may act as authorized persons under relevant Commonwealth legislation. Several memorandums of understanding (MOUs) have been signed between AFP, ACS and state/territory police forces directly related to maritime law enforcement. However, these MOUs are limited to the achievement of specific objectives, such as the interdiction of narcotics and "boat people," and exclude other criminal activities. As these MOU exist in order to coordinate the approaches of state and Commonwealth agencies for a specific purpose, they are not a complete solution to the problems of coordination. There are also volunteer coast guard organizations across the country that can provide a local response to boating accidents, often using the personal craft of the members.

Management and planning

Allocation of activities to operational elements is managed and planned through a series of interdepartmental committees. Two of these operate at a national level in Canberra whilst a third is a series of committees around regional areas. The Operations and Program Advisory Committee (OPAC) develops and reviews the program of surveillance flights and reviews significant issues. After April 2000, its procedures were modified to encourage participating agencies to make greater use of it as a consultative forum. The Planning Advisory Sub-Committee (PASC) determines the long-term requirements for support capacity, including a rolling three-month sailing plan. The Regional Operations and Program Advisory Committee (ROPAC) has the same responsibilities as OPAC but meets in the regional locations, where the bodies attending can include state agencies.

The way that each agency relates to the overall system varies significantly. This is largely a reflection of their separate and often very different responsibilities. This can be seen as creating different "classes" of agencies with widely varying expectations. A review of Coastwatch conducted by the Auditor General[14] developed a typology to demonstrate this phenomenon and distinguish between "major" and "minor" clients.[15] The categorization of different agencies depends largely on whether the agency has a long-term ongoing tasking for Coastwatch[16] or whether it is more a matter of a response to a particular situation[17] and on whether they are strategically or tactically driven. What becomes clear from this review is that:

- there is no one agency that is involved across the full range of maritime surveillance and enforcement activities; and
- there are some agencies that depend upon the continuing operation of the system and others that require it only when specific events occur.

The current system is therefore far from integrated. No single agency has a complete overview of the entire maritime enforcement and compliance system. This includes Coastwatch. Although it has the most complete intelligence overview, and has been able to consolidate the advice of client agencies, Coastwatch does not conduct risk assessments within the various portfolio areas of its clients. This was a task that was specifically left to the individual client agencies when the current system was established in 1988.

Functional areas

Maritime security and border protection

Maritime security has become an increasingly significant area in the last decade as Australia seeks to protect itself from terrorism and reduce the number of illegal immigrants arriving on Australia's shores. The *Customs Act 1901* (Cth) provides powers to board and search vessels[18] and to seize goods.[19] These powers are exercisable in Australian territorial waters and now extended to the contiguous zone under the *Border Protection Amendment Act 1999* (Cth). This legislation also enhances powers under the *Fisheries Management Act 1991* (Cth) and the *Migration Act 1958* (Cth). Powers under the *Migration Act* to deal with suspected illegal entry vessels (SIEVs) have been amended by the *Border Protection (Validation and Enforcement Powers) Act 2001* (Cth). Under the *Quarantine Act 1908* (Cth) officers have the right to board and inspect vessels and installations in Australian waters. Quarantine officers are further permitted to take control of prescribed goods[20] and order crew members of the vessel to undertake medical examination.[21]

The *Maritime Transport Security Act 2003* (Cth) (MTSA) gives effect to Australia's implementation and interpretation of the ISPS Code. Under the new legislation, the Australian government will regulate the security arrangements of Australian ports, port facilities and Australian-owned or flagged ships. It extends ISPS provisions to all ships employed on interstate voyages but not to ones employed on intrastate voyages. The legislation also establishes robust compliance checking of foreign ships and is to be extended to cover offshore oil and gas installations.

Illegal immigration

In response to increased numbers of illegal immigrants arriving in Australia at the turn of the twenty-first century,[22] the federal government adopted new legislation to improve border security, to reduce the number of people entitled to visas, and to increase Commonwealth powers to detain and deport. Under the *Migration Amendment (Excision from Migration Zone) Act 2001* (Cth) certain Australian external territories were excised from the migration zone for purposes related to unauthorized arrivals. This included the external territories of Ashmore and Cartier Islands, Christmas Island, Cocos (Keeling) Islands, the Coral Sea Islands, and certain sea and resource installations. The *Migration Amendment (Excision from Migration Zone) (Consequential Provisions Act) 2001* (Cth) allowed for the detention and removal of unauthorized arrivals in the excision zone and included powers to remove a person to another country where their claims, if any, for refugee status might be handled. The measure was aimed at deterring further movement from, or the bypassing of, other safe countries. The *Border Protection (Validation and Enforcement Powers) Act 2001* (Cth) introduced minimum penalties for people smuggling of five years for a first conviction and eight years for a second conviction. The legislation provided additional statutory authority for future action in relation to vessels carrying unauthorized arrivals and the unauthorized arrivals themselves.

Shipping

The *Navigation Act 1912* (Cth) provides the legislative basis for many of the Commonwealth's responsibilities for the regulation of shipping and navigation, including ship safety, the coasting trade, employment of seafarers, and shipboard aspects of the protection of the marine environment. It also regulates wrecks and salvage operations, tonnage measurement of ships, and the survey, inspection and certification of ships. A vast array of other Commonwealth legislation touches on the regulation of shipping, including the *Protection of the Sea (Prevention of Pollution from Ships) Act 1983* (Cth) and the *Environment Protection (Sea Dumping) Act 1981* (Cth).[23]

The Shipping and Navigation Agreement under OCS establishes the division of jurisdiction between the Commonwealth and the states/territories in relation to shipping.[24] Accordingly, the states and the Northern Territory are responsible for trading ships on intrastate voyages, fishing vessels, pleasure craft and inland waterway vessels, and the Commonwealth has responsibility for:

- trading vessels on an interstate or international voyage;
- fishing vessels and fishing fleet support vessels on an overseas voyage;
- ships belonging to the Commonwealth or a Commonwealth authority; and
- offshore industry mobile units and vessels, other than those confined to one state or territory.

Fishing

The principal pieces of legislation for the management of Commonwealth fisheries are the *Fisheries Management Act 1991* (FMA) and the *Fisheries Administration Act 1991* (FAA) which establish relevant management institutions and mechanisms. The FMA contains the provisions providing for the making of arrangements to manage a fishery in accordance with either Commonwealth or state law, and enabling joint authority management. It also contains provisions for surveillance and enforcement of Commonwealth fisheries laws. The FAA establishes the institutional administrative machinery responsible for managing Commonwealth-managed fisheries, other than joint authorities. The FMA grants extensive enforcement powers to AFMA officers, including stopping, boarding and searching vessels, seizure and powers of arrest. Unlike other regulatory mechanisms, the enforcement provisions apply generally without the need to be imposed by a specific management regime.

Crimes at sea

The *Crimes at Sea Act 2000* (Cth) provides the framework of criminal law applicable to Australian marine areas and to Australian-flag ships, offshore installations and foreign vessels under certain circumstances. Australian law applies according to coastal state jurisdiction. The law that is actually applicable will depend on the nature of the offence and the maritime zone in which it occurred. AFP has a major responsibility for the enforcement of national criminal legislation in Australian waters and the investigation of crimes at sea. However, it has difficulty in unilaterally discharging its responsibilities because "despite having the requisite investigative expertise, it lacks sufficient marine resources and, most importantly, presence in the national maritime environment."[25]

The *Crimes at Sea Act* is supported by an inter-government agreement (IGA) that provides a general framework for inter-agency coordination. However, the IGA lacks the specific coordinating mechanisms, such as those in national plans to counter terrorism, natural disasters and major oil spills, and does not indicate an appropriate enforcement agency. Past practice would indicate that state and federal police might use their own resources to deal with criminal activities close to shore and to act in concert with ADF in respect of criminal activities further from shore.

While the *Crimes at Sea Act* has simplified enforcement by reducing the risk of jurisdictional disputes and providing a legislative basis for cooperative or joint Commonwealth-state/territory law enforcement operations, there is still the problem created by the number of agencies that might be involved in these operations. In the words of an AFP officer engaged in maritime enforcement, "law enforcement within Australia's maritime jurisdiction continues to be fragmented, poorly coordinated and generally reliant upon ad hoc arrangements at the operational level."[26]

Marine pollution

Commonwealth, state/territory law regulating pollution from shipping is predominantly based upon Australia's various international obligations under the framework created by the 1982 United Nations Convention on the Law of the Sea (LOS Convention)[27] and further developed under specific instruments such as MARPOL[28] and the London Convention.[29] While the Commonwealth retains principal constitutional power in this field, the states also have complementary legislation applying in state waters.[30] The

Protection of the Sea (Powers of Intervention) Act 1981 (Cth) gives effect to the 1969 International Convention relating to Intervention on the High Seas in Cases of Oil Pollution Casualties.[31]

Marine protected areas

The process of establishing and managing Commonwealth marine protected areas (MPAs) falls under the *Environment Protection and Biodiversity Conservation Act 1999* (Cth). This Act operationalizes an environmental assessment and approval regime for actions likely to have an impact on matters of national environmental significance. It applies to areas under Australian jurisdiction, including Commonwealth and state and territory marine waters.[32] What is allowable in a MPA depends on its purpose and how the area is managed. Actions that might be prohibited include mining operations, certain other commercial activities, and those affecting native species and heritage. In some cases, virtually all human activity is excluded, or fishing and the removal of sea life may not be permitted. In other cases, recreational and commercial activities, such as fishing, tourism or exploitation, may occur providing the natural attributes of the MPA are not disturbed. Other areas have seasonal restrictions on activities such as changes to shipping routes to reduce impacts on migrating whales. The effectiveness of an MPA will depend on the arrangements made for enforcement and compliance but, with the exception of the Great Barrier Reef Marine Park, relatively little consideration seems to have been given to these requirements so far.

Miscellaneous areas of regulation

Offshore installations are primarily regulated by the *Sea Installations Act 1987* (Cth). This Act ensures that sea installations in Commonwealth waters are operated with regard to the safety of the people using them as well as the environment. The *Historic Shipwrecks Act 1976* (Cth) is designed to protect shipwrecks. Actions near a shipwreck that may have an impact on the shipwreck are prohibited. The *Submarine Cables and Pipelines Protection Act 1963* (Cth) applies to cables and pipelines beneath the high seas and the seas of the EEZ. The Act fulfills Australia's international obligations with regard to cables and pipelines under the LOS Convention by making it an offence to break or damage wilfully or with culpable negligence, a submarine telegraph or telephone cable pipeline or a submarine high-voltage power cable.

Assets for maritime enforcement and compliance

Aerial surveillance

Civil aviation contractors provide most of the aerial surveillance effort for maritime enforcement and compliance in Australia. They provide a fully operational system to Coastwatch with aircraft, crew, operational support, maintenance and related services. Coastwatch closely monitors contractor performance, operations and training. The fixed-wing aerial surveillance operator is currently Surveillance Australia, an Adelaide-based operation bought by the British company, FR Aviation, in 1999. It provides aircraft for surveillance missions under a nine-year contract originally valued at AUD$300 million. Reef Helicopters of Cairns provide rotary-winged surveillance and air transport in the Torres Strait area. This arrangement was extended in late 2004 until 2007, to allow for the evaluation of options for the next generation of surveillance technology, scheduled for operational introduction in the latter year.

Current fixed-wing aircraft on contract to Coastwatch are:

- five Bombardier Dash 8-200 – radar and electro-optic sensors; patrols up to 100 nm beyond the AFZ, radius of action up to 600 nm;
- three Reims F406 – radar and night vision equipment; medium range seaward operations;
- six Britten-Norman Islander – visual littoral search from Exmouth northabout to Brisbane; and
- one Shrike Commander – visual search, based at Broome.

The helicopters used are:

- one Bell 412EP – electro-optical vision equipment; and
- one Bell Longranger IV – visual surveillance and special purpose transportation, no surveillance devices fitted.

In addition, RAAF provides some 250 hours per annum of P3C Orion aircraft surveillance, primarily in waters from Perth southabout to Newcastle. Since the contracted aircraft are normally deployed in the north of Australia, this is the major surveillance capability over southern waters.

Surface assets

Since the first attempt in 1978 to organize a coastal surveillance system, RAN *Fremantle* class patrol boats and ACS sea-going vessels have provided response resources for maritime border protection. These continue as the basic assets but there have been some changes lately, most caused by specific policy decisions. The current surface response force consists of:

- 1,800 boat days per annum from the RAN *Fremantle* class patrol boat fleet with ten vessels based in Darwin and five in Cairns;
- eight *Bay* class ACVs managed by NMU and deployed at various locations around Australia providing 2,400 boat days per annum; and
- *Oceanic Viking*, a 105 meter 9,000 tonne ice-strengthened vessel,

operational from November 2004 under a two-year lease managed by NMU, for patrolling Australia's Southern Ocean EEZs.

Although the emphasis on operations in the Southern Ocean has changed, this list of assets remains much as it was before the incident involving MV *Tampa* in August–September 2001. This event marked a substantial change of policy on illegal entry of people by sea. On 3 September 2001, the federal government announced a naval deployment to intercept and force back to Indonesia an increasing number of vessels carrying people seeking to enter Australian territory unlawfully but with the purpose of claiming asylum. The deployment consisted of three frigates with embarked helicopters, a support vessel and four P3C Orion aircraft. This decision marked the first time that RAN combat vessels had been directly used for maritime border protection by laying a blockade. At a later stage of the operation, HMAS *Manoora*, transiting the area with people removed from the *Tampa*, also became involved. A high level of ADF involvement continued as *Operation Relex*, at least until the end of 2001 when the demands of participating in the anti-terrorist campaign in Afghanistan began to draw vessels away.

As an interim arrangement, the ocean-going survey ships, HMAS *Leeuwin* and HMAS *Melville*, were painted gray, given additional crew (including Army personnel) for boarding operations, and deployed on patrol and enforcement duties in the northwest. These vessels have advantages of range and sea-keeping qualities over the *Fremantle* class patrol boats. Their deployment confirms the gap that exists in Australia's current capabilities for maritime enforcement between guided missile frigates on the one hand and patrol boats on the other.

However, the one significant capability enhancement is an increased capacity to patrol the Southern Ocean, particularly the Heard and McDonald Islands EEZ. This had been done intermittently since 1997 by RAN and from 2000 by a chartered commercial vessel, initially managed by AFMA and, from mid-2003, by NMU. Strategic developments limited RAN's capacity to deploy to the Southern Ocean after 2001, whilst the commercial charter allowed for only two cruises a year (for a total of around 100 days deployed) and had no capacity to impose a non-consensual intervention on apparent breaches of Australian law.

With the value of the Southern Ocean fishery being heightened after 2001, the government decided in 2004 to increase both presence and capability in the Southern Ocean. From the end of 2004 *Oceanic Viking* has been deployed for 200 to 250 days per annum for a two-year trial allocated AUD$89.2 million. It is managed by ACS and maintained by P&O Maritime Services. This vessel is better suited to conditions in the sub-Antarctic but, more significantly, is the first ACS vessel to be armed in peacetime. It carries two 0.5" machine guns (the same armament as on Canadian fisheries vessels) and can launch an armed boarding party.

Costs

Costs are a central issue in judging the efficiency of maritime enforcement and compliance. Unfortunately, these are not known with any accuracy by the public as a comprehensive costing has never been released. Figures are available on the surveillance component of border protection, but even the usefulness and clarity of these numbers has been subject to criticism.[33] They omit agencies that are funded to enforce relevant legislation. Nonetheless, some attempt can be made to provide an estimate of the cost of maritime border protection, depending on the assumptions that are made about its components. The costs of the surveillance component of the system for 2003–04 were estimated to total AUD$257 million comprising $127 million for ACS costs and $129 million for Defence.[34]

However, this is not even the full cost of the surveillance and response components. The costs presented in the ACS budget for civil coastal surveillance cover the costs of Coastwatch, the value of Defence activities devoted to this task, and the transfer of funding from DIMIA to Coastwatch for Dash 8 surveillance hours allocated since 1999. From 2004–05, over a period of four financial years, the $114 million appropriated to DIMIA for surveillance will be transferred to ACS, resulting in a more realistic estimate for maritime surveillance costs in the 2004–05 budget of $308 million. Furthermore, this accounting includes only an allocation of NMU costs against the surveillance task. Other NMU activities, for instance against people smuggling, will be costed against other entries in the ACS budget. During 2004–05, NMU appropriations will include new funding of $44 million for an enhanced Southern Ocean patrol capacity. Including these and related additional elements, such as AusSAR, and Torres Strait border protection, a more realistic costing for the surveillance and response would exceed $400 million. In most definitions of maritime surveillance and enforcement, the full cost of AMSA and AFMA would also be attributed as relevant costs. In that case, a more comprehensive cost of maritime border protection would bring the total cost to around AUD$500 million.[35]

Implications of *Australia's Oceans Policy*

Goals and objectives

The first goal of *Australia's Ocean Policy* is: "To exercise and protect Australia's rights and jurisdiction over offshore areas, including offshore resources."[36] The priority accorded to sovereignty and sovereignty protection was a conscious decision reflecting the basic responsibility of government for national security. Thus the Policy commits government to providing effective surveillance and enforcement within Australia's marine jurisdiction. This is fundamental to protecting Australia's national interests at sea. In turn, *Australia's Oceans Policy* commits the Commonwealth government to a number of ocean-specific surveillance and enforcement measures, including measures as varied as the review and rationalization of effort involved in and the capacity for surveillance and enforcement, possible alternatives to traditional surveillance and enforcement techniques, cooperative multilateral and bilateral activities to reduce incursions into Australian waters, and the development of improved marine intelligence networks and cooperation between Commonwealth and state agencies in surveillance and enforcement actions.[37]

A holistic view of Australia's maritime interests, including ways and means of protecting national interests at sea, is adopted throughout *Australia's Oceans Policy*. In a section on surveillance and enforcement, the Policy notes that the challenges are:

- to ensure that there is an effective and efficient surveillance capacity for Australia's marine jurisdictions; and
- to ensure effective enforcement of national legislation throughout Australia's marine jurisdictions.[38]

The principles of integrated oceans management incorporated in *Australia's Oceans Policy* should also be reflected in national arrangements for oceans management, including maritime enforcement and compliance. The fact that they are not, and that the administrative and legislative frameworks for the Australian marine environment remain extremely complex, throws doubt on the capacity to implement regional marine plans under *Australia's Oceans Policy*. Enforcement and compliance will be important elements contributing to successful implementation, and it is important that attention is given to integrated arrangements.

Regional marine planning

Not a lot of attention has been paid so far to integrated enforcement and compliance in the development of regional marine planning under the *Oceans Policy*. On the basis of a series of community surveys and group workshops conducted as part of the development of the South-east Regional Marine Plan (SERMP), the community and cultural values assessment conducted by the National Oceans Office found strong support for "more policing of the resources of the Region."[39] Specifically, the community group workshops identified "more and better enforcement (and policing) mechanisms, particularly in relation to fishing and introduced marine pests"[40] as an important aspiration of marine-focused community interest groups. However, this may have been mainly in the context of illegal fishing by foreign vessels because "overseas fishers fishing in Australian waters" had been identified earlier as an important issue for the community groups.[41]

A major implication of RMP for integrated enforcement and compliance will be the need to achieve some consistency in arrangements around Australia. While the arrangements for coordinating maritime surveillance and enforcement are generally complicated with a multiplicity of state and federal authorities involved, this is particularly so in southern regions. Unlike northern Australia, there is no regional structure of Coastwatch in southern regions for managing regional surveillance. All coordination for these regions is done through the central Coastwatch structure in Canberra. To some extent Coastwatch will be "out of touch" with state authorities and regional requirements. Another problem with achieving coordination and a consistent approach to enforcement and compliance is that arrangements for managing coastal and marine activities vary from one state or territory to another.

South-east Regional Marine Plan

The *Uses in the Region Assessment Report* for the SERMP assessed the surveillance requirements of the region as follows:

> The current national surveillance issues – those of illegal immigrants and illegal drugs and quarantine – have not been and are unlikely to be, major issues for the South-east Marine Region. Such activity is more likely to continue as a major issue off northern Australia. However, the predicted increase in shipping activity and volumes of cargo will pose its own onshore surveillance issues and create greater threats for illegal drugs and increase the requirements for the monitoring of pollution from ships. Illegal fishing activity is also likely to at least be a potential issue along the outer edges of the Region until effective high seas management and real-time monitoring of flag of convenience vessels is in place.[42]

Coastwatch does not consider the South-east Marine Region (SEMR) as an area of high or emerging threat.[43] Theoretically, there is no reason why people smugglers, prospective terrorists or drug smugglers should not attempt to land illicit cargoes in southern areas, particularly if the surveillance effort in these waters is perceived to be less intensive. Illegal, unregulated and unreported (IUU) fishing activity has also taken place in the SEMR, including understating fish catch, unlicensed fishing, fraud in the abalone industry and operations within Australia's EEZ by foreign fishing vessels.[44] In April 1998, the (then) Deputy Commissioner of the AFP expressed the view that while Australia's maritime surveillance is adequate, the ability to respond to surveillance, to intercept and detain, to board and search, to enforce laws and to effect sovereignty is entirely inadequate.[45] While this view was a general description of the situation throughout Australia's maritime jurisdiction, it is particularly applicable to the SEMR where response capabilities are few and far between.

The SERMP was released during 2004.[46] It includes an action 3.10.1 to:

"Investigate the enforcement and compliance challenges and opportunities associated with the increasing use of spatial management of marine resources in the Region."[47] By addressing only resource issues, this action is rather less than a satisfactory response to achieving integrated enforcement and compliance across the range of illegal activities that might occur in the region.

Northern Region

The RMP for the Northern Marine Region (NMR) is now well advanced. The NMR comprises waters off Arnhem Land, the Gulf of Carpentaria and the Torres Strait. The Torres Strait poses particular problems for maritime enforcement and compliance. It is as near as Australia comes to a land boundary with foreign countries. Several Australian islands in Torres Strait are only 15 minutes by boat from Papua New Guinea (PNG), whilst a trip from PNG to Cape York, the nearest point on the mainland, takes several hours in a small outboard-powered dinghy. The West Papua border with Indonesia lies less than 160 kilometers from Boigu Island, one of the northernmost of the Australian islands in the Strait. There are over 40,000 people movements per annum across Torres Strait.[48] The AFP has identified smuggling of cannabis into Australia and arms out of Australia across the Strait as among the main law enforcement problems in the region.[49]

Policy considerations

Policy development for maritime enforcement and compliance reflects a history of unwillingness to respond to any but pressing current problems. The policies adopted have generally ameliorated the worst problems and have avoided large expenditures. However, they have not addressed structural weaknesses, which have created a persistent risk of fragmentation and remain a barrier to integration. The major exception to this rule was the establishment of Coastwatch in 1988 as an agency with financial control over the civil maritime surveillance program and, hence, the ability to take ultimate control of operational procedures. All subsequent strengthening of the system has built on this reactive approach.

Absence of a central authority

A major barrier to integration is that no single agency has the core role of and thereby, legislative authority for, overall law enforcement in areas of national maritime jurisdiction. There is no agency with the responsibility of reporting on the overall effectiveness and efficiency of maritime enforcement and compliance. There is no body designated as the lead agency for developing legislation on maritime enforcement and compliance and certainly none with the role of simplifying the historical tangle of existing law. To use organizational jargon, there is no agency that "owns" the problem. Coastwatch itself has no legislative basis, which has meant that, in the past, questions of its own structure and independence have been placed second to those of a client agency.[50] The current system is one of "distributed responsibility" in that many agencies have responsibility for distinct components but no one agency has the power to assess and manage the overall performance of the whole system. The existing arrangements presume that there is a partnership between the participating agencies and that cooperation will produce optimal outcomes. Yet, just as important are the great differences in the nature of the responsibilities of the constituent agencies. Most of the time inter-agency cooperation works, but there is also potential for the system to fail. The consequences of agency differences have the capacity to undermine effective coordination of operations and to prevent the development of optimal central objectives for the system as a whole.

Intelligence

Breakdowns in inter-agency coordination with intelligence and evaluation most consistently explain failures of the system. Landings by boats carrying unlawful entrants at Montague Sound in 1992, on the Coburg Peninsular in 1998 and on the East Coast in 1999[51] all followed failures to pass on important intelligence between agencies. The development of a centralized intelligence function within the National Surveillance Centre of Coastwatch (opened in January 2000) was supposed to have removed such problems. Yet the problems of lack of coordination reappeared spectacularly in October 2001 with the confusion over the veracity of pictures in what has become known as the "children in the water" incident. The government claimed that children had been thrown overboard from a vessel carrying asylum seekers but later it was established that they had not been. The incident was highly politicized, and this distorted some of the earlier handling of the intelligence.[52]

Effective maritime surveillance and enforcement requires comprehensive knowledge of what is happening in offshore areas. It is only by knowing what is happening in these areas that a state can properly respond to maritime incidents. Following events of 11 September 2001, the concept of maritime domain awareness has become central to the maintenance of maritime homeland security in the United States.[53] This concept recognizes that, under existing arrangements, no single agency has the capability independently to create a comprehensive picture of the maritime domain and that an architecture that readily supports information sharing and collaboration is crucial.[54]

Search and rescue

SAR is an activity that is closely related to maritime enforcement and compliance.[55] It requires similar assets and command and control procedures.

AusSAR provides the national center for coordination of Australia's civil SAR activities combining air, land and maritime requirements. In each state and territory, the police are the SAR authority with responsibility for marine SAR operations for persons or ships in waters within the limits of ports of a state or territory and in respect of pleasure craft and fishing vessels. ADF is responsible for SAR for all military and visiting military forces and will also provide assets to assist in civil SAR operations as requested by AusSAR. SAR in Australian waters requires a high level of cooperation and coordination, including with volunteer rescue organizations.

An incident that attracted considerable adverse publicity for both state and Commonwealth SAR authorities occurred in April 2001 when the 10-meter fishing boat *Margaret J* disappeared with three men onboard on a fishing trip to Hunter Island off Tasmania's northwest corner.[56] The sunken craft was located a few days after the search was commenced but no trace of the crew was found until the boat's life raft and one body were found three weeks later on an island off Tasmania's northeast corner. This was some 125 nm east of the official search zone and almost exactly where local fishermen had predicted any survivors and a life raft would have drifted. Bitter recriminations followed between the authorities involved, Tasmania police at the state level and AusSAR at the federal level, over issues of coordination and responsibility, with much public criticism of the effectiveness of the search and the quantity of resources employed. These criticisms included allegations that both the state and Commonwealth governments had reduced spending on SAR services.[57]

Adjusting to change

One of the characteristics of the Australian system of maritime surveillance and enforcement has been the slow pace of improvement. In a system of distributed responsibility, caution and lack of common purpose amongst the members usually leads to change occurring only when it is strongly in the interest of an individual agency or is shown as necessary by some major breakdown of the system. There is little interest in changes, and indeed few mechanisms to promote changes that might improve the system as a whole. An equally significant issue arising from the same cause is that, as the system is currently structured, a problem that is not the responsibility of one of the constituent agencies will not be seen to exist. Agencies will not take on a task that does not fall explicitly within their own well-defined responsibilities. Current practices of public sector management and general under-funding of public services reinforce this attitude. Slow adjustment to change leads to difficulty in recognizing the emergence of new threats especially with the current unwillingness to "own" non-portfolio problems.

Crisis-driven change

Throughout its history in Australia, civil maritime surveillance and enforcement has been driven by politics. A structure of distributed responsibility seldom lends itself to pro-active policy development. Most recently, it has been evident with the establishment of OTS and the new Joint Offshore Protection Command that links civil maritime surveillance with military maritime surveillance.

Most changes to existing arrangements have been made in reaction to the political outcry caused by some particular development. The best way to describe the development of the Australian regime for maritime enforcement and compliance is that it is crisis-driven incrementalism. This has produced gradual evolution and, generally, improved operational performance of the system. Yet it continues to maintain a very real risk of confusion of policy focus, decisions of short-lived utility and risk of hijack. It remains too easy for single agency agendas or outside bodies to take control of policy. This appears to be what happened during the *Tampa* incident. DIMIA appears to have forced its interests to the fore in what was initially a search and rescue operation.[58] Indeed, the most significant factor during the blockade of boat people and the subsequent confusion over what events took place is the almost complete invisibility of Coastwatch during the entire period. Under current arrangements, the only option for the government is to continue using the navy to blockade any future influx of boat people. Yet the only proposal in the government's 2001 election manifesto relating to maritime response capacity was to double the number of sea days available from ACS *Bay* class vessels. Whilst this represents a more efficient use of previously under-utilized assets, the *Bay* class vessels are even less relevant to current policy on boat people than the patrol boats.

Conclusion

Australia faces major challenges with achieving integrated maritime enforcement and compliance. However, pressures for changing current arrangements will not stop. Potential law and order problems at sea include drug smuggling, people smuggling, crimes at sea, illegal fishing and maritime terrorism. The present complicated system for dealing with these problems is an unsatisfactory answer to the challenges. While this is a general problem throughout Australia's maritime jurisdiction, it is particularly acute in southern regions in view of the number of state agencies involved, the relative lack of Commonwealth assets in these regions and the lack of experience of agencies at working together. A satisfactory system appears to have emerged in northern regions, but these depend on informal working relationships between the agencies involved and are not necessarily indicative of an institutionalized integrated approach to maritime enforcement and compliance. With the implementation of *Australia's Oceans Policy*, the conservation objectives of RMP will require monitoring and establish the need for additional enforcement activity at sea.

Australia has traditionally drawn a careful line between civil maritime surveillance and military maritime surveillance although this is now changing with the development of the Joint Offshore Protection Command. The distinction between traditional security against military threats and non-traditional economic, resource, human and environmental security is becoming blurred. Increasingly, military assets are being used in support of constabulary roles, particularly in Australia in recent years against the threat of people smuggling, and the threat of terrorism is accelerating this trend. A distinction between military and civilian responsibilities in maritime surveillance and enforcement has proven to be a luxury that Australia can no longer afford. It may have made sense when the civil area of interest was mainly along the littoral, but makes no sense now that the civil surveillance area is much larger, concepts of security more intertwined, and surveillance and intelligence systems more technologically advanced and expensive.

If Australia is to resort to more than crisis-driven incrementalism to meet these pressures and to introduce integrated maritime enforcement and compliance, a significant reorganization of government functions is required. There is no hierarchical system for Commonwealth public administration and no Commonwealth department has the power to control the way that another department meets its administrative responsibilities. The only effective way to do this is to persuade a minister of the need for change. The problem for maritime enforcement and compliance is that there is no single minister to influence. Many ministers have responsibility for bits of the area but they will only take an integrated view if there is political reason sufficient to focus their attention. A possible solution would be to appoint a minister with responsibility for overall policy and legislative development in maritime enforcement and compliance (perhaps with responsibility for oceans issues generally) and to support him or her with an agency that has the requisite legislative authority and capabilities. Despite the benefits of integrated maritime enforcement and compliance and the limitations of existing arrangements for maritime surveillance and enforcement in Australia, we are unlikely to see a more integrated system in the foreseeable future.

Notes

1 A list of previous reviews can be found at: Auditor General *Coastwatch Australian Customs Service* (Audit Report No. 38 1999–2000) (Australian National Audit Office, Canberra: April 2000) at 111. Online at: www.anao.gov.au (accessed 15 January 2004).

2 See Anthony Bergin and Marcus Haward "Australia's new oceans policy" (1999) 14 *The International Journal of Marine and Coastal Law* 387–98; Donald R. Rothwell and Stuart B. Kaye "A legal framework for integrated oceans and coastal management in Australia" (2001) 18 *Environmental and Planning Law Journal* 278–92.

3 See, for example, Derek Woolner "Australian coastal surveillance: Changing policy pressures" (2001) 119 *Maritime Studies* 2.
4 The total length of Australia's coastline is 59,736 kilometers if the coastline of islands is taken into account. Geoscience Australia "Coastline Lengths." Online at: www.auslig.gov.au/facts/dimensions/coastlin.htm (accessed 15 January 2004).
5 Geoscience Australia "Australia's Oceans and Seas." Online at: www.auslig.gov.au/facts/dimensions/oceans.htm (accessed 15 January 2004).
6 National Oceans Office *Resources – Using the Ocean* (National Oceans Office, Hobart: 2002) at 133, Table 27.
7 See the discussion in Donald R. Rothwell and Marcus Haward "Federal and international perspectives on Australia's maritime claims" (1996) 20 *Marine Policy* 29–46.
8 Hugh Smith "The Use of Armed Forces in Law Enforcement" in Doug MacKinnon and Dick Sherwood (eds) *Policing Australia's Offshore Zones – Problems and Prospects* (Centre for Maritime Policy, University of Wollongong, Wollongong: 1997) 74–97.
9 1978 Treaty between Australia and the Independent State of Papua New Guinea concerning the Maritime Boundaries between the Two Countries, including the Area known as Torres Strait, and Related Matters, done in Sydney, 18 December 1978, in force 15 February 1985 [1985] ATS No. 5.
10 Greg McLeod "Keeping Australia's Seas Safe" (unpublished paper submitted for a Masters of Public Policy and Administration, Charles Sturt University, 2002) at 26, Table 2 (on file).
11 Ibid., at 44.
12 Major fleet units (frigates, tankers and amphibious ships) are involved from time to time in maritime enforcement, particularly in the Southern Ocean and the northeast Indian Ocean around Christmas and Ashmore Islands.
13 Prime Minister of Australia "Strengthening Offshore Maritime Security" Media Release, 15 December 2004.
14 Auditor-General, note 1 at 112
15 Ibid., at Appendix 3, 112.
16 This applies in the case of AFMA and ACS.
17 As is the case with DIMIA and the AFP.
18 *Customs Act 1903* (Cth), ss 184A, 185, 185A, 187.
19 Ibid., s. 185.
20 *Quarantine Act 1908* (Cth), s. 70B.
21 Ibid., s. 78.
22 Between 1998–99 and 2000–01, there was an increase of 450 percent in flows of unauthorized arrivals by boat to Australia; see Department of Immigration (Australia). Online at: www.immi.gov.au/legislation/refugee/ (accessed 15 January 2004).
23 National Oceans Office *Ocean Management – The Legal Framework* (National Oceans Office, Hobart: 2002) at 14.
24 Ibid.
25 McLeod, note 10 at 50.
26 Ibid., at 37.
27 1833 UNTS 396.
28 1340 UNTS 61; as implemented in Australia by the *Protection of the Sea (Prevention of Pollution from Ships) Act 1983* (Cth) and related state legislation.
29 1972 Convention for Prevention of Marine Pollution by Dumping from Ships and Aircraft 932 UNTS 3, and a 1996 Protocol (1997) 26 *International Legal Materials* 7; as implemented in Australia by the *Environment Protection (Sea Dumping) Act 1981* (Cth).

30 See generally Michael White *Marine Pollution Laws of the Australasian Region* (Federation Press, Annandale, NSW: 1994).
31 970 UNTS 211.
32 For an assessment of the implications of the *Environment Protection and Biodiversity Conservation Act 1999* (Cth) Act for offshore marine management, see Rothwell and Kaye, note 2.
33 Joint Committee on Public Accounts and Audit (JCPAA) *Review of Coastwatch* (Parliament of Australia, Canberra: August 2001) 33ff.
34 Figures drawn from various entries within the ACS *Annual Report 2003–04*. Online at: www.customs.gov.au/site/page.cfm?u=4283 (accessed 16 January 2005).
35 Calculated by adding in AMSA's 2003–04 projected spending of AUD$35 million and AFMA's 2003–04 budget estimate of AUD$65 million, including the cost of AusSAR.
36 Commonwealth of Australia *Australia's Oceans Policy* Volume 1 (Environment Australia, Canberra: 1998) at 4.
37 Commonwealth of Australia *Australia's Oceans Policy* Volume 2 (Environment Australia, Canberra: 1998) Section 5.3 at 40–2.
38 Ibid., at 40.
39 National Oceans Office *Communities Connecting with the Ocean* (National Oceans Office, Hobart: 2002) at ii.
40 Ibid., at 15.
41 Ibid.
42 National Oceans Office, note 6 at 157.
43 Ibid., at 132.
44 Senator Warwick Parer (Minister for Resources and Energy) "Government takes action on orange roughy" *Media Release* DPIE 97/237P (13 November 1997).
45 Douglas R. Mackinnon and James Hinkley "Report on Critical Issues Workshop: The regulation and enforcement of crime in Australia's maritime zones" (1999) 104 *Maritime Studies* 11.
46 National Oceans Office *South-east Regional Marine Plan* (National Oceans Office, Hobart: 2004).
47 Ibid., at 57.
48 JCPAA, note 33 at 114.
49 JCPAA, note 33, Volume 3 – AFP, Submission No. 43, S585 and S588.
50 See, for instance, Auditor General, note 1 at 41; JCPAA, note 33, Volume 2, Lofty Mason, Submission No. 31, S302–303; Anthony Bergin "Taking surveillance littorally" 14 April 1999 *Australian Financial Review*; Anthony Bergin "Presenting Australian coast's case for the defence" 1 July 1999 *Australian Financial Review*; and Sam Bateman and Anthony Bergin "Conference Summation" in Doug Mackinnon and Dick Sherwood (eds) *Policing Australia's Offshore Zones Problems and Prospects* (Centre for Maritime Policy, University of Wollongong: 1997) 294–9.
51 On 17 May 1999, the small Panamanian-registered cargo vessel *Kayuen* was intercepted while attempting to sail into Jervis Bay on the south coast of New South Wales and escorted to Port Kembla where 69 illegal Chinese immigrants were found hidden onboard. It appears that this ship traveled across the north of Papua New Guinea and down the east coast of Australia well outside the Great Barrier Reef with the intention of transferring the illegal immigrants to a small craft off Port Hacking, south of Sydney. This incident came only six weeks after a similar one when 60 illegal Chinese immigrants landed undetected at Scotts Head (NSW) on 10 April 1999.
52 Patrick Weller *Don't Tell the Prime Minister* (Scribe Short Books, Melbourne:

2002) at 41. For further background, see David Marr and Marian Wilkinson *Dark Victory* (Allen & Unwin, Crows Nest, NSW: 2003).

53 Coast Guard Homeland Security Strategic Task Force *Maritime Homeland Security The Way Forward: Final Report* (US Coast Guard, Washington: 16 October 2001) 2–7.

54 Ibid., at 2–3.

55 This was highlighted during the MV *Tampa* incident in August 2003. For discussion of search and rescue and law of the sea issues raised by this incident, see Donald R. Rothwell "The law of the sea and the MV *Tampa* incident: Reconciling maritime principles with coastal state sovereignty" (2002) 13 *Public Law Review* 118–27.

56 Details of this incident are from Ben Nichols "Abandoned search for Tasmanian fishermen reveals impact of cost-cutting." Online at: www.wsws.org/articles/2001/may2001/fish-m28.shtml (accessed 15 January 2004).

57 Ibid.

58 Marr and Wilkinson, note 52 at 14–29.

Part III

The precautionary principle/approach

6 Canada and the precautionary principle/approach in ocean and coastal management

Wading and wandering in tricky currents

David L. VanderZwaag, Susanna D. Fuller and Ransom A. Myers

Introduction: the tricky currents of precaution

The precautionary principle or approach, while firmly grounded in international environmental law,[1] fisheries law[2] and in common sense,[3] may be likened to a life raft swirling in tricky currents. While the principle has great potential to save lives and salvage the environment by requiring anticipatory precautionary and preventative measures in the face of scientific uncertainty[4] and by having a core notion of placing the burden of proof on those who propose change,[5] it is buffeted by political and practical implementation challenges.[6] Debate continues over terminology with some preferring the term precautionary approach because of its less onerous legal connotations.[7] Exactly what should trigger precautionary action remains controversial with the Rio Declaration on Environment and Development suggesting a threshold of "serious or irreversible" harm.[8] The extent the precautionary approach should be driven by "sound science" and risk assessment[9] versus social values and public perceptions is a further area of contention.[10] The Rio Declaration calls on states to apply the precautionary approach. This leaves open the question as to which persons and institutions within states should be made responsible for making precautionary determinations and judgments.[11]

Tensions continue over how extreme precautionary measures should be.[12] Extreme measures include outright bans, phase-outs for risky chemicals or technologies, and reversals in the burden of proof where the proponents of development activities would not be allowed to proceed unless they demonstrate lack of significant harm or some other standard of safety/acceptability.[13] Less extreme measures include, among others, requiring pollution prevention plans as a precondition to licensing polluting activities,[14] broadly applying environmental impact assessment including alternatives assessment,[15] and ensuring strict or absolute liability approaches for pollution damage.[16]

In the fisheries management field a broad array of precautionary measures are also available. Proponents of new fisheries might be required to demonstrate no significant ecological damage or to meet some other legal litmus before license approval. Other measures include: ensuring that all discards are quantified and incorporated into estimates of fishing mortality;[17] ensuring that fish spawn at least once, so that all fish contribute to the population before being harvested;[18] limiting a fishery by the catch of a non-target species (e.g. the pollock fishery in Alaska is shutdown once a specific amount of halibut is caught);[19] conducting environmental assessments of fishing gear and managing according to the level of harm each gear risks within the concept of ecosystem management; developing gear zoning legislation;[20] eco-labeling of fish thereby providing a market-based incentive for precautionary management and fishing practices;[21] and following adaptive management processes where decisions can be flexible based on the outcome of regulated measures.

In light of the controversial and elusive nature of the precautionary approach, the eventual "firming up" of precaution will likely depend largely on national law and policy efforts. This chapter summarizes, through a four-part format, Canadian initiatives and efforts to implement the precautionary approach. The first section describes Canada's general steps to adopt the precautionary approach including the 1998 *Canada-wide Accord on Environmental Harmonization*[22] and the Government of Canada's *Framework for Application of Precaution in Science-based Decision Making about Risk* issued in 2003[23] and caselaw developments relating to precaution.[24] The section concludes with a brief look at the limited embracing of precaution in environmental impact assessment review and strategic planning processes. The second section reviews Canada's efforts to address marine pollution – ocean dumping, land-based, vessel-source and seabed activities – in light of precaution. The third section examines Canadian experiences with implementing precaution in the field of living marine resource management including fisheries, aquaculture and biodiversity protection. The final section highlights Canada's rather non-precautionary responses to the threats of climate change.

Canadian general steps and wanderings

Environmental harmonization accord

Adopted on 29 January 1998 by the Canadian Council of Ministers of the Environment (with the exception of Quebec), the *Canada-wide Accord on Environmental Harmonization* pledges governments to cooperate in establishing consistent environmental measures, in preventing inter-jurisdictional disputes and to apply common environmental management principles. Besides recognizing such principles as polluter pays, pollution prevention, public participation and Aboriginal rights, the Accord expressly adopts the precautionary principle:

> [W]here there are threats of serious or irreversible environmental damage, lack of full scientific certainty shall not be used as a reason for postponing cost-effective measures to prevent environmental degradation . . .[25]

Sub-agreements developed under the Accord[26] might be described as "cautious wadings" as the precautionary principle is not strongly embraced. The *Sub-agreement on Canada-wide Environmental Standards*,[27] while expressly repeating commitment to the precautionary principle,[28] dilutes the commitment through various qualifications, such as the statements that standards to be developed will be based on "sound science"[29] and that environmental measures will be determined in a sustainable development context recognizing "socio-economic considerations."[30] The *Sub-agreement on Environmental Impact Assessment*,[31] aimed at avoiding duplication in project assessments through designation of a lead party responsible for administering the assessment process, avoids express mention of the precautionary principle in favor of highlighting other principles – effectiveness, transparency and public accountability, and efficiency and certainty.[32]

Government of Canada's framework document

In 2003, the Government of Canada released a *Framework for Application of Precaution*, prepared through a multi-departmental approach, to suggest guiding principles for operationalizing precaution.[33] The Framework proposes ten "guiding principles". The first five involve general principles of application, for example, recognizing the legitimacy for decisions to be guided by a society's chosen level of protection against risk, suggesting sound scientific information must be the basis for applying the precautionary approach, and noting the importance of increased transparency, accountability and public involvement.

The latter five principles focus on precautionary measures and suggest precautionary measures should be:

- subject to reconsiderations based on the evolution of science, technology and society's chosen level of protection;
- proportional to the potential severity of the risk being addressed and to society's chosen level of protection;
- non-discriminatory and consistent with measures taken in similar circumstances;
- cost-effective, with the goal of generating an overall net benefit for society, at least cost and efficiency in the choice of measures; and
- least trade restrictive.

The Framework's three most fundamental limitations are: attempting to furl the sails of precaution through a "sound science" limitation, failing clearly to address the burden of proof in decision making and neglecting the important approach of alternatives assessment. Perhaps the most important limitation is the attempt in Principle 4.3 to restrict application of the precautionary approach to situations where there is adequate scientific information and evaluation. Principle 4.3 reads, "Sound scientific information and its evaluation *must be* the basis for applying the precautionary approach . . ." (emphasis added). The accompanying commentary reinforces the need for risk assessment: "Scientific data relevant to the risk must be evaluated through a sound, credible, transparent and inclusive mechanism leading to a conclusion that expresses the possibility of occurrence of harm and the magnitude of that harm . . ."

Such reliance on scientific justification and risk assessments is in line with international trade law restrictions which continue to be controversial.[34] There continue to be strong arguments that national values and policy judgments about acceptable risk should not be trumped by trade interests.[35]

A second major limitation is the restricted treatment of the burden of proof. The Framework fails to recommend as a guiding principle the general burden of proof reversal. Instead, in commentary for Principle 4.3, the Framework restricts the burden of proof to the issue of who should bear the burden of producing scientific information. The Framework suggests the responsibility for providing the scientific information base should generally rest with the party taking action associated with the potential of serious harm, but that burden may shift depending on who in a concrete scenario would be in the best position to provide the information base.

A third limitation is neglecting the important approach of alternatives assessment. While risk assessment may be a useful tool in understanding risk, an increasing number of academic writers are urging greater use of "alternatives assessment."[36] Under such an approach, decision-makers are encouraged to consider alternative technologies, locations, timings and scales with the objective of identifying least environmentally intrusive options.

Canadian caselaw

Although caselaw treatment of the precautionary approach in Canada has not been extensive,[37] the Supreme Court of Canada has opened the net for citizens and environmental groups displeased with decisions inadequately considering the precautionary approach to seek judicial review. In *Spraytech* v. *Town of Hudson*,[38] the Court recognized the precautionary principle is part of international law[39] and relied on the principle for justifying a broad interpretation of provincial statutory authority allowing towns to regulate pesticides through by-laws under the rubric of preventive action. The Court, in upholding the Town of Hudson's by-law restricting non-essential uses of pesticides, restated the potential importance of international legal principles in not only statutory interpretation but also in assessing the reasonableness of discretionary administrative decisions. Justice L'Heureux-Dube re-emphasized the wave-making language from *Baker* v. *Canada (Minister of Citizenship & Immigration)*:[40] "the values reflected in international human rights law may help to inform the contextual approach to statutory interpretation and judicial review."[41]

A key case, *Ecology Action Centre Society* v. *Attorney General of Canada*,[42] was the first Canadian case to raise the precautionary principle in the context of fisheries and marine environmental protection. The Ecology Action Centre, located in Halifax, Nova Scotia, and represented by legal counsel from the Sierra Legal Defence Fund, challenged the legality of the Regional Director-General's Variation Order under the *Fisheries Act*[43] allowing draggers to fish on Georges Bank. The applicant argued that harmful alteration of fish habitat by draggers is not authorized pursuant to the *Fisheries Act* and should be subjected to environmental impact assessment under the *Canadian Environmental Assessment Act*.[44] In August 2004, the judge ruled against the applicant, citing that the act of fishing could not be considered an undertaking or activity as specified in Section 35(2) of the Act. [45] Of note in this case is the fact that the science pertaining to the effects of trawling on the sea floor was not contested.

The case of *Brighton* v. *Nova Scotia (Minister of Agriculture and Fisheries)*[46] dealt with a citizen challenge, partly based on the precautionary approach, to a ministerial decision in favor of licensing a finfish net cage aquaculture farm in Northwest Cove, Nova Scotia. There a group of concerned citizens (appellants) argued the Minister had failed to err on the side of caution in light of so many unanswered questions regarding environmental consequences of the proposal. The appellants referred to the precautionary approach called for under Canada's *Oceans Act* in section 30 and the preamble.[47] While Justice MacDonald agreed the Minister was under a duty to proceed cautiously, whether legislatively directed or not, he found the Minister in fact had proceeded cautiously in light of stringent license conditions and ongoing monitoring requirements.

The case demonstrates some of the potential difficulties those challenging administrative decisions in light of the precautionary approach may face. Where natural resource legislation provides broad licensing discretion and even favors economic development, judges are likely to be highly deferential.[48] Courts may also be hesitant to address reversal in the burden of proof given traditional faith in bureaucratic expertise.

Environmental assessment and strategic planning

While environmental impact assessment processes may be viewed as inherently precautionary by supporting anticipatory and preventive planning,[49] laws and practices at both the provincial and federal levels have not strongly embraced the precautionary approach. Provincial environmental assessment laws do not make explicit reference to the precautionary principle.[50] The *Canadian Environmental Assessment Act (CEAA)*[51] may be criticized for various "non-precautionary" aspects including: the largely self-assessment approach where federal departments/agencies that are the proponent, funder, regulator or grantor of a land interest are responsible for assessments and final decisions;[52] the limitation to project proposals and failure to include assessment of government policies, programs and plans;[53] the limits of public participation particularly at the screening stage of review[54] where public comment is discretionary and participant funding is not ensured;[55] and the lack of decision criteria.[56] CEAA has also been criticized for not requiring alternatives to proposed projects to be addressed.[57]

Amendments to CEAA assented to on 11 June 2003, while not remedying the key deficiencies,[58] do at least insert precaution as an overall purpose of the Act:[59]

> 4(1) The purposes of this Act are
> (a) to ensure that projects are considered in a careful and precautionary manner before federal authorities take action in connection with them, in order to ensure that such projects do not cause significant adverse environmental effects . . .

A major gap in environmental assessment application has been in relation to fish harvesting. Federal environmental assessment review has not been applied to the issuing of fishing licenses, including destructive forms of bottom trawling.[60]

While environmental assessment legislation has not explicitly mandated the precautionary approach, various review panels have in practice addressed precaution either because of participant arguments or by direction in terms of reference. For example, in the joint federal–provincial assessment of Sable natural gas projects off Nova Scotia, some intervenors argued the precautionary principle should be used to impose zero-discharge limits on oil-based or synthetic drilling muds and for produced water.[61] The panel rejected such extreme versions of precaution and was content with recommending various measures relating to offshore wastes including implementation of an environmental effects monitoring program, further exploration of alternatives to oil-based muds and implementation of environmentally superior waste treatment methodologies if they became available during the life of the project.[62]

A further general route towards sorting out the implications of precaution might be through strategic planning efforts which to date have not involved detailed precautionary commitments. Canada's *Ocean Act*[63] calls for the development of a national oceans management strategy based upon the principles of sustainable development, integrated management and the precautionary approach.[64] However, *Canada's Oceans Strategy*,[65] released in 2002, simply reaffirms the Government of Canada's commitment to promoting the wide application of the precautionary approach to marine resource management with additional general commitments to promote ecosystem-based management, to establish marine protected areas, to improve understanding of the marine environment and to give priority to maintaining ecosystem health and integrity.[66] Potential for future flux in precautionary approach application is indicated: "*Canada's Oceans Strategy* will be governed by the ongoing policy work being undertaken by the Government of Canada."[67]

Through amendments to the *Auditor General Act*[68] in 1995, various federal departments and agencies, including the Department of Fisheries and Oceans, have been required to table sustainable development strategies with Parliament.[69] DFO's *Sustainable Development Strategy 2005–2007*[70] does not expressly discuss commitment to the precautionary approach, while the need for smart regulation is mentioned several times.[71]

Wadings and wanderings in marine pollution control

As the following discussion describes, Canada has most strongly embraced the precautionary approach in the field of ocean dumping, but is largely wading with rather general and weak commitments in the areas of land-based marine pollution control and the regulation of vessel-source pollution and seabed activities.

Ocean dumping

Pursuant to the *Canadian Environmental Protection Act, 1999*,[72] Canada has adopted the "reverse listing" approach to ocean dumping in line with the 1996 Protocol to the London Convention. No person is allowed to dispose of waste or other matter in waters under Canadian jurisdiction unless done in accordance with a Canadian permit, and only substances listed on a "safe list" set out in Schedule 5 may be disposed of at sea.[73] Schedule 5 includes dredged material; fish wastes; ships, aircraft, platforms or other structures;[74] inert, inorganic geological material; uncontaminated organic matter of natural origin; and bulky substances primarily composed of iron, steel, concrete or other similar matter.[75] As an additional precautionary measure, applicants for ocean disposal are required to undertake a waste prevention audit exploring the feasibility of reducing or preventing wastes through such techniques as process modification, input substitution and closed-loop recycling.[76]

Land-based marine pollution

While an assessment of Canadian precautionary approaches to land-based marine pollution is complicated by the overlap of provincial and federal jurisdictional controls,[77] a partial picture of Canadian precautionary efforts may be gleaned from looking at the three main federal statutes governing land-based pollution – the *Canadian Environmental Protection Act, 1999* (CEPA, 1999), the *Fisheries Act* and the *Pest Control Products Act*[78] – and Canada's *National Programme of Action for the Protection of the Marine Environment for Land-based Activities* (NPA).[79]

Canadian Environmental Protection Act, 1999

Through CEPA, 1999 the Government of Canada has waded into the waters of precaution through four specific references to the precautionary principle. The Preamble recites a Rio Declaration version of precaution.[80] Second, the legislation also imposes an administrative duty on the Government of Canada to follow the precautionary principle and emphasizes pollution prevention in implementing the Act.[81] For example, the Minister of the Environment, in deciding whether to authorize the manufacture or import of new chemical substances to Canada, would be obliged to take a precautionary approach.[82] Third, the Act also requires a National Advisory Committee to use the precautionary principle in giving advice and recommendations,[83] for example on proposed regulations.[84] Fourth, the Act requires the Ministers of Environment and Health, when conducting and interpreting the results of toxicity assessments, to apply "a weight of evidence approach and the precautionary principle."[85]

CEPA, 1999 might be described as taking a wandering approach to precaution because the Act is not consistent with a strong embrace of the precautionary principle as applied to ocean dumping. The Act, like the previous CEPA adopted in 1988, leaves over 23,000 chemicals on the market,[86] and will continue a reactive substance-by-substance toxicity assessment approach before regulatory actions are taken.[87] While the federal Ministers of Health and the Environment are required to establish a Virtual Elimination List, setting lowest measurable levels for anthropogenic toxic substances that are persistent and bioaccumulative, they are granted discretion to not actually require virtual elimination in light of "relevant social, economic or technical matters."[88] While the Act requires the precautionary principle to be followed as an administrative duty, the Act does not provide explicit guidance as to risk management implications.[89]

The Canadian Environmental Protection Act is exceedingly weak, partly due to constitutional considerations, in general powers given to the federal government to control land-based marine pollution beyond toxic chemicals. The Act only allows the federal government to establish objectives, guidelines and codes of practice for land-based marine pollution.[90] The Minister of the Environment is only allowed to impose pollution prevention plans for listed toxic substances or international air or water pollutants.[91]

Fisheries Act

Legal wanderings toward precaution are especially evident in the main federal lever to control marine pollution, the federal *Fisheries Act*.[92] Although the Act predates precautionary principle development in international law, the Act does contain provisions somewhat in line with precaution by a general prohibition on deposits of deleterious substances into water frequented by fish and a general prohibition against harmful alterations of fish habitat.[93] However, six general sets of regulations[94] under the Act allow considerable pollution discharges for specific industries and adopt rather crude, non-precautionary environmental standards, such as pollution limits based on production capacity[95] and acute toxicity testing for effluents.[96]

Pest Control Products Act

A modernized *Pest Control Products Act*, receiving Royal Assent on 12 December 2002, does go some distance in supporting precaution. The Act sets out the primary objective of preventing unacceptable risks to people and the environment from the use of pest control products.[97] The Act places the burden of proof on the person requesting registration of pesticides to demonstrate to the Minister of Health that the risks and value of the pesticide are acceptable.[98]

However, the Act has been seriously criticized for its numerous shortcomings including a failure to strongly embrace the precautionary approach.[99] The Act does not even give a preambular "honorable mention" to precaution and unlike CEPA, 1999 does not entrench precaution as a general administrative duty. The Act marginalizes the precautionary principle by only requiring the principle to be taken into account in the course of re-evaluation or special review[100] of a pest control product.[101] The Act does not give legal force to the substitution principle that would require older pesticides to be replaced with newer, less toxic products and non-chemical alternatives.[102] The Act does not define acceptable or unacceptable risk.[103] The Act fails to give the Pest Management Regulatory Agency a clear statutory mandate.[104] The Act also avoids imposing a legislative ban on the use of pesticides for "cosmetic use," partly because of constitutional limitation concerns.[105]

National Programme of Action for Land-based Activities

Canada's *National Programme of Action for the Protection of the Marine Environment from Land-based Activities* (NPA), released in June 2000, is a helpful document in highlighting the main environmental problems in various regions (Pacific, Arctic, Southern Quebec/St. Lawrence, Atlantic) and in establishing national priorities for action. For example, inadequate sewage treatment is listed as a high priority because of substantial environmental and economic effects shared by various regions. Shellfish harvesting closures from bacterial contamination on the Pacific coast are reported at nearly 1,000 km^2 (up from 710 km^2 in 1989).[106] The harvesting of shellfish has been permanently or temporarily prohibited in the Quebec region because of bacterial contamination in nearly half of all shellfish areas.[107] In 1986, 35 percent of the classified shellfish growing area in Atlantic Canada was closed to harvesting of shellfish because of fecal bacteria.[108]

However, the *National Programme of Action* might be described as weak in addressing precaution. The NPA document gives "lip service" to the precautionary approach but does not explore what the implications of precaution might be.[109] The NPA was not accompanied by a specific implementation budget but has depended on the vagaries of funding at national, provincial and municipal levels.[110]

Vessel-source pollution

Although Canada's legal framework for controlling pollution from ships has precautionary aspects, such as imposing strict liability on shipowners for oil pollution damage[111] and prohibiting discharges of garbage from ships,[112] the framework is not strongly precautionary. The *Canada Shipping Act*[113] does not mention precaution as a guiding principle. *Oil Pollution Prevention Regulations*[114] adopt the compromise standards of the MARPOL Convention[115] with considerable pollution allowed. For example, oil tankers are authorized to discharge oily mixtures from cargo spaces if they are more than 50 nautical miles from the nearest land and the instantaneous rate of discharge of oil in the effluent does not exceed 30 liters per nautical mile.[116] *Pollutant Substances Pollution Prevention Regulations*,[117] rather than imposing a strict precautionary "reverse listing" where ships would only be allowed to discharge substances listed on a "safe list", lists hundreds of substances (including many heavy metals and pesticides) that are prohibited from discharge.[118]

Perhaps the least precautionary area of vessel-source pollution control is Canada's approach to controlling ballast water releases from ships. The 2002 audit of the Commissioner of the Environment and Sustainable Development (CESD) called upon Environment Canada to address the threat of invasive species properly, and specifically noted Transport Canada's failure to regulate ballast water dumping and the past and potential socio-economic ramifications of this neglect.[119] With regard to the precautionary principle, the CESD concluded:

> The precautionary principle, pollution prevention, and the concept of "polluter pays" have been part of Canada's environmental policies for more than a decade. The federal government is not applying them to manage invasive species that threaten our environment.[120]

A subsequent report in May 2003 by the House of Commons Standing Committee on Fisheries and Oceans was also critical of federal efforts to address invasive species.[121] Environment Canada was criticized for neglecting invasive species issues and concentrating its efforts instead on getting a new *Species at Risk Act* enacted.[122] The report recommended that the Minister of Fisheries and Oceans expedite the development and implementation of ballast water management regulations.[123]

An Invasive Alien Species Strategy for Canada, issued in September 2004,[124] noted the ongoing limited capacities of departments and agencies to address invasive species and the inadequacy of existing legislation, policies and programs.[125] The Strategy lamented that less than one percent of marine container shipments are inspected for hitchhiking alien species at Canadian ports.[126] The Strategy was quite general and vague as to what legal responses should be:

> As appropriate, and where feasible, federal departments and agencies and their provincial and territorial counterparts will develop legal and regulatory tools and amend existing legislation and regulations to strengthen measures to prevent, detect, respond, and manage invasive alien species.[127]

Seabed activities

The legal framework governing offshore petroleum exploration and development in Canada, the main type of seabed activity,[128] does not strongly embrace a precautionary approach. None of the multiple statutes governing environmental aspects of offshore petroleum activities mentions the precautionary principle.[129] No specific regulatory requirements have been forged for chemicals used in offshore drilling and production activities. Flexible guidelines have been issued.[130] Management of seismic surveys and threats of seismic sounds from air guns on marine species has been controversial in light of scientific uncertainties[131] and lack of agreement on how precautionary mitigative measures should be.[132]

The precautionary principle is minimally acknowledged in the oil and gas regulatory field. For example, the memorandum of understanding (MOU) between the Canada–Nova Scotia Offshore Petroleum Board and the Department of Fisheries and Oceans[133] for coordinating management roles over offshore petroleum activities recognizes the precautionary approach as one of the principles of cooperation.[134] A similar MOU between the Canada–Nova Scotia Offshore Petroleum Board and Environment Canada also refers to the precautionary principle.[135]

In British Columbia, a Scientific Review Panel was appointed by the British Columbia Minister of Energy and Mines on 19 October 2001 to review scientific, technological and regulatory issues associated with possible lifting of an offshore oil and gas moratorium in place since the early 1970s.[136] Reporting in January 2002, the Panel recognized the precautionary approach but did not extensively discuss the law and policy implications.

The Panel endorsed the 1998 Wingspread Statement on the Precautionary Principle, which advocated the proponent of an activity, rather than the public, should bear the burden of proof.[137] However, the Panel also emphasized the observation in the 2001 Lowell Statement on Science and the Precautionary Principle that "The goal of precaution is to prevent harm, not to prevent progress."[138] The Panel noted the divisions among specialists regarding how precautionary responses should be where there is a credible risk of substantial environmental damage.[139] The Panel recommended that before actual exploration activities take place, a quantitative risk analysis be undertaken along with a thorough cost–benefit analysis.[140]

Living marine resource management and precaution

Fisheries

Various Canadian fisheries policy documents have acknowledged the need for adopting the precautionary approach in fisheries management, but at quite a general level without detailed suggestions for reforms. For example, *A Policy Framework for the Management of Fisheries on Canada's Atlantic Coast*,[141] resulting from multi-stakeholder consultations on future directions for fisheries management, calls for the development and adoption of a comprehensive risk management framework incorporating precaution. The Framework indicates that conservation-oriented decisions will be sought through establishment of target reference points (where resource benefits can be obtained on a sustainable basis), setting limit reference points (where unacceptable risk of serious or irreversible harm may occur), and determining in advance what corrective actions will be taken when limit reference points are approached.[142] The Framework document leaves considerable uncertainty as to what future actions will be. For example, the document states specific actions in Phase II of the Policy Review may include, among others, "establishing parameters for applying precaution within the decision-making framework, including defining negative outcomes and risk tolerances, setting limit reference points for stocks and determining corrective actions."[143]

The Department of Fisheries and Ocean's *Policy Framework for Conservation of Wild Salmon*,[144] although not adopting the precautionary approach as a guiding principle,[145] does call for precaution to be followed in future decision making relating to conservation of wild salmon. Under the precautionary umbrella, the Policy proposes categorizing wild salmon Conservation Units into one of three status zones (green, amber or red) with social and economic factors dominating management decisions in the green zone and biological considerations being paramount in the red zone.[146]

A Policy for Selective Fishing in Canada's Pacific Fisheries[147] adopts the precautionary approach in Principle 1. The Policy establishes conservation of Pacific fisheries as the primary objective and pledges that productivity of the resource will not be compromised because of short-term factors or considerations.

A DFO policy for determining when new fisheries can be initiated[148] also recognizes the precautionary approach. The policy suggests placing the burden on fisheries proponents to provide a reasonable scientific basis for fisheries management through stock assessment information.

Practical efforts to invoke precaution in Canadian fisheries management, however, have been limited.[149] Considerable energies have been focused on determining reference points for selected fish stocks.[150] While fisheries closures have been used effectively, establishment of marine protected areas has lagged. Over-reliance on quota management has occurred and at times pre-cautionary scientific advice ignored. In some cases, the precautionary approach has been invoked without adequate knowledge of ecosystem effects.

Fisheries closures and lag in marine protected areas

Specific areas, determined to be essential for spawning or juvenile habitat, can be closed seasonally or permanently under the *Fisheries Act*. Prior to the adoption of the FAO Code of Conduct for Responsible Fisheries or the ratification of the UN Agreement on Straddling Stocks, the federal Department of Fisheries and Oceans already closed certain areas to fishing. An example is the "Haddock Box" area, on the Scotian Shelf, which has been closed since 1987 to protect juvenile haddock, and thereby allowing the stock to rebuild.[151] More recently, fisheries closures of areas encompassing habitat structure, including the closure of 426 square kilometers in the Northeast Channel between Georges and Browns Bank and 15 square kilometers in the Laurentian Channel for deep-sea coral conservation[152] and areas of the Hecate Strait in the Pacific to protect ancient sponge reefs,[153] signify that the federal government is beginning to recognize the importance of habitat for fisheries conservation. While closed areas offer advantages for seafloor habitat, juvenile species and spawning grounds, their implementation tends to displace fishing effort, which has been a concern in further area closures.

The *Oceans Act* authorizes establishment of marine protected areas (MPAs) as a measure to protect biodiversity and sustain fish populations.[154] However, Canadian efforts to designate and protect such areas have been extremely slow. MPA establishment may not be popular with fishers, and the criteria for protection and selection of areas, as well as specific objectives for protection, have yet to be agreed upon.[155] Only two MPAs have been designated under the *Oceans Act*, the Endeavor Hot Vents off British Columbia[156] and the Gully submarine canyon in the western North Atlantic.[157] DFO lists 12 other Areas of Interest for future designation.[158]

The *Canada National Marine Conservation Areas Act*, receiving Royal Assent on 13 June 2002,[159] also recognizes the precautionary principle but designation of areas is at a political commitment stage. The Act, besides mentioning the precautionary principle in the Preamble,[160] requires the development of management plans within five years after MPAs are established. Primary considerations in development and modification of plans are to be the "principles of ecosystem management and the precautionary principle."[161] In the Speech from the Throne on 30 September 2002, the Government of Canada announced a commitment to create five new national marine conservation areas over the next five years.[162]

Over-reliance on quota management and ignoring of precautionary scientific advice

Canada's experiences with fish stock collapse, such as the crisis in the East Coast groundfishery since 1992, help demonstrate the limitations with the primary method of fisheries management, the setting of total allowable catches (TACs) based on biological reference points. Simply managing according to reference points[163] does not lead to sustainable fisheries, due to the uncertainty in reference point estimation and general lack of specific management objectives.[164] Estimates of fishing mortality are biased and do not include discards[165] or indirect mortality.[166] Fisheries regulations encourage large amounts of discarding, and political decisions often overrule precautionary scientific advice in fisheries management.[167]

An example of the non-precautionary approach spawned by over-reliance on setting TACs and failure to follow scientific precautionary advice may be seen in the events surrounding the collapse of commercial fisheries for Atlantic cod (*Gadus morhua*), especially off Newfoundland. Massive social disruption, including loss of work for some 40,000 fisheries workers has occurred, and the Canadian government has spent approximately CDN$4 billion on special fisheries relief programs.[168] Fundamental to this collapse was the fact that fishing mortality was consistently much higher than the recommended level, and no effective measures were taken to ensure that young fish were not discarded or that spawning biomass was maintained at adequate levels.[169]

Following the 1992 moratorium on directed fishing, numerous measures were taken with little regard for precaution. Catches continued through by-catch from other groundfish species, sentinel fisheries that began in 1995 and recreational/food fisheries.[170] Exploitation rates calculated from the results of tagging experiments conducted through the sentinel fishery estimated fishing mortality to be between 10–30 percent of the total fish biomass[171] despite the ban on directed fishing. In 2000, the Fisheries Resources Conservation Council (FRCC) recommended a TAC not to exceed 7,000 tonnes,[172] to be taken in the recreational, sentinel and by-catch fisheries. In 2002, the FRCC recommended a TAC of 5,600 tonnes, including all fisheries and by-catch, even though there was known to be an estimated overshoot of the 2001 quota by 1,900 tonnes.[173]

The failure of the cod to return to even a fraction of historical levels prompted the federal Minister of Fisheries and Oceans to declare a complete closure of the cod fishery in the Gulf of St. Lawrence and off northeast Newfoundland and Labrador on 24 April 2003.[174] The Newfoundland and Labrador population of Northern cod was declared endangered by the Committee on the Status of Endangered Wildlife in Canada (COSEWIC) in May 2003.[175] If measures are to be taken in the name of precaution, they must be acted upon before a species declines to the point of being classified as endangered. Indeed, the application of the precautionary approach is to ensure that species do not decline to such low abundances.[176] The extremely low abundances of groundfish on the east coast of Canada renders previously determined reference points inadequate for further management of these stocks.

One of the emerging issues in marine conservation is effects of bottom trawling on target populations and the sea floor.[177] While Canada advocates conservation of marine resources, during his address to the United Nations General Assembly in November 2004 on the Sustainable Fisheries Resolution[178] the Minister of Fisheries, Geoff Regan, maintained that no fishing gear is inherently destructive.[179] Canada is one of the few industrialized fishing nations that has yet to recognize fully the differential effects of fishing gear on the sea floor and continues to be quite unsupportive[180] of high seas conservation initiatives related to bottom trawling.[181] This issue is tightly linked to precaution, as there is little scientific information regarding the effects of bottom trawling on the high seas and information on fish stocks is lacking, making it difficult to calculate fishing effort that would ensure maximum sustainable yield (MSY).

Invoking precaution with inadequate knowledge

There is also a danger of invoking the precautionary approach without adequate knowledge about ecosystem effects, as shown by two Canadian examples. In 1999, the FRCC,[182] acting in the interests of conserving groundfish stocks in the northwest Atlantic, recommended the following: "In applying the precautionary approach to groundfish management, action must be taken immediately in order to improve opportunities for the conservation and recovery of cod ... We strongly suggest that the seal herds be reduced by up to 50 percent of their current population levels."[183]

Here, precaution was suggested, as a large scale culling of seals, with no real knowledge of the effect of this population on the cod fishery. Such a suggestion does not take into account the complexities of marine food webs nor does it adhere to an ecosystem approach. However, fishermen and managers alike have often blamed the burgeoning seal population for the continued low biomass of cod. More recent findings suggest that seal populations may in fact have a positive effect on cod stocks in the Gulf of St. Lawrence.[184]

In response to the FRCC recommendation, the Minister appointed a panel to investigate both scientific and management objectives for the seal populations. The results of this panel showed that there were no reference points for seal population management, nor were there specific management objectives.[185]

In a report to the Northwest Atlantic Fisheries Organization (NAFO),[186] Canada proposed that in the interests of protecting juvenile groundfish species on the Southern Grand Banks, fishing for Greenland halibut (turbot) be restricted to below 700 meters. This would be a precautionary measure taken to ensure that juveniles are not subject to dangerously high fishing mortality. The difficulty with this management suggestion (which was in fact rejected by NAFO) is that there was no consideration given to the effects of increasing fishing pressure in deeper waters. Making a decision based on what is known about the by-catch of juveniles, without considering the risk involved in limiting a fishery to depths where little is known, is an example of improperly invoking precaution.

Need for a broader precautionary approach in fisheries

With a focus on the scientific basis for precaution through reference points and control limits, Canada has lagged behind in strongly implementing precaution in support of sustainable fisheries[187] and a broader approach to precaution needs to be considered. A broader approach would emphasize the collection of ecological data, multidisciplinary inputs into management decisions, increased communication between fishers and managers, and a commitment to preservation of fishing communities.[188]

Aquaculture

The precautionary principle has not strongly infiltrated the field of aquaculture management in Canada. The various memoranda of understandings, whereby the federal government and provinces have agreed to jurisdictional arrangements for controlling marine aquaculture developments, do not mention the key principles of sustainable development including the precautionary principle.[189] No provincial aquaculture legislation has expressly incorporated reference to the principle.[190] The Department of Fisheries and Oceans, Office of Sustainable Aquaculture, has developed an *Aquaculture Policy Framework*,[191] but the policy document is quite general as to how the precautionary approach might apply to aquaculture activities:

> DFO's use of the precautionary approach in the context of aquaculture development will be informed by the *Oceans Act* and federal direction regarding risk management, including the application of the precautionary approach.[192]

The 2004 Report of the Commissioner of the Environment and Development assessing the adequacy of salmon aquaculture management by Fisheries and Oceans Canada was very critical.[193] The Report noted significant gaps in scientific knowledge about the potential environmental effects of salmon aquaculture, lack of credible siting criteria for aquaculture sites, limited monitoring of aquaculture operations and lack of progress in dealing with the deposit of deleterious substances from fish farms.[194]

The details of how the precautionary approach might affect the regulation of genetically modified (GM) aquaculture products remain to be worked out. *DFO's Aquaculture Policy Framework* does not specifically address the issue of genetic modification of fish but appears supportive of technological innovations.[195] A Royal Society of Canada Expert Panel Report on the Future of Food Biotechnology in Canada,[196] undertaken on behalf of Health Canada, the Canadian Food Inspection Agency and Environment Canada, recommended "it would be prudent and precautionary to impose a moratorium on the rearing of GM fish in aquatic facilities" because of the paucity of scientific information pertaining to genetic interactions between cultured and wild fish.[197] The Department of Fisheries and Oceans, as part of the action plan of the Government of Canada in response to the Royal Society of Canada Report on Food Biotechnology, pledged to develop specific regulations governing transgenic aquatic organisms.[198] The action plan indicates DFO agreement with the need to keep reproductively capable transgenic fish and transgenic aquatic organisms in secure land-based facilities.[199] In the seventh progress report on Action Plan implementation, Health Canada and the Canadian Food Inspection Agency, in collaboration with DFO and Environment Canada, were reported as leading the portion of the project, *Transforming the Horizontal Regulatory Governance of Biotechnology in Canada*, involving the development of common regulatory governance principles.[200]

A *National Code on Introductions and Transfers of Aquatic Organisms*,[201] prepared by a federal–provincial Task Group on Introduction and Transfers, has been issued for assessing proposals to move aquatic organisms intentionally from one water body to another, including for aquaculture purposes.[202] The Code seeks to establish a standard set of risk assessment and approval procedures and recognizes the precautionary approach as a guiding principle. It notes that if a risk assessment outcome is uncertain, priority should be given to conserving the productive capacity of the native resource.[203]

Biodiversity (species and habitat) protection

Unlike Australia,[204] Canada has not adopted broad biodiversity protection legislation, but has enacted a *Species at Risk Act* that demonstrates again a rather wandering approach to precaution.[205] The Act embraces the precautionary principle in various ways including a preambulary reference[206] and placing a legal duty on ministers[207] required to develop recovery strategies and action plans for endangered or threatened species to consider the precautionary principle.[208] However, the legislation wanders away from a strong precautionary approach in various ways, including leaving the actual listing of species at risk largely to political discretion.[209] The Act also leaves wide discretion for the Minister of Fisheries and Oceans to issue "incidental harm" permits allowing activities to proceed even though they may affect a listed species, its residence or its critical habitat.[210] Various prohibitions, including those against harming individuals of listed endangered/threatened wildlife species and damaging the residences of such species, will not apply to activities permitted by a recovery strategy or action plan and also authorized under an Act of Parliament.[211]

Canada has yet to develop detailed regulatory requirements for whale watching in accord with the precautionary approach. Existing *Marine Mammal Regulations*[212] under the *Fisheries Act* provide only a general obligation for persons not to disturb a marine mammal.[213] A recent report has documented the growing scientific concerns over the effects of whale watching vessels on whale behavior and life processes, and has urged more precautionary controls be imposed through regulation, for example, limiting the number of vessels near whales, establishing distance and speed restrictions, and covering the duration of time any one vessel can spend in contact with an animal or group of marine mammals.[214] The Department of Fisheries and Oceans is soliciting views regarding possible amendments to the *Marine Mammal Regulations* and intends to develop regulatory proposals for public discussion.[215]

Climate change and precaution

Perhaps Canada has displayed the most serious case of policy wanderings in responding to potential threats of climate change. The significant environmental risks associated with climate change, such as melting of sea ice in the Arctic and potential disastrous effects on wildlife, such as polar bears and ice-dependent seals,[216] would seem to raise the "classic case" of the need for strong precautionary actions.[217] However, Canada has wandered away from firm responses. The UN Framework Convention on Climate Change,[218] ratified by Canada in 1992,[219] endorses a weak "cost-effective" policy towards precaution as advocated by the United States:[220]

> The Parties should take precautionary measures to anticipate, prevent or minimize the causes of climate change and mitigate its adverse effects. Where there are threats of serious or irreversible damage, lack of full scientific certainty should not be used as a reason for postponing such measures, taking into account that policies and measures to deal with climate change should be cost-effective so as to ensure global benefits at the lowest possible cost. To achieve this, such policies and measures should take into account different socio-economic contexts, be comprehensive, cover all relevant sources, sinks and reservoirs of greenhouse gases and adaptation, and comprise all economic sectors . . .[221]

Although adopting a commitment under the Kyoto Protocol[222] to reduce greenhouse gas emission by 6 percent from 1990 levels during the commitment period 2008–12, Canada has been criticized for trying to "weasel out" of emission reduction commitments. In climate change negotiations, Canada was a leading advocate for counting carbon dioxide soaked up by forests and soils (carbon sinks) against emission targets.[223] Former Federal Environment Minister, David Anderson, fought hard to gain international acceptance for Canada receiving Kyoto credits for exporting clean-energy exports to the United States of natural gas and hydro-electric power.[224]

While Canada did ratify the Kyoto Protocol in late 2002,[225] how Canada will implement Kyoto reduction commitments is still very much "up in the air." *A Climate Change Plan for Canada*[226] set only very general directions for action in seven areas.[227] A further plan, *Moving Forward on Climate Change: A Plan for Honouring the Kyoto Commitment*,[228] released on 13 April 2005,[229] largely sets out only broad response parameters. For example, for the about 700 companies considered to be large final emitters (LFEs), including companies in the mining, manufacturing, oil and gas, and thermal electricity sectors, a system of addressing their greenhouse emissions is proposed that is largely market-based. While LFE companies could choose to invest for in-house emission reductions, they could also comply with emission reduction targets by various other compliance options including the purchase of emission reductions from other LFE companies and the purchase of "green" international credits recognized by Canada.[230] The Plan, while proposing LFE regulations to be developed in 2005 to, among other things, set rules for domestic and international trading, does not spell out details and pledges that the government will consult with Canadians on how the *Canadian Environmental Protection Act, 1999* might be used to implement the LFE system.[231]

The Plan also relies heavily on incentives and voluntary approaches. For example, one of the most criticized parts of the Plan is the decision to leave automobile industry reduction commitments to a "soft" memorandum of understanding.[232] The industry has agreed to reduce greenhouse gas emissions in 2010 by 5.3 megatonnes from the light duty vehicle sector (cars and light duty trucks) through such means as advanced emission technologies and more alternative fuel and hybrid vehicles.[233]

Conclusion

To date, Canada has waded and wandered in the tricky currents of precaution. Only cautious steps have been taken to incorporate the precautionary principle into federal legislation while the provinces and territories have hardly tested the legal waters. With the exception of ocean dumping control, Canada has largely wandered towards weak versions of precaution by emphasizing the need for "sound science" and cost-effectiveness and giving primacy to economic growth.

Given political, social and cultural differences in Canada, the precautionary principle will likely provide an ongoing touchstone for discourses about ecosystem protection and broader public interests.[234] As noted by Stanley Fish, legal principles are not neutral but provide catalytic sources for interpretive arguments:

> Principles don't by themselves aggravate or produce anything; principles never appear "by themselves" but are deployed and configured by partisan agents in particular situations. Principles, in short, are part of the arsenal or equipment of prudence, not an alternative to it.[235]

Tensions are certain to continue over application of the precautionary principle. The extent to which the precautionary approach should be constrained by scientific research and traditional risk assessment is one area of contention.[236] The appropriate ethical viewpoint directing precautionary action is another issue with various competing philosophies, such as deep ecology favoring strong environmental rights and utilitarian approaches embracing cost–benefit and risk–benefit analysis.[237]

Key questions remain. What institutional innovations are needed and politically realistic at the national and international levels to support precautionary decision making?[238] To what extent will the parameters of the precautionary principle be dictated through bureaucratic fiats and judicial decisions rather than determined through discursive processes?[239] Navigation through the tricky currents of precaution is likely to be a long and rough voyage.[240]

Acknowledgments

Professor VanderZwaag, the lead author, would like to acknowledge the Social Sciences and Humanities Research Council (SSHRC) for supporting his research on the roles of sustainable development principles and human rights norms in controlling toxic chemicals and the research assistance of Melanie MacLellan, Anastasia Makrigiannis and Emma Butt, Dalhousie Law School. Further research support was provided by AquaNet, a Centres of Excellence Network for Aquaculture in Canada, based at Memorial University, and funded by the Natural Sciences and Engineering Research Council of Canada and the SSHRC through Industry Canada. This chapter is an updated and revised version of a paper published under the same title in (2002–03) 34 *Ottawa Law Review* 117–58.

Notes

1 Almost every recent international environmental agreement and declaration has included a version of the precautionary principle/approach. For a partial listing, see Carolyn Raffensperger and Joel A. Tickner (eds) *Protecting Public Health & the Environment: Implementing the Precautionary Principle* (Island Press, Washington, DC: 1999) Appendix B.

2 For example, a precautionary approach is called for in the FAO Code of Conduct for Responsible Fisheries (UN Food and Agriculture Organization, Rome: 1995) and the 1995 UN Convention on Straddling Fish Stocks and Highly Migratory Fish Stocks, 34 *International Legal Materials* 1542 (1995). See, generally, S.M. Garcia "The precautionary principle: Its implications in capture fisheries management" (1994) 22 *Ocean and Coastal Management* 99; Justin Cooke and Michael Earle "Towards a precautionary approach to fisheries management" (1993) 2 *Review of European Community and International Environmental Law* 252.

3 Common sense sayings include: "an ounce of prevention is worth a pound of cure," "a stitch in time saves nine" and "if in doubt don't pump it out." David VanderZwaag "The precautionary principle in environmental law and policy: Elusive rhetoric and first embraces" (1998) 8 *Journal of Environmental Law and Practice* 355 at 358.

4 See Arie Trouwborst *Evolution and Status of the Precautionary Principle in International Law* (Kluwer Law International, The Hague: 2002) at 10–11.

5 Andrew Jordan and Timothy O'Riordan "The Precautionary Principle in Contemporary Environmental Policy and Politics" in Raffensperger and Tickner, note 1, chapter 1 at 24.

6 For recent reviews of implementation challenges, see David VanderZwaag "The precautionary principle and marine environmental protection: Slippery shores, rough seas, and rising normative tides" (2002) 33 *Ocean Development and International Law* 165; Christopher D. Stone "Is there a precautionary principle?" (2001) 31 *Environmental Law Reporter* 10790.

7 For example, the United Nations Food and Agriculture Organization (FAO) has preferred the term "approach" as it is "weaker" in meaning allowing considerations of cost-effectiveness and local capabilities. See *The Precautionary Approach to Fisheries with Reference to Straddling Fish Stocks and Highly Migratory Fish Stocks*, United Nations Conference on Straddling Fish Stocks and Highly Migratory Fish Stocks, New York, 14–31 March 1994. U.N. Doc. A/CONF. 164/INF/8. This paper hereinafter will refer to both the precautionary approach and principle.

8 *United Nations Conference on Environment and Development: Rio Declaration on Environment and Development*, 14 June 1992, 31 *International Legal Materials* 874 at 879 (Rio Declaration). Principle 15 of the Rio Declaration states:

> In order to protect the environment, the precautionary approach shall be widely applied by States according to their capabilities. Where there are threats of serious or irreversible damage, lack of full scientific certainty shall not be used as a reason for postponing cost-effective measures to prevent environmental degradation.

9 What is meant by risk assessment may be the source of some debate. For a discussion of the need to move from traditional risk assessment seeking to quantify threats towards multidisciplinary risk assessment garnering wisdom from the natural and social sciences, see Nicolas de Sadeleer *Environmental Principles in an Age of Risk: From Political Slogans to Legal Rules* (Oxford University Press, Oxford: 2002) 180–95.

10 The role of risk assessment in risk management under uncertainty is a special area of divisiveness. For discussions, see Paul C. Lin-Easton "It's time for environmentalists to think small – real small: A call for the involvement of environmental lawyers in developing precautionary policies for molecular nanotechnology" (2001) 14 *Georgetown International Environmental Law Review* 107 at 128–9; Kenneth R. Foster, Paolo Vecchia and Michael H. Repacholi "Risk management: Science and the precautionary principle" (2000) 288 *Science*

978–9; Wybe Th. Douma "The precautionary principle in the European Union" (2000) 9 *Review of European Community and International Environmental Law* 132 at 142.

11 For a recent call for an independent regulatory authority to broaden precautionary discourse about new biotechnology, see William Leiss and Michael Tyshenko "Some Aspects of the 'New Biotechnology' and its Regulation in Canada" in Debora L. VanNijnatten and Robert Boardman (eds) *Canadian Environmental Policy: Context and Cases*, 2nd edition (Oxford University Press Canada, Don Mills, Ontario: 2002) chapter 17.

12 For reviews of the spectrum of regulatory responses and strategies, see John S. Applegate "The precautionary preference: An American perspective on the precautionary principle" (2000) 6 *Human and Ecological Risk Assessment* 413 at 415–16; Joel A. Tickner "A Map Toward Precautionary Decision Making" in Raffensperger and Tickner, note 1 at 171–2; Adrian Deville and Ronnie Harding *Applying the Precautionary Principle* (The Federation Press, Sydney: 1997).

13 See Carl F. Cranor "Asymetric information, the precautionary principle, and burdens of proof" in Raffensperger and Tickner, note 1, chapter 4 at 93–4.

14 VanderZwaag, note 6 at 168.

15 Alternatives assessment involves public examination of a full range of alternatives to a potentially damaging human activity and includes the fundamental question of whether a potentially hazardous activity is necessary and what less hazardous options are available. See Mary O'Brien "Alternatives assessment: Part of operationalizing and institutionalizing the precautionary principle" in Raffensperger and Tickner, note 1, chapter 12 at 208.

16 See Bruce Pardy "Applying the precautionary principle to private persons: Should it affect civil and criminal liability?" (2002) 43 *Les Cahiers de Droit* 63.

17 See Heather Breeze *Conservation Lost at Sea: Discarding and Highgrading in the Scotia-Fundy Groundfishery in 1998* (Ecology Action Centre and Conservation Council of New Brunswick, Halifax: 1998) 15; and Ransom A. Myers, Susanna D. Fuller and Daniel G. Kehler "A fisheries management strategy robust to ignorance: Rotational harvest in the presence of indirect fishing mortality" (2000) 57(12) *Canadian Journal of Fisheries and Aquatic Sciences* 2357.

18 See Gordon Mertz and Ransom A. Myers "A simplified formulation for fish production" (1998) 55 *Canadian Journal of Fisheries and Aquatic Sciences* 478.

19 International Pacific Halibut Commission, *The Pacific Halibut: Biology, Fishery and Management* Technical Report 40 (IPHC, Seattle: 1998).

20 The National Oceanographic and Atmospheric Association in the United States has conducted gear assessments. Michael C. Barnette "A review of the fishing gear utilized within the Southeast Region and their potential impacts on essential fish habitat" (2001) *National Oceanographic and Atmospheric Association Technical Memorandum NMFS-SEFSC-449* 62.

21 The Marine Stewardship Council has developed criteria for certification of fisheries. This endeavor began as a partnership between World Wildlife Fund and Unilever, one of the world's largest fish companies. No fisheries were certified in Canada as of May 2002 and six fisheries were certified globally. Online at: www.msc.org/ (accessed 24 January 2004).

22 Canadian Council of Ministers of the Environmental. Online at: www.ccme.ca/assets/pdf/ cws_accord_env_harmonization.pdf (accessed 24 January 2004) [hereinafter Harmonization Accord].

23 Government of Canada, Privy Council Office *A Framework for the Application of Precaution in Science-based Decision Making about Risk.* Online at:

www.pro_bcp.gc.ca/docs/Publications/precaution/precaution_e.pdf (accessed 2 December 2003).

24 For a general review of caselaw developments relating to precaution see Elizabeth Fisher "Is the precautionary principle justiciable?" (2001) 13 *Journal of Environmental Law* 315.

25 Harmonization Accord, note 22 at principle 2.

26 Three sub-agreements were adopted in January 1998 in the areas of inspections, standards and environmental assessment.

27 Online at: www.ccme.ca/assets/pdf/cws_envstandards_subagreement.pdf (accessed 24 January 2004).

28 Ibid., at principle 3.1.3.

29 Ibid., at principle 3.1.2.

30 Ibid., at principle 3.1.7.

31 Online at: www.ccme.ca/assets/pdf/envtlassesssubagr_e.pdf (accessed 24 January 2004).

32 Ibid., at section 3 (principles).

33 Note 23. Based on the previously released *A Canadian Perspective on the Precautionary Approach/Principle: Discussion Document*, the Framework was prepared by Agriculture and Agrifood Canada, Canadian Environmental Assessment Agency, Canadian Food Inspection Agency, Department of Fisheries and Oceans, Department of Foreign Affairs and International Trade, Environment Canada, Finance Canada, Health Canada, Industry Canada, Justice Canada, Natural Resources Canada, Privy Council Office, Transport Canada and Treasury Board Secretariat.

34 See David Winickoff, Sheila Jasonoff, Lawrence Bursch, Robin Grove-White and Brian Wynne "Adjudicating the GM food wars: Science, risk and democracy in world trade law" (2005) *Yale Journal of International Law* 81–123.

35 Ibid., at 85.

36 See Mary O'Brien *Making Better Environmental Decisions: An Alternative To Risk Assessment* (MIT Press, Cambridge: 2000).

37 For a good overview of cases, see J. Aboucher "Implementation of the Precautionary Principle in Canada" in Timothy O'Riordan, James Cameron and Andrew Jordan (eds) *Reinterpreting the Precautionary Principle* (Cameron, London: 2001) chapter 10 at 245–51.

38 Indexed as *114957 Canada Ltée (Spraytech, Societé d'arrosage)* v. *Hudson (Ville)*, [2001] 2 S.C.R. 241. For reviews of the decision, see Howard Epstein "Case comment: *Spraytech* v. *Town of Hudson*" (2001) 19 *Maritime Provinces Law Reports* (3d) 1; Gibran Van Ert *Using International Law in Canadian Courts* (Kluwer Law International, The Hague: 2002) at 225–7. For a discussion of the lower court rulings and the legal issues raised, see John Swaigen "The *Hudson* case: Municipal powers to regulate pesticides confirmed by Quebec Courts" (2000) 34 *Canadian Environmental Law Reports* (N.S.) 162.

39 The Court indicated there might be sufficient state practice to establish the precautionary principle as a principle of customary international law.

40 [1999] 2 S.C.R. 817.

41 Ibid., at 861.

42 See Ecology Action Centre and Sierra Legal Defence Fund, Media Release "Lawsuit launched to protect fish habitat from destructive draggers" (4 July 2001). Online at: www.sierralegal.org/m_archive/2001/pr01_07_04.html (accessed 24 January 2004).

43 R.S.C. 1985, c. F-14.

44 S.C. 1992, c. 37.

45 (2004) 9 C.E.L.R. (3d) 161 (F.C.T.D.).

46 [2002] N.S.J. No. 298 (S.C.) (QL).

47 Ibid., at para. 44.

48 In *Brighton*, Justice MacDonald suggested at least reasonableness simpliciter if not patent unreasonableness was the appropriate standard of review. For a discussion of the current approach to substantive review and possible future directions, see William Lahey and Diana Ginn "After the revolution: Being pragmatic and functional in Canada's trial courts and courts of appeal" (2002) 25 *Dalhousie Law Journal* 259.

49 See Warwick Gullett "The precautionary principle in Australia: Policy, law & potential precautionary EIAs" (2000) 11 *Risk: Health, Safety and Environment* 93 at 117.

50 Nova Scotia's *Environment Act*, S.N.S. 1994–95, c. 1, does adopt the precautionary principle in its purpose section but not in the substantive provisions governing the environmental impact assessment process itself in Part IV.

51 S.C. 1992, c. 37.

52 Ibid., s. 5.

53 A 1999 Cabinet Directive on the Environmental Assessment of Policy Plan and Program Proposals has adopted a flexible approach to such assessment with public participation not ensured and limited coverage to plans, policies and programs raising significant environmental concern and subject to Cabinet or Ministerial approval. The Directive has been described as lacking transparency and being widely ignored. See Robert B. Gibson "The major deficiencies remain: A review of the provisions and limitations of Bill C-19, an Act to Amend the *Canadian Environmental Assessment Act*" (2001) 11 *Journal of Environmental Law and Practice* 83 at 101.

54 While CEAA provides for other types of assessment, namely, comprehensive studies, panel reviews and possibly mediation, some 99 percent of assessments are limited to screening review. For example, during the 1999–2000 period, 5,662 screenings were initiated compared with nine comprehensive studies and three panels. Canadian Environmental Assessment Agency *Performance Report for the Period Ending March 21, 2000* (Minister of Public Works and Government Sources Canada, Ottawa: 2000) 31.

55 Pursuant to section 58(1.1) of CEAA, participant funding is limited to mediations and review panel assessments.

56 Section 37 of CEAA leaves wide discretion for decision-makers to allow questionable projects to proceed by open-textured determinations such as "not likely to cause significant adverse environmental effects" and "mitigation measures" but the most serious "loophole" may be the authority for responsible authorities to even allow projects likely to cause significant adverse effects to proceed if "they can be justified in the circumstances" (CEAA, s. 37(1)(a)(ii)). Justified is not defined under the Act, and although pursuant to section 58(1)(a) the Minister of Environment is authorized to develop criteria for determining justification, no such criteria have been developed.

57 See Shauna Finlay "Sustainable Development and the *Canadian Environmental Assessment Act*" (1999) 8 *Journal of Environmental Law and Practice* 377 at 385–6. Section 16 of CEAA provides responsible authorities discretion to require consideration of the need for the project and alternatives to the project. For comprehensive studies, review panel reviews and mediations, an assessment of *alternative means* of carrying out the project is required not a comprehensive alternatives assessment.

58 For a critique and a report seeking to lay a foundation for further review of CEAA, see House of Commons Standing Committee on Environment and Sustainable Development, *Sustainable Development and Environmental Assessment:*

Beyond Bill C-9 (June 2003). Online at: www.parl.gc.ca/InfoComDoc/37/2/ENVI/Studies/Reports/envirp02-e.htm (accessed 24 January 2004).

59 *An Act to amend the Canadian Environment Assessment Act*, S.C. 2003, c. 9. An especially strong case can be made in the case of wildlife species at risk that the onus of proof should be on the proponent to demonstrate the lack of significant adverse effects on wildlife at risk or biological diversity. See Canadian Wildlife Service *Environmental Assessment Best Practice Guide for Wildlife at Risk in Canada* 1st edn. (27 February 2004) at Guideline 10.

60 House of Commons Standing Committee, note 58 at 8.

61 Canadian Environmental Assessment Agency, Nova Scotia Department of the Environment, National Energy Board, Natural Resources Canada, Nova Scotia Department of Natural Resources, Canada-Nova Scotia Offshore Petroleum Board *The Joint Public Review Panel Report: Sable Gas Projects* (October 1997) 31–2.

62 Ibid., at 33.

63 S.C. 1996, c. 31.

64 Ibid., ss. 29, 30.

65 Government of Canada *Canada's Oceans Strategy: Our Oceans, Our Future* (Fisheries and Oceans Canada, Oceans Directorate, Ottawa: 2002).

66 Ibid., at 11–12.

67 Ibid., at 11.

68 R.S.C. 1985, c. A-17 as amended by S.C. 1995, c. 43.

69 The first strategies were due in 1997, and the legislation requires departmental sustainable development strategies to be updated at least every three years.

70 Online at: www.dfo-mpo.gc.ca/sds-sdd2005-06/Index_e.htm (accessed 13 April 2005).

71 The federal government has encouraged a Smart Regulation agenda, which has raised concerns among environmentalists regarding the potential for cost-efficiency and business competitiveness, to roll back regulatory commitments to protect the environment. See External Advisory Committee on Smart Regulation *Smart Regulation: A Regulatory Strategy for Canada* (September 2004). Online at: www.pco-bcp.gc.ca/smartreg-regint/en/08/summ.html (accessed 13 April 2005).

72 S.C. 1999, c. 33.

73 Ibid., s. 125.

74 Floating debris or other material that can create marine pollution must be removed to the maximum extent possible and the substances must not pose a serious obstacle to fishing or navigation after disposal. Ibid., Sch. 5(3).

75 Such substances must not have a significant adverse effect, other than a physical effect, on the sea or seabed and such dumping is to occur in locations where disposal at sea is the only practicable option and without posing a serious obstacle to fishing or navigation. Ibid., Sch. 5(6).

76 Ibid., Sch. 6(2)(3). A permit to dispose of waste is to be refused if opportunities exist to reuse, recycle or treat the waste without undue risks to human health or the environment or disproportionate costs. Ibid., at Sch. 6(6).

77 For a discussion of the numerous uncertainties and complexities over federal and provincial jurisdiction to control environmental matters, since Canada's *Constitution Act, 1867* fails to specifically allocate jurisdiction over the environment, see Marie-Ann Bowden "Jurisdictional Issues" in Elaine L. Hughes, Alastair R. Lucas and William A. Tilleman (eds) *Environmental Law and Policy* (3rd edn.) (Emond Montgomery Publications Ltd., Toronto: 2003) chapter 2.

78 S.C. 2002, c. 28. Not yet in force at the time of writing.

79 The NPA was finalized in June 2000. Online at: www.npa-pan.ca/npa/index_e.htm#26 (accessed 24 January 2004).

80 The Preamble states:

> Whereas the Government of Canada is committed to implementing the precautionary principle that where there are threats of serious or irreversible damage, lack of full scientific certainty shall not be used as a reason for postponing cost-effective measures to prevent environmental degradation.

81 S.C. 1999, c. 33, s. 2 provides:

> In the administration of this Act, the Government of Canada shall . . . exercise its powers in a manner that protects the environment and human health, applies the precautionary principle that, where there are threats of serious or irreversible damage, lack of full scientific certainty shall not be used as a reason for postponing cost-effective measures to prevent environmental degradation and promotes and reinforces enforceable pollution prevention approaches . . .

82 Section 84 allows the Minister of Environment to permit or prohibit the manufacture or import of new substances suspected of being toxic or capable of becoming toxic, and the general administrative duty to follow the precautionary principle should guide such decisions.

83 CEPA, 1999, s. 6(1.1).

84 Section 95(3) requires the Minister of Environment to provide an opportunity to the Committee for giving advice on proposed regulations.

85 Section 76.1. For a discussion of some uncertainty about what a weight of evidence approach means in light of a lack of definition under the Act, see Abouchar, note 37 at 239–40.

86 The chemicals allowed to remain on the market are listed on the Domestic Substances List, an inventory of over 23,000 substances manufactured in or imported into Canada on a commercial scale between January 1984 and December 1986. Environment Canada *CEPA Annual Report: April 1998 to March 1999* (Minister of Public Works and Government Services, Ottawa: 1999) at 13.

87 Section 74. However, CEPA, 1999 in section 73 requires the Ministers of Health and Environment, within seven years of the Act receiving Royal Assent, to categorize the over 23,000 substances on the Domestic Substances List on the basis of available information into those presenting the greatest potential for human exposure in Canada and those which are persistent and bioaccumulative.

88 Section 65(3).

89 For a further discussion of the unpredictability surrounding inclusion of the precautionary principle in CEPA, see Marcia Valiante "Legal Foundations of Canadian Environmental Policy: Underlining our Values in a Shifting Landscape" in Van Nijnatten and Boardman, note 11, chapter 1.

90 Sections 120–1.

91 Section 56.

92 R.S.C. 1985, c. F-14.

93 Ibid., ss. 36(3) and 35.

94 The regulations are: *Chlor-Akali Mercury Liquid Effluent Regulations*, C.R.C., c. 811; *Meat and Poultry Products Plant Liquid Effluent Regulations*, C.R.C., c. 818; *Metal Mining Liquid Effluent Regulations*, C.R.C., c. 819; *Petroleum Refinery Liquid Effluent Regulations*, C.R.C., c. 828; *Potato Processing Plant Liquid Effluent Regulations*, C.R.C., c. 829; and *Pulp and Paper Effluent Regulations*, SOR/92-269.

95 For example, the *Potato Processing Plant Liquid Effluent Regulations* authorize

deposits of biochemical oxygen demanding matter (BOD) and total suspended solids (TSS) based on the amount of raw potatoes processed, with potato chip plants authorized to deposit 1.5 kg/tonne of BOD and 2.1 kg/tonne of TSS on a daily basis.

96 The *Pulp and Paper Effluent Regulations* designate acutely lethal effluent as a prescribed "deleterious substance" under the *Fisheries Act*, and acutely lethal is defined in relation to effluent as that which at 100 percent concentration kills more than 50 percent of the rainbow trout subjected to it during a 96-hour test period.

97 S.C. 2002, c. 28.

98 Ibid., s. 7(6).

99 See World Wildlife Fund and Canadian Environmental Law Association, *Bill C-53 Briefing Note.* Online at: www.cela.ca/toxics/C-53briefingnote.pdf (accessed 24 January 2004).

100 Section 17(1) requires the Minister of Health to initiate a special review of a pesticide registration if the Minister has reasonable grounds to believe that the health or the environmental risks of the product are, or its value is, unacceptable.

101 Section 20(1) provides:

> The Minister may cancel or amend the registration of a pest control product if ... in the course of a re-evaluation or special review, the Minister has reasonable grounds to believe that the cancellation or amendment is necessary to deal with a situation that endangers human health or safety or the environment, taking into account the precautionary principle set out as subsection (2).

102 See comments of Mrs. Karen Kraft Sloan (North York Lib.), House of Commons Debates, 15 April 2002, *Hansard* Vol. 137, No. 168; and David R. Boyd *Unnatural Law: Rethinking Canadian Environmental Law and Policy* (UBC Press, Vancouver: 2003) at 123.

103 See comments of Hon. Charles Caccia (Davenport Lib.), House of Commons Debates, 9 April 2002, *Hansard* Vol. 137, No. 164. Section 7(7) of the Act may be viewed as adopting a weak version of precaution since the Minister of Health is mandated to apply a "scientifically based approach" in determining acceptable risks. The Minister is also required to apply margins of safety for various vulnerable sub-groups including pregnant women, children and seniors. For a review of the "margin of safety" approach in United States' environmental laws, see Frank B. Gosse "Paradoxical perils of the precautionary principle" (1996) 53 *Washington and Lee Law Review* 851 at 855.

104 The Agency was created within Health Canada in 1995 to assist the Minister of Health in administering the Act.

105 The then Minister of Health, Anne McLellan, in beginning second reading debate, stated:

> One does have to remember that federal authority for the Pest Control Products Act relies primarily on the criminal law power . . . To include in this legislation a ban on the use of pesticides for what people refer to as cosmetic use would be exposing individuals to criminal prosecution for engaging in an activity which has not been proven to constitute an unacceptable risk. Such a measure I would submit would be beyond the proper scope of the criminal law power. (House of Commons Debates, 21 March 2002, *Hansard*, Vol. 137, No. 161)

106 NPA, note 79 at 37.

107 Ibid., at 67.
108 Ibid., at 88.
109 The document states: "Canada's NPA is based upon the principles of sustainable development, integrated management and the precautionary approach". Ibid., at 4.
110 Chapter 1 of the NPA suggests a lack of "political fervor":

> Many programs are already in place, or are being actively developed, to protect the marine environment. The NPA takes into account the priorities and actions of these existing programmes and recognizes the cost-effectiveness of building upon them (Ibid., at 1).

111 *Marine Liability Act*, S.C. 2001, c. 6, s. 51.
112 *Garbage Pollution Prevention Regulations*, C.R.C. 1978, c. 1424.
113 R.S.C. 1985, c. S-9. A revamped *Canada Shipping Act, 2001*, S.C. 2001, c. 26, received Royal Assent on 1 November 2001 but many provisions are not yet in force and will depend on further orders of the Governor in Council pursuant to s. 334. The revised Act does not mention the precautionary principle.
114 SOR/93-3 [hereinafter OPPR].
115 1973 Convention on the Prevention of Pollution from Ships, *International Environment Reporter* 21:2301; 1978 Protocol *International Environment Reporter* 21:2401.
116 OPPR, note 114, at s. 34.
117 C.R.C. 1978, c. 1458; as amended by SOR/83-347; SOR/2002-276.
118 Ibid., s. 4.
119 See *Report of the Commissioner of the Environment and Sustainable Development to the House of Commons* (2002). Online at: www.oag-bvg.gc.ca/domino/reports.nsf/html/c2002menu_e.html (accessed 24 January 2004) Chapter 4.
120 Ibid., at para. 4.104.
121 *Aquatic Invasive Species: Uninvited Guests, Fourth Report of the Standing Committee on Fisheries and Oceans* (May 2003). Online at: www.parl.gc.ca/InfocomDoc/37/2/FOPO/Studies/Reports/foporp04/05-hon-e.htm (accessed 24 January 2004).
122 Ibid., at 22–3.
123 Ibid., at 27.
124 Online at: www.cbin.ec.gc.ca/primers/ias_invasives.cfm?lang=e (accessed 8 April 2005).
125 Ibid., at 13.
126 Ibid., at 14.
127 Ibid., at 23. It should be noted that a Canadian Action Plan to Address the Threat of Aquatic Invasive Species is projected to be finalized by September 2005. Ibid., at 28. At the time of writing, Transport Canada was preparing *Ballast Water Control and Management Regulations*.
128 This chapter does not specifically address the control of offshore mineral exploration/exploitation where general legislative controls may apply, such as the prohibition of deleterious deposits under the *Fisheries Act*.
129 A complicated regulatory scheme, addressing the oil and gas sector, involves offshore petroleum board leadership in regulating activities offshore Newfoundland and Nova Scotia and responsibility of the National Energy Board for regulating operations in the rest of Canada's offshore. Key pieces of legislation include: *Canada–Newfoundland Atlantic Accord Implementation Act*, S.C. 1987, c. 3; *Canada–Newfoundland Atlantic Accord Implementation (Newfoundland) Act*, R.S.N. 1990, c. C-2; *Canada–Nova Scotia Offshore Petroleum Resources Accord Implementation Act*, S.C. 1988, c. 2; *Canada–Nova Scotia Offshore Petroleum Resources Accord Implementation (Nova Scotia) Act*, S.N.S. 1987, c. 3; and *Canada*

Oil and Gas Operations Act, R.S.C. 1985, c. O-7. For useful overviews of the legislative and regulatory complexities, see Van Penick "Legal framework in the Canadian offshore" (2001) 24 *Dalhousie Law Journal* 1; and Angus Taylor and Jim Dickey "Regulatory regime: Canada–Newfoundland/Nova Scotia Offshore Petroleum Board issues" (2002) 24 *Dalhousie Law Journal* 51.

130 National Energy Board, Canada–Newfoundland Offshore Petroleum Board and Canada–Nova Scotia Offshore Petroleum Board *Guidelines Respecting the Selection of Chemicals Intended to be Used in Conjunction with Offshore Drilling and Production Activities* (January 1999). Online at: www.cnsopb.ns.ca/Environment/evironment.html (accessed 24 January 2004).

131 See Fisheries and Oceans Canada *Review of Scientific Information on Impacts of Seismic Sound on Fish, Invertebrates, Marine Turtles and Marine Mammals* Habitat Status Report 2004/002 (September 2004). The Review notes that, based upon available evidence, it can be concluded "that seismic sounds in the marine environment were neither completely without consequences nor are they certain to result in serious or irreversible harm to the environment." Ibid., at 1.

132 A set of proposed mitigation measures was issued for discussion purposes on 19 February 2005 by the Government of Canada and the Provinces of British Columbia, Newfoundland and Labrador and Nova Scotia. See Fisheries and Oceans Canada "Mitigation of Seismic Noise in the Marine Environment: Statement of Canadian Practice." Online at: www.dfo-mpo.gc.ca/canwaters-eauxcan/infocentre/media/seismic-sismique/statement_e.aps (accessed 13 April 2005).

133 Online at: www.cnsopb.ns.ca/Environment/evironment.html (accessed 24 January 2004).

134 Section 4.3 recognizes the precautionary approach:

> Both Parties promote the wide application of the precautionary approach to the conservation management and exploitation of marine resources in order to protect these resources and preserve the marine environment. The uncertain and incomplete nature of science relating to the environment invokes the precautionary approach where it is necessary to exercise caution in adopting safe minimal standards for all development. When there are threats of serious or irreversible damage, lack of full scientific certainty will not be used as a reason for postponing measures to prevent environmental degradation. (Ibid.)

135 Online at: www.cnsopb.ns.ca/Environment/evironment.html (accessed 24 January 2004).

136 See Douglas M. Johnston *B.C. Offshore Development Issues: The 2002 Dunsmuir Symposium Report* (Maritime Awards Society of Canada, Victoria: 2002) at 1.

137 *British Columbia Offshore Hydrocarbon Development: Report of the Scientific Review Panel* (January 15, 2002) at ii (Executive Summary). Online at: www.offshoreoilandgas.gov.bc.ca/reports/scientific-review-panel/ (accessed 24 January 2004).

138 Ibid., at ii and Appendix 20.

139 Ibid., at 46.

140 Ibid.

141 Online at: www.dfo-mpo.gc.ca.afpr-rppa (accessed 14 April 2005).

142 Ibid., at s. 3.2.1.

143 Ibid.

144 The Policy Framework was not yet finalized at the time of writing but a final policy is expected to be announced by 30 June 2005. *A Policy Framework for Conservation of Wild Pacific Salmon* (Fisheries and Oceans Canada, Vancouver: December 2004) at 39.

145 Three guiding principles are set out: conservation of wild salmon and their habitat is the first priority in resource management decision making; resource management decisions will be made in an open, transparent and inclusive manner; and biological, social and economic benefits and costs will be balanced. Ibid., at 12.
146 Ibid., at 17–18.
147 Fisheries and Oceans Canada, Ottawa: January 2001. Online at: www-comm.pac.dfo-mpo.gc.ca/publications/selectivep_e.pdf (accessed 24 January 2004).
148 Department of Fisheries and Oceans *New Emerging Fisheries Policy* (September 2001). Online at: www.dfo-mpo.gc.ca/communic/fish_man/nefp_e.htm (accessed 24 January 2004).
149 For example, the Department of Fisheries and Oceans has initiated an Objective-Based Fisheries Management (OBFM) pilot through the Scotia-Fundy Groundfish Management Plan for 2002–07. One of the three general objectives of the Plan is to ensure fishing for groundfish does not cause reductions in resource productivity or modifications to ecosystem structure/function that are difficult or impossible to reverse by adopting a precautionary approach to management. Fisheries and Oceans Canada *Groundfish Management Plan: Scotia-Fundy Fisheries Maritimes Region, April 1, 2002–March 31, 2007* (March 2002). A response by the Department of Fisheries and Oceans to the recent federal discussion paper on the precautionary approach admits that there is little tangible evidence of the implementation of precaution in fisheries management. The authors advise setting specific biological limits to overfishing for all stocks, rather than wait for an integrated approach to precautionary management. See P.A. Shelton and J.C. Rice "Limits to overfishing: Reference points in the context of the Canadian perspective on the precautionary approach" 2002/084 (Canadian Science Advisory Secretariat, Department of Fisheries and Oceans, Ottawa: 2002). Online at: www.dfo-mpo.gc.ca/csas/Csas/English/Research_Years/2002/2002_084e.htm (accessed 24 January 2004).
150 See, for example, *Proceedings of the National Meeting on Applying the Precautionary Approach in Fisheries Management* 2004/003 (Canadian Science Advisory Secretariat, Fisheries and Oceans Canada, Ottawa: 2004).
151 See Ken T. Frank, Nancy L. Shackell and Jim E. Simon "An evaluation of the Emerald/Western Bank juvenile haddock closed area" (2000) 57 *ICES Journal of Marine Science* 1023.
152 See Fisheries and Oceans Canada, "Minister Thibault announces 2002 Georges Bank Groundfish Management Plan" News Release NR-MAR-02-10E (21 June 2002). Online at: www.pac.dfo-mpo.gc.ca/English/release/p-releas/2002/nr044e-htm (accessed 13 April 2005); Fisheries and Oceans Canada, "Closure to Protect Deep Water Coral Reef" News Release NR-MAR-04-14E (13 August 2004). Online at: www/mar.dfo-mpo.gc.ca/communications/maritimes/news04e/NR-MAR-04-14E.html (accessed 13 April 2005).
153 See Department of Fisheries and Oceans "Groundfish trawlers help to protect unique sponge reefs in British Columbia." Press Release NR-PR-02-044E (July 18, 2002). Online at: www-comm.pac.dfo-mpo.gc.ca/pages/release/p-releas/2002/nr044_e.htm (accessed 24 January 2004).
154 S.C. 1996, c. 31, s. 35(2) provides: "For the purposes of integrated management plans referred to in sections 31 and 32, the Minister will lead and coordinate the development and implementation of national system of marine protected areas on behalf of the Government of Canada."
155 See Glen S. Jamieson and Colin O. Levings "Marine protected areas in Canada – Implications for both conservation and fisheries management" (2001) 58

Canadian Journal of Fisheries and Aquatic Sciences 138. Although there are currently some 198 protected areas with a marine component under federal and provincial jurisdiction in Canada, the majority of these are wildlife reserves or bird sanctuaries.

156 *Endeavour Hydrothermal Vents Marine Protected Area Regulations*, SOR/2003-87.

157 *Gully Marine Protected Area Regulations*, SOR/2004-112.

158 Fisheries and Oceans Canada "Marine Protected Areas of Interest." Online at: www.dfo-mpo.gc.ca/canwaters-eauxcan/oceans/mps-zpm/mpa_.e.asp (accessed 13 April 2005).

159 S.C. 2002, c. 18.

160 The Preamble states:

> Whereas the Government of Canada is committed to adopting the precautionary principle in the conservation and management of the marine environment so that, where there are threats of environmental damage, lack of scientific certainty is not used as a reason for postponing preventive measures. (Ibid.)

161 Ibid., at s. 9(3).

162 "The Canada We Want" Speech from the Throne to open the Second Session of the Thirty-Seventh Parliament of Canada (30 October 2002). Online at: www.pco-bcp.gc.ca/sft-ddt/hnav/hnav08_e.htm (accessed 24 January 2004).

163 The 1995 UN Straddling Fish Stocks and Highly Migratory Fish Stocks Agreement specifies that F_{MSY} the fishing mortality that can produce maximum sustainable yield (MSY) as a limit point that cannot be exceeded. $B_{MSY,}$ or the biomass that can sustain MSY given $F_{MSY,}$ is suggested as a target for re-building overfished stocks.

164 See Laura J. Richards and Jean-Jacques McGuire "Recent international agreements and the precautionary approach: New directions for fisheries management science" (1998) 55 *Canadian Journal of Fisheries and Aquatic Sciences* 1545.

165 See Breeze, note 17.

166 See Myers *et al.*, note 17.

167 Despite scientific advice that suggested the collapse of the cod stocks, political motivations prevented precautionary management decisions. See Jeffrey A. Hutchings, Carl Walters and Richard L. Haedrich "Is scientific inquiry incompatible with government information control?" (1997) 54(5) *Canadian Journal of Fisheries and Aquatic Sciences* 1198.

168 See Michael Harris *Lament for an Ocean: The Collapse of the Atlantic Cod Fishery – A True Crime Story* (McLelland and Stewart, Toronto: 1998).

169 See Ransom A. Myers, Jeff A. Hutchings and Nicholas J. Barrowman "Why do fish stocks collapse? The example of cod in Canada" (1997) 7 *Ecological Applications* 91.

170 See Department of Fisheries and Oceans *Northern (2J+3KL) Cod Stock Status Update* (DFO Science Stock Status Report A2-01, Department of Fisheries and Oceans, St. John's: April 2002). Online at: www.dfo-mpo.gc.ca/ca/csas/ (accessed 24 January 2004).

171 Ibid.

172 FRCC *2000/2001 Conservation Requirements for Georges Bank Groundfish Stocks and 2J3KL Cod: Report to the Minister of Fisheries and Oceans* FRCC 00.R.4 (Minister of Public Works and Government Services Canada, Ottawa: May 2000) at 24.

173 Letter from Fred Woodman, chairman FCCC, to the Minister of Fisheries and

Oceans (22 May 2002). Online at: www.frcc-ccrh.ca/htm (accessed 24 January 2004).

174 See CBC News, "Thibault shuts down cod fisheries" (14 April 2003). Online at: www.stjohns.cbc.ca/regional/servlet/View?filename=nf_closure_20030424 (accessed 24 January 2004).

175 COSEWIC, May 2003 Species Assessments. Online at: www.sararegistry.gc.ca/status/showDocument_e.cfm?id=126 (accessed 24 January 2004).

176 For a discussion of the still nascent national framework for precautionary fisheries management, see Peter A. Shelton, Jake C. Rice, Denis Rivard, Ghislain A. Chouinard and Alain Fréchet *Recent Progress on the Implementation of the Precautionary Approach on Canadian Cod Stocks Leading to the Re-introduction of the Moratorium* (International Council for the Exploration of the Sea, Theme Session on Reference Point Approaches to Management with the Precautionary Approach CM 2003/Y:15, ICES, Copenhagen: 2003).

177 At the Third World Conservation Congress in Bangkok, Thailand in November 2004, a resolution was passed on the Conservation and Sustainable Management of High Seas Biodiversity (RESWCC3064). The resolution calls upon the IUCN to facilitate compliance of states party to the LOS Convention, the UN Straddling Fish Stocks and Highly Migratory Fish Stocks Agreement, the Convention on Biological Diversity, the FAO Compliance Agreement and the Convention on Migratory Species.

178 UNGA Res. 59/25.

179 Online at: www.dfo-mpo.gc.ca/media/speech/2004/20041116_e.htm (accessed 15 February 2005).

180 At the 10th meeting of the Subsidiary Body on Scientific, Technical and Technological Advice (SBSTTA) to the Convention on Biological Diversity in February 2005, Canada, the United States and Iceland blocked any specific references to the high seas and bottom trawling in the setting of these goals.

181 Canada is a party to the Convention of Biological Diversity (CBD). In 1995, the CBD agreed to a work program on marine and coastal biodiversity, the Jakarta Mandate. However, the program of work did not include many time-bound or quantitative targets. When the 2002 World Summit on Sustainable Development in Johannesburg agreed to try to reduce and halt the loss of biodiversity by 2010, participants realized that several additional targets and goals on marine and coastal biodiversity were necessary to achieve this goal.

182 FRCC *1999 Conservation Requirements for the Gulf of St. Lawrence Groundfish and Codfish Stocks in Divisions 2GH and 3PS and Science Priorities: Report to the Minister of Fisheries and Oceans Canada* FRCC.99.R.1 (Minister of Public Works and Government Services Canada, Ottawa: April 1999). Online at: www.frcc-ccrn.ca/eindex.htm (accessed 24 January 2004).

183 Ibid., at 11.

184 See D.P. Swain and A.F. Sinclair "Pelagic fishes and the cod recruitment dilemma in the Northwest Atlantic" (2000) 57 *Canadian Journal of Fisheries and Aquatic Sciences* 1321.

185 See Department of Fisheries and Oceans *Understanding Seals and Sealing in Canada: Report of the Eminent Panel on Seal Management* (Fall 2001). Online at: www.dfo-mpo.gc.ca/seal-phoque/reports-rapports/expert/repsm-rgegp_e.htm (accessed 24 January 2004). The Panel's report assisted DFO in developing a multi-year management strategy for seals. Fisheries Resource Management – Atlantic, Fisheries and Oceans Canada *Atlantic Seal Hunt 2003–2005 Management Plan* (DFO, Ottawa: 2003). Online at: www.dfo-mpo.gc.ca/seal-phoque/report-rapport_e.htm (accessed 24 January 2004).

186 See D.W. Kulka "Distribution of Greenland halibut and by-catch species that overlap the 200-mile limit spatially and temporally and in relation to depth: Effect of depth restrictions on the fishery" (2001) NAFO SCR Doc 01/40 (Northwest Atlantic Fisheries Organization, Dartmouth, NS: 2001).

187 There are a few exceptions, for example, where an inherently precautionary harvesting technique has allowed for a sustainable fishery as in the snow crab (*Chionectes opilio*) fishery. Only males are harvested as they are larger and females can escape from the traps. Coupled with limited entry access to the fishery, the snow crab populations may be sustainably harvested, as in the Gulf of St. Lawrence.

188 See Ray Hilborn, Jean-Jacques Mcguire, Ana N. Parma and Andrew A. Rosenburg "The precautionary approach and risk management: Can they increase the probability of successes in fishery management?" (2002) 58(1) *Canadian Journal of Fisheries and Aquatic Sciences* 99. The authors make the case that precaution must be used not only towards sustainable fish stocks, but towards sustaining fishing communities as well. They advocate a process of portfolio management, where a fisher has several licenses and can thereby have an effective multi-species fishery that will allow for declines in one or more species without causing a major economic decline in the community. Also see, Tony Charles "The precautionary approach and 'burden of proof' challenges in fishery management" (2002) 70(2) *Bulletin of Marine Science* 683 at 692–3.

189 The MOUs with British Columbia, New Brunswick, Nova Scotia and Newfoundland delegate aquaculture licensing to the provincial level while the MOU with Prince Edward Island leaves licensing responsibility with the federal government. For a review, see David VanderZwaag, Gloria Chao and Mark Covan "Canadian aquaculture and the principles of sustainable development: Gauging the law and policy tides and charting a course" (2002) 28 *Queen's Law Journal* 279 at 299–301 and (2003) 28 *Queen's Law Journal* 529 at 532–4.

190 For example, Nova Scotia's *Fisheries and Coastal Resources Act*, S.N.S. 1996, c. 25, in the purpose section (s. 2) emphasizes the need to increase aquaculture production and to optimize aquaculture processing and only mentions community involvement as a principle of sustainable development.

191 Fisheries and Oceans Canada, *DFO's Aquaculture Policy Framework* (Department of Fisheries and Oceans, Ottawa: 2002).

192 Ibid., at 24. The Standing Senate Committee on Fisheries has recommended that DFO define the precautionary approach as it pertains to aquaculture and issue a written public statement on how the precautionary approach is being applied to the aquaculture sector. See Report of the Standing Senate Committee on Fisheries, *Aquaculture in Canada's Atlantic and Pacific Regions* (June 2001) Recommendation 14. Online at: www.parl.gc.ca/37/1/parlbus/commbus/senate/com-e/fish-e/rep-e/repintjun01-e.htm (accessed 1 December 2003).

193 Online at: www.oag-bug.gc.ca/domino/reports.nsf/html/c20041005ce.html (accessed 12 April 2005) Chapter 5.

194 Ibid.

195 The Policy states: "Government policy for aquaculture must recognize the significant potential for innovation in the aquaculture sector and benefits such innovation will yield in a variety of disciplines." Note 191 at 9.

196 The Royal Society of Canada *Elements of Precaution: Recommendations for the Regulation of Food Biotechnology in Canada* (January 2001). Online at: www.rsc.ca/foodbiotechnology/indexEN.html (accessed 24 January 2004).

197 Ibid., at 167.

198 See Action Plan for the Government of Canada in response to the Royal Society of Canada Expert Panel Report, *Elements of Precaution: Recommendations for the Regulation of Food Biotechnology in Canada* (23 November 2001) at 28. Online at: www.hc-sc.gc.ca/english/protection/royalsociety/index.htm (accessed 24 January 2004).

199 Action Plan, ibid., at 26. For a review of Canada's approach to regulating genetically modified foods in light of precaution, see Canadian Biotechnology Advisory Committee *Improving the Regulation of Genetically Modified Foods and Other Novel Foods in Canada: Report to the Government of Canada Biotechnology Ministerial Coordinating Committee* (August 2002) at 23–36. The Committee noted the current paradigm for regulatory decisions pertaining to health and the environment is based on scientific evaluations and risk assessments and recommended further study and analysis to identify ways to address social and ethical issues related to biotechnology. Ibid., at 45–7.

200 Online at: www.hc-sc.gc.ca/english/protection/royalsociety/rsc-src_02_05/index.html (accessed 14 April 2005). For a further review of Canadian issues relating to biotechnology in the aquatic sector, see Douglas J.B. Moodie "The cautious Frankenfish: Environmental protection and other Canadian regulatory issues relating to transgenic fish" (2004) 7 *Macquarie Journal of International and Comparative Environmental Law* 49.

201 (January 2002). Online at: www.dfo-mpo.gc.ca/science/aquaculture/code/prelim_e.htm (accessed 24 January 2004).

202 The Code states that it does not cover issues relating to aquarium fish, baitfish, live fish for the food trade and transgenic aquatic organisms. Ibid., at para. 1.1.4.

203 Ibid., at para. 2.2.8.

204 *Environment Protection and Biodiversity Conservation Act 1999* (Cth), No. 91, 1999.

205 S.C. 2002, c. 29.

206 The Preamble states:

> [T]he Government of Canada is committed to conserving biological diversity and to the principle that, if there are threats of serious or irreversible damage to wildlife species, cost-effective measures to prevent the reduction or loss of species should not be postponed for a lack of scientific certainty . . . (Ibid.)

207 The Minister required to lead protective efforts may vary with the Minister of Fisheries and Oceans being responsible for aquatic species. Ibid., at s. 2(1) (defining "competent minister").

208 Section 38 provides:

> In preparing a recovery strategy, action plan or management plan, the competent minister must consider the commitment of the Government of Canada to conserving biological diversity and to the principle that, if there are threats of serious or irreversible damage to listed wildlife species, cost-effective measures to prevent the reduction or loss of the species should not be postponed for a lack of full scientific certainty.

209 Section 27 leaves listing largely to the discretion of the Governor in Council through the power to amend the List of Wildlife Species at Risk. The potential listing of marine fish species has in fact become very controversial and an extended consultation process has been followed for several species including the Atlantic cod (Newfoundland and Labrador population) listed as endangered by the Committee on the Status of Endangered Wildlife in Canada.

Fisheries and Oceans Canada "Extended Listing Process for 12 Aquatic Species: The Species at Risk Act (SARA)" Media Room-Backgrounder (April 2004). Online at: www.dfo-mpo.gc.ca/media/backgrou/2004sara_c.htm (accessed 10 February 2005).

210 Section 73. Issuance of permits would be subject to a determination by the Minister that the activity will not jeopardize the survival or recovery of the species. Ibid., s. 73(3)(c). Lack of a definition of "jeopardize" and provision for public participation in the determination process raise concerns about whether there are adequate "checks" on ministerial discretion. In 2004, the Minister issued approximately 9,600 incidental harm permits to commercial fishers allowing the taking of threatened northern and spotted wolfish. Canada, Species at Risk Public Registry *Notice of Permit*. Online at: www.sararegistry.gc.ca/agreements/ViewPermit_e.cfm?id=47 (accessed 14 April 2005).

211 Section 83(4). For a discussion of how this could be a "huge door" for allowing substantial interferences with listed species, see David L. VanderZwaag and Jeffrey A. Hutchings "Canada's marine species at risk: Science and law at the helm, but a sea of uncertainties" (2005) 36 *Ocean Development & International Law* 219 at 232.

212 SOR/1993-56.

213 Ibid., at s.7.

214 See Jon Lien "The conservation basis for the regulation of whale watching in Canada by the Department of Fisheries and Oceans: A precautionary approach" (2001) *Canadian Technical Report of Fisheries and Aquatic Sciences* 2363. Online at: www.dfo-mpo.gc.ca/Library/259973.pdf (accessed 24 January 2004).

215 See Fisheries and Oceans Canada website. Online at: www.dfo-mpo.gc.ca/communic/lien/intro_e.htm (accessed 24 January 2004).

216 See Susan Joy Hassol *Impacts of a Warming Arctic: Arctic Climate Impact Assessment (ACIA)* (Cambridge University Press, Cambridge: 2004) at 8.

217 David Freestone "The Precautionary Principle" in Robin R. Churchill and David Freestone (eds) *International Law and Global Climate Change* (Graham & Trotman/Martinus Nijhoff, London: 1991) chapter 2 at 38. Effects of climate change on the Pacific and Atlantic coasts may also be serious, including decline in salmon stocks due to warmer water temperatures, inundation of wetlands, beaches and other sensitive coastal ecosystems, damage to coastal infrastructure, and increases in toxic algae blooms. See Government of Canada "Regional impacts: Climate change in British Columbia." Online at: www.climatechange.gc.ca/english/issues/how_will/fed_bc.shtml/ (accessed 24 January 2004) and Government of Canada "Regional impacts: Climate change in Nova Scotia." Online at: www.climatechange.gc.ca/english/issues/how_will/fed_novascotia.shtml (accessed 24 January 2004).

218 May 1992. (1992) 31 *International Legal Materials* 849.

219 Can. T.S. 1994 No. 7 (ratified by Canada on 4 December 1992; in force for Canada 21 March 1994).

220 See James Cameron and Juli Abouchar "The precautionary principle: A fundamental principle of law and policy for the protection of the global environment" (1991) 14 *B.C. International and Comparative Law Review* 1 at 12.

221 Note 213 at art. 3(3).

222 10 December 1997. (1998) 37 *International Legal Materials* 22.

223 Cathy Wilkinson and Stephanie Cairns *Negotiating the Climate: Canada and the International Politics of Global Warming* (David Suzuki Foundation, Vancouver: November 2000).

224 See, for example, Harry Sterling "Kyoto: Lost in rhetorical smog" *The Globe and Mail* (8 May 2002) A21.

225 For legal arguments in support of the federal ratification decision despite substantial provincial opposition, see D.M. McRae and J.H. Currie "Treaty-making and treaty implementation: The Kyoto Protocol" (2003) 29(2) *Canadian Council on International Bulletin* 1.

226 The document was adopted in November 2002. Online at: www.climatechange.gc.ca/plan_for_canada/pldn/index.html (accessed 3 December 2003).

227 The action areas included: transportation, housing and commercial/institutional buildings, large industrial emitters, renewable energy and cleaner fossil fuel, small and medium-sized enterprises and fugitive emissions, agriculture, forestry and landfills, and the international market.

228 Government of Canada (2005). Online at: www.climatechange.gc.ca (accessed 14 April 2005).

229 Government of Canada "Canada Launches Project Green by Releasing a Plan to Honour its Climate Change Commitment" News Release (13 April 2005). Online at: www.climatechange.gc.ca/english/newsroom/2005/plan05_NR.asp (accessed 14 April 2005).

230 Heavy criticisms have arisen over letting the "biggest polluters" buy "hot air" from other countries rather than reducing emissions in Canada. See Jeff Sallot "Opposition tears into Liberals' Kyoto Plan" *The Globe and Mail* (14 April 2005) A6.

231 Note 228 at 18.

232 Natural Resources Canada *Memorandum of Understanding between the Government of Canada and the Canadian Automotive Industry Respecting Automobile Greenhouse Gas Emissions* (5 April 2005). Online at: www.nrean-rncan.gc.ca/media/mous/2005/20050405_e.htm (accessed 14 April 2005).

233 Note 228 at 18.

234 See Jaye Ellis "The Straddling Stocks Agreement and the precautionary principle as interpretive device and rule of law" (2001) 32 *Ocean Development & International Law* 289.

235 Stanley Fish *The Trouble with Principle* (Harvard University Press, Cambridge, MA: 1999) at 232.

236 See Bill Durodié "Plastic Panics: European Risk Regulation in the Aftermath of BSE" in Julian Morris (ed.) *Rethinking Risk and the Precautionary Principle* (Butterworths – Heinemann, Oxford: 2000) chapter 8 at 161.

237 See Indur M. Goklany *The Precautionary Principle: A Critical Appraisal of Environmental Risk Assessment* (CATO Institute, Washington, DC: 2001).

238 For example, a global regulatory body or discussion forum might be established for addressing unresolved conflicts over biotechnology. The latter has been suggested by Murphy. See Sean D. Murphy "Biotechnology and international law" (2001) 42 *Harvard International Law Journal* 47. On the institutional challenges surrounding climate change, see Bruce Yundle "The precautionary principle as a force for global political centralization: A case-study of the Kyoto Protocol" in Morris, note 236 at chapter 9.

239 On the need for discursive procedures to foster social learning and to air differing values and interests, see Andy Stirling "The Precautionary Principle in Science and Technology" in O'Riordan, Cameron and Jordan, note 37 at chapter 3.

240 As noted by Michael M'Gonigle, the implementation of precaution is not a "discreet task" but involves development of an ecological society. See R. Michael M'Gonigle "The Political Economy of Precaution" in Raffensperger and Tickner, note 1, chapter 7 at 142.

7 Australia and the precautionary principle

Moving from international principles to domestic and local implementation

Lorne K. Kriwoken, Liza D. Fallon and Donald R. Rothwell

Introduction: the tricky currents of precaution

The precautionary principle has been widely accepted in policy directives and legislation at all levels of government in Australia. It has also been specifically recognized as policy guidance for Australian coastal and ocean planning and management. For example, the Commonwealth's *Guiding Principles for Management of Coastal Resources* define the precautionary approach as:

1 If there is high risk of serious or irreversible adverse impacts from a use of a resource, that use should be permitted only if those impacts can be mitigated or there are overwhelming grounds for proceeding in the national interest.
2 If a use is assessed as having a low risk of causing serious or irreversible adverse impacts, or there is insufficient information with which to assess fully and with certainty the magnitude and nature of impacts, decision-making should proceed in a conservative and cautious manner. The absence of scientific certainty should not be a reason for postponing measures to prevent or mitigate negative impacts.[1]

Whilst the acceptance of the precautionary principle is widespread, the operational details of precaution, in many cases, remain to be tested. Implementation of the precautionary principle therefore remains a significant challenge for all levels of government in Australia, and not just at federal or state/territorial levels, but also for local and municipal governments.

The overall aim of this chapter is to explore the application of the precautionary principle in the context of the Australian coastal and ocean environment. It commences with relevant background to the precautionary principle in Australia through the development of ecologically sustainable development, the *Intergovernmental Agreement on the Environment* and *Australia's Ocean Policy*. Relevant legislative instruments are reviewed and judicial and administration applications of the principle are assessed. Case studies of two coastal and ocean sectors are undertaken and an assessment made of how the precautionary principle has been applied in each instance. Concluding comments are made as to the state of the precautionary principle in Australia.

Australian general steps and wanderings

Ecologically sustainable development

In June 1990, the Hawke Labor Government released a discussion paper on ecologically sustainable development (ESD) that identified a range of environmental problems (air, water, land degradation, loss of species, inadequate waste control, and Australia's contribution to stratospheric ozone depletion and climate change).[2] After two years of work the process produced nine sectoral (agriculture, energy use, energy production, fisheries, forest use, manufacturing, mining, tourism and transport) and two intersectoral reports containing a multitude of consensus-based recommendations for consideration by Australian federal, state and territory governments.[3]

The 1992 *National Strategy for Ecologically Sustainable Development* included three core objectives and a set of guiding principles. The latter recognized the precautionary principle. This set the stage for the Council of Australian Governments to form an Intergovernmental Committee for Ecologically Sustainable Development (ICESD) in 1992, pursuant to its adoption of an *Intergovernmental Agreement on the Environment* (IGAE).[4] The ESD process and the ESD Steering Committee that monitored the implementation of the *National Strategy for Ecologically Sustainable Development* influenced the mandate of ICESD. ICESD published the *Summary Report on Implementation of the Ecologically Sustainable Development Strategy* in December 1993,[5] the *Forward Agenda* in 1994,[6] and the *1993–1995 Report on Implementation of the ESD Strategy* in July 1996.[7] However, these reports tend to be descriptive of achievements under the ESD Steering Committee's own strategy, and it is therefore difficult to assess the effectiveness of implementation merely from a reading of the reports. It is especially difficult to say what is directly related to the ESD process and what would have occurred irrespective of its existence. In 1998, ICESD was wound up and future monitoring of the implementation of the ESD Strategy, if any, is likely to rest with the federal Department of the Environment and Heritage.

Intergovernmental Agreement on the Environment

In February 1992, through the IGAE, all levels of government throughout Australia (Commonwealth, state and local) agreed to follow the precautionary principle as part of a commitment to ecologically sustainable development. Where there are threats of serious or irreversible environmental damage, the IGAE agreed that a lack of full scientific certainty should not be used as a reason for postponing measures to prevent environmental degradation. In the application of the precautionary principle, public and private decisions should be guided by careful evaluation to avoid, wherever practicable, serious or irreversible damage to the environment and an assessment of risk-weighted consequences of various options.[8]

Australia's Oceans Policy

Australia's Oceans Policy was released in 1998. It recognized the need for understanding and protecting of biodiversity, promotion of ecologically sustainable development, encouraging equitable, efficient and economic utilization of resources, job creation, and establishing broad principles and planning and management approaches in an effort to achieve the goal of ecologically sustainable development for Australia's oceans.[9] *Australia's Oceans Policy* adopts *Principles for Ecologically Sustainable Development*, recognizing that ocean health and integrity is fundamental to ecologically sustainable development. Specific acknowledgement is made to the precautionary principle:

> Incomplete information should not be used as a reason for postponing precautionary measures intended to prevent serious or irreversible environmental degradation of the oceans.[10]

A key task for the National Oceans Office, established to implement and give effect to *Australia's Oceans Policy*, has been to interpret and apply principles of ecologically sustainable development and the precautionary principle.

Legislative instruments supporting the precautionary principle

The precautionary principle is now widely accepted in Australian Commonwealth legislation.[11] Examples of specific federal legislative requirements for the application of the precautionary principle include:

- *Environment Protection and Biodiversity Conservation Act 1999*
- *Environment, Sport and Territories Legislation Amendment Act 1995*
- *Fisheries Administration Act 1991*[12]
- *Fisheries Management Act 1991*[13]
- *Fisheries Legislation Amendment Act 1997*[14]
- *Great Barrier Reef Marine Park Act 1975*[15]
- *Hazardous Waste (Regulation of Exports and Imports) (Waigani Convention) Regulations 1999*
- *National Environment Protection Council Act 1994.*[16]

Of these instruments, the *Environment Protection and Biodiversity Conservation Act 1999* (EPBC Act) (Cth) represents the most significant attempt to reform Commonwealth environmental law in Australia since the introduction of the *Environmental Protection (Impact of Proposals) Act 1974* (Cth).[17] The principles of ecologically sustainable development are referred to in the objectives of the Act. The objectives of the Act also include a definition of the precautionary principle as part of the principles of ecologically sustainable development:

> if there are threats of serious or irreversible environmental damage, lack of full scientific certainty should not be used as a reason for postponing measures to prevent environmental degradation . . .[18]

The EPBC Act emphasizes the protection of those aspects of the environment that are "matters of national environmental significance." Six of the seven matters of national environmental significance identified by the Council of Australian Governments Agreement are direct triggers that invoke the EPBC Act, including World Heritage properties, Ramsar listed wetlands, listed threatened species and communities, listed migratory species, protection of the environment from nuclear actions, and the marine environment.

Of significance under the Act is the obligation placed upon the Minister of Environment and Heritage to consider the precautionary principle in decision making. The EPBC Act states:

> *Taking account of precautionary principle:*
>
> (1) The Minister must take account of the precautionary principle in making a decision listed in the Table in subsection (3), to the extent he or she can do so consistently with the other provisions of this Act.
>
> *Precautionary principle:*
>
> (2) The precautionary principle is that lack of full scientific certainty should not be used as a reason for postponing a measure to prevent degradation of the environment where there are threats of serious or irreversible environmental damage.[19]

Under the EPBC Act there are 14 decisions for which the precautionary principle must be taken into account. These decisions involve determining whether a proposed action is a controlled action, and whether to approve the taking of an action. Decisions also concern the granting of permits and making/adopting/varying management plans for various protected areas.[20]

The EPBC Act remains a relatively new initiative, and there is considerable scope in its potential application to a range of marine activities in support of *Australia's Oceans Policy*.[21] One area of potential application, especially in giving effect to the precautionary approach, is with respect to fisheries in both Commonwealth and state/territorial waters. While the focus of the Act is on matters of national environmental importance, the potential exists in the Act's extension to the marine environment and thereby fisheries within those waters, to fall under the application of the precautionary principle.[22] The consequential interaction of the EPBC Act with its strong environmental principles and the *Fisheries Management Act 1999* (Cth) has the potential to raise some significant issues due to the different management approaches of each Act.[23]

Other illustrations of the precautionary principle being implemented federally can be found in the *Environmental Protection (Sea Dumping) Act 1981* (Cth), which regulates ocean dumping in Australian waters consistent with the 1996 Protocol to the 1972 London Convention.[24] The Act has been substantially adjusted in light of the 1996 Protocol, which while still not in force has been given effect under Australian law, and accordingly implements an important international initiative supporting the precautionary principle.

The *Hazardous Waste (Regulation of Exports and Imports) Regulations 1999* (Cth) are intended to give effect to the 1995 *Convention to Ban the Importation into Forum Island Countries of Hazardous and Radioactive Wastes and to Control the Transboundary Movement and Management of Hazardous Wastes within the South Pacific Region* (Waigani Convention).[25] The precautionary principle occurs once in these Regulations. Other areas of Commonwealth responsibility where the precautionary principle is recognized are fisheries, marine pollution and seabed activities.[26]

State and territorial legislation recognizing the precautionary principle

At a state and territorial level in Australia there are also examples of where the precautionary principle has been given effect. The *Protection of the Environment Administration Act 1991* (NSW) commits to the principles of inter-generational equity and the conservation of biological diversity and has adopted the precautionary principle as an objective. The principle is worded in a traditional fashion. If there are threats of serious or irreversible environmental damage, lack of full scientific certainty should not be used as a reason for postponing measures to prevent environmental degradation. Under the Act, the New South Wales Environment Protection Authority must have regard to the need to maintain ecologically sustainable development, including the precautionary principle. The requirement that the principle be applied is also set out in a number of federal natural resource policies that inform state policies including the *National Strategy for Rangeland Management*, the *National Strategy for the Conservation of Australia's Biological*

Diversity, the *Decade of Landcare*, the *National Water Quality Management Strategy* and the *National Forests Policy Statement.*[27] The *Environment Protection Act 1993* (SA) also provides a commitment:

> to apply a precautionary approach to the assessment of risk of environmental harm and ensure that all aspects of environmental quality affected by pollution and waste (including ecosystem sustainability and valued environmental attributes) are considered in decisions relating to the environment . . .

In Tasmania, a more comprehensive approach has been taken in the Tasmanian Resource Management and Planning System (RMPS) adopted in 1994. RMPS consists of an integrated policy, statutory and legislative framework for resource management.[28] The cornerstone of the RMPS is the *State Projects and Policies Act 1993* (Tas), which provides for an integrated assessment of state policies and sets out the principles and objectives of sustainable development.[29] The Tasmanian *State Coastal Policy* was developed under the RMPS framework in 1996.[30] The *Living Marine Resources Management Act 1995* (Tas) and the *Marine Farming Planning Act 1995* (Tas) emphasize a precautionary approach to development and use a "reverse onus of proof" in enforcement and compliance.

Judicial and administrative application of the precautionary principle

During the past decade, Australian courts at all levels have had occasion to consider aspects of the precautionary principle, both as doctrine of the common law[31] and as endorsed by statute. The leading case is *Leatch* v. *National Parks and Wildlife Service.*[32] Considering the status of the precautionary principle in New South Wales law, and the related impact of the 1992 Convention on Biological Diversity,[33] Justice Stein considered that, even though there was no specific reference to the precautionary principle in the *National Parks and Wildlife Act 1974* (NSW), or to ecologically sustainable development, the legislation allowed the decision-maker to take into account any matter that could be relevant, including the precautionary principle. The judge found that a consideration of the state of knowledge or uncertainty regarding a species, the potential for serious or irreversible harm and the adoption of a cautious approach, were consistent with the subject, scope and purpose of the legislation in question. Subsequent decisions of the New South Wales Land and Environment Court have endorsed and applied the *Leatch* decision, though none have further advanced the debate on application of the principle.[34]

The application of the precautionary principle has since been confirmed in nearly all Australian states and territories and is now firmly established as an appropriate legal consideration. What remains evident, however, is an uncertainty as to not only what the content of the principle is, but also how the courts are to apply it.[35] For example, in the South Australian case of the *Tuna Boat Owners Association of SA Inc* v. *Development Assessment Commission*,[36] following an appeal from the Environment Resources and Development Court,[37] the Supreme Court of South Australia stated in reference to the application of the principle:

> There can be no hard and fast rules about what is required in a case such as this. Everything will depend upon the circumstances of the particular case, especially the level of knowledge about the impacts of the particular proposal.[38]

A significant difficulty for the development of the jurisprudence concerning the precautionary principle has been the lack of superior court authority. In the leading Federal Court case, *Friends of Hinchinbrook Society Inc* v. *Minister for Environment and Others*,[39] Justice Sackville sought to limit its application to mean interpreting the statute in a common sense cautious way, fully considering available evidence as to environmental impact. To date, there has been no serious consideration of the principle by the High Court of Australia. However, there have been emerging signs of the Court's willingness to look to international law principles when interpreting the common law and in ensuring that in the case of ambiguity, statutes are interpreted consistently with international law.[40]

A significant administrative recognition of the application of the precautionary principle was apparent in 1996 when a major mining company, North Limited, submitted a development application for the Lake Cowal Gold Project. This involved the development of an open pit gold mine, but was refused consent by the New South Wales government on the grounds that the unknown risks to the significant environment of Lake Cowal, a national estate listed wetland, could only be avoided by refusing the proposal. The company has since revised its risk assessment and resubmitted its environment impact assessment.

Case studies on the precautionary principle

Aquaculture

Aquaculture involves the farming of aquatic organisms and includes breeding, hatching, rearing and the cultivation of fish, mollusks, crustaceans and aquatic plants for sale.[41] It is a growing sector of the seafood industry that has undergone significant expansion in Australia over the past decade.[42] There are over 2,000 aquaculture licensees in Australia and fewer than 100 large-scale commercial producers that account collectively for the majority of production and employment.[43] While over 40 species are being produced, this figure is broken down (by order of value) into pearls (comprising of between 30 and 50 percent of total aquaculture), salmon/trout (16 percent), prawns (8 percent), oysters (8 percent) and tuna (more recently 33 percent).

Atlantic salmon (*Salmo salar*) aquaculture was introduced to the island state of Tasmania from Norway in 1985 and quickly became the second most important aquaculture harvest before being overtaken by tuna fattening in 1998.[44] This industry was valued at AUD$85 million in 1999–2000 and it contributes over 83 percent of the region's total aquaculture production.[45] The fastest growing sector in the Australian aquaculture industry is Southern bluefin tuna (*Thunnus maccoyii*), which began in 1991 to add value to a severely diminished tuna catch quota. In 1999, tuna aquaculture was worth approximately AUD$130 million although it has reached its limit of expansion, constrained by the total allowable catch quota of 5,265 tonnes per annum set by the Commonwealth government.[46]

Given the success and growth of the aquaculture industry in Australia, environmental issues and access to suitable sites in marine waters are important considerations.[47] Suitable sites are highly contested. Although 85 percent of Australian aquaculture production is in coastal state waters, the opportunities for expansion are more likely to be in offshore marine waters. This expansion, and the concern for developing an environmentally sustainable industry, have resulted in Commonwealth government, aquaculture industry and research body representatives drafting national and state management regimes that deal with issues such as disease resistance, immunity stress, fish nutrition and feed development for freshwater and marine fish, water quality, reproduction, and genetic variation and modification of oysters, prawns and fish.

Management of marine aquaculture in Australia

The Commonwealth government has no specific statutory responsibility for aquaculture management in Australia, although (along with state and territory governments) it does have a role in ensuring the industry is ecologically sustainable. At the federal level, the *Environment Protection and Biodiversity Conservation Act 1999* (Cth) and the *Wildlife Projection (Regulation of Exports and Imports) Act 1982* (WP(REI) Act) (Cth) specifically promote ecologically sustainable development. Under the EPBC Act, the Commonwealth government has increased its assessment of marine aquaculture operations proposed in sensitive areas, and any aquaculture development that is likely to have a significant impact on a matter of national environmental significance requires the approval of the Environment Minister. The primary objective of the WP(REI) Act is to enable Australia to comply with its obligations under the 1973 Convention on International Trade in Endangered Species of Wild Fauna and Flora (CITES)[48] by regulating the export and import of certain animal and plant products. From December 2003, exports of all aquaculture species will only be allowed if it can be demonstrated that they have been farmed in an ecologically sustainable way and have not had a detrimental effect on matters of environmental significance.

Aquaculture is a primary production activity under the authority of regional state governments.[49] Marine farming is licensed under state legislation and licenses include environmental conditions to ensure that these activities do not result in unacceptable impact on the marine environment.[50] Australia also has an important responsibility in the planning and management of marine aquaculture operations, particularly in the coordination of policy including quarantine, disease control, product quality, labeling, funding and taxation.

Developments in Australian aquaculture management

In March 1994, the Commonwealth government released the *National Strategy on Aquaculture in Australia*. The Strategy was developed to provide a vision for "an environmentally sustainable aquaculture industry" that generated economic benefit for Australia. However, the Strategy makes no mention of the precautionary principle and although it states that the "industry must be sensitive to community and government concerns for the protection of the environment and operate in an ecologically sustainable manner," no direction is provided as to how this may be achieved.[51]

The implementation of the Strategy was expected to be effective once the lead agencies developed and implemented action plans relating to each of the major issues. In 1997, a *Strategy Implementation Review* was undertaken to document its achievements and future action required to fulfill the Strategy's aims.[52] Table 7.1 summarizes the major achievements and priorities identified in the Implementation Review, including the need for more work to be undertaken on ecologically sustainable development with regard to the marine aquaculture industry, an environmental framework, increased research and development, and increased resource and market access.

The Implementation Review concluded that the "continued growth and development of a strong, environmentally sustainable aquaculture industry depend on the timely implementation of the priorities for future action."[54] Again, no specific mention is made to the precautionary principle. However, the Implementation Review recognized that aquaculture can affect the environment, and that appropriate management practices and technologies and management plans, codes of practice and risk assessments of the possible interaction between cultured animals and wild fish stocks were required if the industry was to be ecologically sustainable. In particular, it called for enhanced and enforced risk assessment protocols for intra- and inter-state movement of aquatic organisms and the importation of aquatic organisms (including aquarium fish, imported fish, fish products and ballast waters) to minimize the accidental import of disease incursions and exotic organisms.

Federal government responses to ecologically sustainable aquaculture practices are also recognized in *Australia's Oceans Policy*,[55] which includes a commitment to further implement the *National Aquaculture Strategy*, to develop objective-based environmental standards and management

Table 7.1 Key goals of the *National Strategy on Aquaculture* and the major achievements and priorities identified in the *Strategic Implementation Review*[53]

Key goal	*Achievement*	*Priorities for future action*
Government framework	Developed state/territory/ regional plans for the aquaculture industry	Relevant authorities to improve coordination of aquaculture management
Environmental management	Revised relevant legislation and license conditions to ensure that aquaculture operates within acceptable social and environmental standards	Develop national framework/guidelines for aquaculture From these guidelines develop industry codes of practice that incorporate a framework for regulatory control
Water and land use planning	Developed resource/ management plans that include aquaculture and recognize industry needs	Designate aquaculture zones in coastal areas and achieve secure long-term tenure of sites and operations
Research and development	Improved research and development coordination	Improve communication between industry and research providers Coordinate research objectives and communicative research outcomes
Education and training	Cooperative Research Centre for Aquaculture financed collaborative research for 23 doctorate students	Develop more short-term specialist/technical courses and workshops, particularly in biology and marketing

measures, to develop consistent national guidelines for environmental standards, and to identify best practices for ecological sustainability, clarity, efficiency and effectiveness. Despite these commitments, *Australia's Oceans Policy* fails to incorporate specific measures relating to the precautionary principle and aquaculture.

The voluntary *Code of Conduct for Australian Aquaculture* was initiated in 1998 through the Australian Aquaculture Forum (renamed the National Aquaculture Council in 2000), a body representing industry, governments and environmental interest groups.[56] The Code aims to maintain ecological and economic sustainability and adopts the following principles: ecological sustainable development; economic viability; long-term protection of the environment to ensure the viability of suitable aquaculture sites; compliance with, and auditing of adherence to, regulation and the Code; resource sharing and consideration of other users of the environment; and research and development. Again, the precautionary principle is not referred to directly in the Code, however, it provides direction to facilitate ecological sustainability and recommends that the aquaculture industry comply with all regulations, respect the rights of others, protect the environment, treat aquatic animals humanely, and promote safety of seafood and other aquatic foods for human consumption. It supports a total catch management approach based on natural resources management and promotes the maintenance of efficient and sustainable stocking densities.

In 1998, Australia also became a member of the Network of Aquaculture Centres in the Asia-Pacific to promote sustainable development. The voluntary *Code of Conduct for a Responsible Seafood Industry* was developed to set standards of behavior to ensure the conservation of marine ecosystems and management of living aquatic resources.[57] In particular, the Code states that operators will encourage the development and operation of aquaculture in a manner, and at a rate, in accordance with ecologically sustainable principles.

In 1999, the Commonwealth government released the *National Aquatic Animal Health Plan* (AQUAPLAN) to provide a strategic direction to animal health issues.[58] It includes eight programs, including protection of Australia's aquatic animal health status, quarantine, surveillance, and methods and protocols to manage emergency aquatic disease outbreaks. During the same period, the Commonwealth government convened a national aquaculture workshop (which included industry and government representatives) because the national aquaculture strategy and subsequent review were considered to be too broad in scope and unclear with respect to time frames.[59] Following this workshop, the *Aquaculture Industry Action Agenda Discussion Paper* was released in 2001. It presents the following vision for the aquaculture industry and states:

> By 2010 a sustainable, vibrant and rapidly growing Australian aquaculture industry will achieve at least $AUD2.5 billion in annual sales by being the world's most globally competitive aquaculture producer.[60]

The Discussion Paper aims to assist the aquaculture industry in developing environmental principles and codes of practice.[61] In addition, it provides time frames, restates that marine aquaculture operations need continual improvement, and refers directly to the *National Strategy for Ecologically Sustainable Development*. However, despite the precautionary principle not being actively mentioned, a number of precautionary "influenced" directions are included, for example:

- To avoid negativity "from coastal communities over their perceived loss of aesthetic and recreational values, as well as concerns expressed over the possible negative impact of aquaculture on the environment . . . one possible response to reduced resource availability and lengthy approval processes in high-demand coastal areas is to undertake aquaculture in coastal areas with fewer users, in off-shore areas or inland."
- The Australian aquaculture industry does not need to use significant amounts of chemicals because it is free from many diseases found in other countries.
- Incursions of endemic and exotic pests, weeds and diseases in the aquatic environment need to be managed to maintain biodiversity.
- The aquaculture industry has expressed concerns over diseases entering Australia through imports of aquaculture products from overseas.[62]

As the *Aquaculture Industry Action Agenda Discussion Paper* points out, if ecologically sustainable development is to be achieved for aquaculture, "the industry needs to go beyond compliance and take steps, such as developing Best Practice Management Practices" to integrate environmental management and the precautionary principle into each stage of their operations.[63] The Discussion Paper is intended to form the basis for new guidelines to regulate and manage the aquaculture industry and provide consistency across jurisdictions.[64]

Industry focus – marine aquaculture in Tasmania

Similarly to other marine areas in Australia, marine aquaculture in Tasmania has expanded rapidly over the past decade and has become one of the state's major industries. As discussed, commercially farmed species in Tasmania's clean marine waters include Atlantic salmon, trout, abalone, scallops, mussels, Pacific oysters (*Crassostrea gigas*) and Rock lobsters (*Jasus edwardsii*).[65]

Marine farms are located in Tasmanian territorial waters within three nautical miles of the coast. In 1999, 145 marine farms were registered in Tasmania occupying some 20 km^2 of coastal water.[66] In Tasmania, prescribed legislation legally implementing marine farm planning initially came under the *Fisheries Act 1959* (Tas), which was amended in 1982 to include marine farming activities.[67] Since then, the *Marine Farming Planning Act 1995* (Tas) has been developed to overcome deficiencies in the previous Act and to meet the rapid growth of marine aquaculture in Tasmania. This new legislation is loosely incorporated into Tasmania's RMPS in an effort to promote sustainable development, fair and orderly use of the environment, integration of marine farm activities with other marine users, public involvement, economic development and the sharing of responsibility between spheres of government, community and industry. This legislative and management scheme has now been confirmed by the *State Coastal Policy Validation Act*

2003 (Tas), which seeks to give effect to the revised Tasmanian *State Coastal Policy*.

The *Marine Farming Planning Act 1995* (Tas) allows for the preparation of Marine Farm Development Plans in an effort to outline the legislative framework, to provide a basis for aquaculture zoning, to develop management plans, and to identify and mitigate environmental impacts resulting from aquaculture operations. These development and zoning plans also "designate areas that are unsuitable for siting marine farms and prescribe limits to the areal coverage of farms in zones where aquaculture will be permitted."[68]

All statutes under the Tasmanian RMPS have sustainable development "objectives" that include the development of state policies under the *State Policy and Projects Act 1993* (Tas). The Tasmanian *State Coastal Policy* specifically states that marine farming must be planned and conducted in accordance with sustainable development principles.[69] In addition, other relevant terrestrial and marine resource management and planning legislation is to be consistent with the objectives, principles and outcomes of the Policy. In particular, section 2.1.5 of the Policy describes the implementation of the precautionary principle to the development of the "coastal zone" in accordance with national ecologically sustainable development principles.

Additionally, the Tasmanian *State of the Environment Report* concluded that more data and research were required on the environmental impacts of marine farming and the precautionary principle must be applied in the planning and managing of marine farming operations.[70] This point is exemplified in the policy and background information to the *Draft Marine Farm Development Plan for the Pipe Clay Lagoon and Georges Bay* as follows:

> There is a general shortage of detailed long-term scientific information providing baseline environmental data on the marine environment. There has also been limited research into the impacts of the marine farming industry in Tasmania . . . however, it is recognized that the time and costs involved in obtaining a comprehensive data set would delay strategies to cope with the expected pressures for growth in the industry over the next few years.[71]

This example demonstrates how the Tasmanian Government and the aquaculture industry absolved their responsibility when applying the precautionary principle as set out in the *Tasmanian State Coastal Policy* by attempting to legitimize "industry pressures" as a valid reason for overlooking the objectives set out in Tasmania's marine and environmental legislation and policies.

Interestingly, the Victorian government utilized the precautionary principle when it rejected the farming of Pacific oysters within its territorial waters. This decision was in part based upon a report detailing the environmental impacts of Pacific oyster farming in Tasmania.[72] However, despite the rejection of this industry by the Victorian government, the Tasmanian aquaculture industry continues to farm this species. Dense aggregations of Pacific oysters in rocky areas limit food and space for native species. Escapes from marine farms have also resulted in feral populations becoming established outside enclosures, and the parasitic copepod (*Mytilocola orientalis*) has been introduced to commercial mussels.[73] Environmental sustainability, the precautionary principle and inter/intra-generational equity may have been compromised with regard to Pacific oyster farming in Tasmanian waters.

Overall, the Tasmanian aquaculture industry has benefited from the introduction of the *Marine Farming Planning Act 1995* (Tas), particularly under section 65 that refers to tenure of security. Tenure ensures that marine farm leases are secured for a period of 30 years and operators can apply for an additional renewal period of ten years. Given that 30 years is a relatively long time when considering the environmental impacts that marine farms can have on the environment, the responsibility of ensuring ecologically sustainable aquaculture operations, the precautionary principle and inter/intra-generational equity for all marine users is placed directly back onto individual marine farmers.

Biodiversity protection – marine protected areas

Consistent with its developing obligations under the 1992 Convention on Biological Diversity, a *Draft National Strategy for the Conservation of Australia's Biological Diversity* was released in March 1992 with the broad goal of protecting biological diversity and maintaining ecological processes and systems. The principles of the Strategy recognized the role of *in situ* conservation and supported the specific use of marine protected areas (MPAs) throughout Australia as a means of supporting marine biodiversity. In the Australian context, a marine protected area is defined as an area of sea (which may include the seabed and subsoil under the sea) established by law for the protection and maintenance of biological diversity and of natural and cultural resources. MPAs are considered an important management tool for promoting marine conservation and management, protecting biodiversity and supporting the sustainable use of marine resources.[74] Approximately 58.5 million ha of terrestrial areas are protected (about 7.6 percent of the Australian mainland). However, the marine environment is very different. Approximately 38.9 million ha is conserved in marine protected areas, or about 3.5 percent of the Australian exclusive economic zone (EEZ). Nevertheless, Australia is still considered a world leader in declaring MPAs for marine conservation and management, with nearly one-quarter of all declared MPAs in the world residing in Australian territory. The administration, management and control of Commonwealth reserves and conservation zones under the EPBC Act is the responsibility of the Director of National Parks, within the Department of the Environment and Heritage.

The precautionary principle is mentioned in an Act that has specific relevance to MPAs. The object of the *Great Barrier Reef Marine Park Act 1975* (Cth) was to "make provision for and in relation to the establishment, control, care and development of a marine park in the Great Barrier Reef Region" in accordance with the provisions of the Act.[75] The precautionary principle is referred to in Part VB of the Act, which deals with management plans. It also occurs in section 39Z, which requires that the Great Barrier Reef Marine Park Authority be informed by the precautionary principle in the preparation of management plans. The precautionary principle is defined as having the same meaning as in section 3(5)(1) of the IGAE which is set out in a schedule to the *National Environment Protection Council Act 1994* (Cth).

New Commonwealth MPAs can be proclaimed in waters from three nautical miles to the 200 nautical mile boundary of the Australian EEZ. The designation of MPAs often straddles the boundaries of state and federal jurisdictions. In these cases, MPAs are jointly managed with the Commonwealth government and the relevant state or territory. States and territories can also declare MPAs within three nautical miles.

Australia's Ocean Policy also supports the increased declaration of MPAs in Commonwealth waters.[76] Consistent with this approach, the South-east Regional Marine Plan gives special emphasis to developing a system of representative MPAs in Commonwealth waters within the region. Noting that in the past, Commonwealth and state governments have tended to develop their own MPAs independently, the Marine Plan envisages enhanced linkages across jurisdictions. Other initiatives identified in the Marine Plan include further identification of MPAs from broad areas of interest within the South-east Region, along with a review of the process and methods used for identifying candidate MPAs.[77]

Since the designation of the first MPA in 1938,[78] significant advances have been made in MPA development. One of the most recent advances has been the promotion of the 1999 *National Representative System of Marine Protected Areas* (NRSMPA) with a goal to "establish and manage a comprehensive, adequate and representative system of MPAs to contribute to the long-term ecological viability of marine and estuarine systems, maintain ecological processes and systems, and protect Australia's biological diversity at all levels."[79] Secondary goals of the NRSMPA are to promote integrated ecosystem management, to manage human activities, to provide for the needs of species and ecological communities, and to provide for the recreational, aesthetic, cultural and economic needs of indigenous and non-indigenous people, where these are compatible with the primary goal.[80] The NRSMPA represents a national system of MPAs that contain representative samples of Australia's marine ecosystems. The system explicitly adopts the following principles: regional framework, comprehensiveness, adequacy, representativeness, highly protected areas, the precautionary principle, consultation, indigenous involvement and integration of decision making. The precautionary principle is specifically referred to with respect to MPAs as follows:

> The absence of scientific certainty should not be a reason for postponing measures to establish MPAs to protect representative ecosystems. If an activity is assessed as having a low risk of causing serious or irreversible adverse impacts, or if there is insufficient information with which to assess fully and with certainty the magnitude and nature of impacts, decision-making should proceed in a conservative and cautious manner.[81]

The NRSMPA uses the 1998 *Interim Marine and Coastal Regionalisation for Australia* (IMCRA)[82] as a framework for the representative designation of MPAs. IMCRA represents a series of maps and descriptions used to identify distinct biological and physical characteristics. Maps are produced at a regional scale (or meso-scale referring to 100s to 1,000s of km) and at a provincial scale (greater than 1,000s of km). By using these two scales, it is possible to plan at a broad ecological level and at more detailed ecosystem, community and species distribution levels. In this way, IMCRA can assist in identifying areas that need further representation and provide for priority setting and delivery of programs to support the NRSMPA.

IMCRA maps have been produced, outlining five distinct categories ranging from no protected areas in an IMCRA bioregion to a category representing greater than 50 percent coverage. The 60 IMCRA bioregions represent only 2.2 million km^2, which is a small portion of Australia's EEZ. What is also evident is that a vast difference exists in representation between bioregions. Twenty-one bioregions have no MPAs, 21 have MPAs with coverage of less than 1 percent and five bioregions have coverage between 1 and 10 percent. MPAs are skewed towards 11 bioregions, representing 92 percent of MPAs. Understandably these bioregions represent the Great Barrier Reef Marine Park (GBRMP) (Queensland), the Great Australian Bight Marine Park (South Australia and Western Australia) and Shark Bay Marine Park and Ningaloo Marine Park (Western Australia). The GBRMP is therefore not representative of the size of most Australian MPAs. Most MPAs are very small, and the low number of large MPAs contributes disproportionately to the total protected area.

The small size and number of MPAs declared at state level is illustrated in Tasmania. The 1990 *Joint Policy for the Establishment and Management of Marine Reserves in Tasmania*[83] identified additional sites for MPA designation. Four new MPAs were declared under the *National Parks and Wildlife Act 1970* (Tas). The fish, however, within these MPAs were protected under the *Living Marine Resources Act 1995* (Tas).[84] In 2001, over ten years later, the Tasmanian government released the *Tasmanian Marine Protected Area Strategy* to establish and manage new MPAs.[85] In 2003, areas previously identified as potential MPAs in the Davey and Twofold Shelf bioregions were finally declared, representing the first substantial designations in over a decade.[86]

In contrast, just off the Tasmanian continental shelf and surrounding Macquarie Island (which is under the jurisdiction of the state of Tasmania), the Commonwealth government has declared significantly sized new MPAs. The Tasmanian Seamounts Reserve was declared on 16 May 1999, recognizing the unique habitat and wildlife of the area and to protect the vulnerable benthic communities of the seamounts from human-induced disturbance.[87] Located 170 km south of Hobart, the MPA covers 37,000 ha and includes some 70 seamounts that are remnants of extinct volcanoes between 200 and 500 meters high and several kilometers across at their base. This distinctive geological feature supports a unique benthic community with at least eight new genera. The unique aspect of this MPA is the way in which it has been zoned. A "managed resource zone" includes the area from the surface to a depth of 500 meters. This zone promotes the long-term protection and maintenance of biological diversity and also provides access to commercial fishing using non-trawling methods, such as the tuna longline fishery. The "highly protected zone" includes the areas from a depth of 100 meters to 500 meters below the seabed. This zone is managed to protect the integrity of the benthic ecosystem and excludes fishing, petroleum or mineral exploration.

Conclusion

Australia has a mixed record in terms of implementing and giving effect to the precautionary principle. There has been considerable governmental effort in defining the principle and seeking to embed it within environmental and resource management policies and strategies. Further, transposition of the precautionary approach into national law and administration has been affected by articulating it as an objective or required consideration in legislation at both federal, state and territory levels. This has, at least, ensured that the precautionary approach is addressed, even if in disparate ways. However, operationalizing the principle has proved a challenge. As this chapter has demonstrated, in certain sectors of coastal and ocean management, such as aquaculture and marine protected areas, progress has been made. However, distinctive federal and state legislative frameworks and policies create challenges to ensuring complementarities in a national approach. Other sectors, such as land-based marine pollution still have considerable hurdles before a true precautionary approach will have been taken.[88] The Australian experience suggests that disciplined interpretations and methodologies for application of the precautionary principle will evolve over time. The implementation of *Australia's Oceans Policy* and the work of the National Oceans Office in developing regional marine plans is another step in this process, which has begun to show promise with the release of the South-east Regional Marine Plan. The Plan places particular emphasis upon the further development of MPAs and implementation of the NRSMPA.[89] It holds the promise of greater coordination of law, management and policy, which for a precautionary approach in a country as large as Australia, will be vital.

Acknowledgments

The authors would like to thank Dr Greg Rose and Dr Sali Bache for assistance on an earlier draft of this chapter.

Notes

1 K. Sainsbury, M. Haward, L.K. Kriwoken, M. Tsamenyi and T. Ward *Australia's Ocean Policy, Oceans Planning & Management: Issues Paper 1 – Multiple Use Management in the Australian Marine Environment: Principles, Definitions and Elements* (Environment Australia, Canberra: 1997) at 17.
2 Commonwealth of Australia *Ecologically Sustainable Development: A Commonwealth Discussion Paper* (Australian Government Publishing Service, Canberra: 1990).
3 Commonwealth of Australia *National Strategy for Ecologically Sustainable Development* (Australian Government Publishing Service, Canberra: 1992).
4 Reproduced as the Schedule to the *National Environment Protection Council Act 1994* (Cth).
5 ICESD *Summary Report on the Implementation of the National Strategy for Ecologically Sustainable Development* (December 1993). Online at: www.deh.gov.au/esd/national/nsesd/summary93/index.html (accessed 1 December 2003); ICESD *Report on the Implementation of the National Strategy for Ecologically Sustainable Development 1993–1995* (Australian Government Publishing Service, Canberra, 1996). Online at: www.deh.gov.au/esd/national/nsesd/summary95/index.html (accessed 1 December 2003).
6 Commonwealth of Australia *ESD: The Forward Agenda* (Commonwealth of Australia, Canberra: 1994).
7 ICESD (1993), note 5.
8 For comment, see C. Barton "The status of the precautionary principle in Australia: Its emergence in legislation and as a common law doctrine" (1998) 22 *Harvard Environmental Law Review* 509 at 524.
9 Commonwealth of Australia *Australia's Oceans Policy: Specific Sectoral Measures* (Environment Australia, Canberra: 1998) 3; R. Reichelt "Introduction and Welcome Address" in *Towards a Regional Marine Plan for the South-east* (National Oceans Advisory Group, National Oceans Office, Hobart: 2000).
10 Commonwealth of Australia *Australia's Ocean Policy* (Environment Australia, Canberra, 1998) 19; further specific details on the precautionary principle are found in Appendix 1 "Policy Guidance for Oceans Planning and Management."
11 R. Harding *Environmental Decision-making: The Roles of Scientists, Engineers and the Public* (Federation Press, Sydney: 1998).
12 ss. 4, 6.
13 ss. 3, 4. The application of the precautionary principle has been considered in *Dixon and Australian Fisheries Management Authority and Executive Director of Fisheries* (2000) AATA (15 June 2000), and *Aika* v. *AFMA* (2003) FCA 248.
14 Schedule 1.
15 s. 39Z.
16 s. 15 (a) and Schedule 1.
17 R. Padgett and L.K. Kriwoken "The Australian Environment Protection and

Biodiversity Conservation Act 1999: What role for the Commonwealth in environmental impact assessment?" (2001) 8 *Australian Journal of Environmental Management* 25–36.

18 *Environmental Protection and Biodiversity Conservation Act 1999* (Cth), s. 3A.

19 Ibid., s. 391.

20 See, in particular, *Environment Protection and Biodiversity Conservation Act 1999* (Cth), ss. 75, 133, 201, 216, 237, 258, 269A, 270A, 270B, 280, 285, 295, 316, 328, 338, 370.

21 In addition to potentially giving effect to parts of *Australia's Oceans Policy*, see D.R. Rothwell and S.B. Kaye "A legal framework for integrated oceans and coastal management in Australia" (2001) 18 *Environmental and Planning Law Journal* 278–92.

22 See *Environment Protection and Biodiversity Conservation Act 1999* (Cth), s. 23.

23 For a precise illustration of the potential impact of the EPBC Act in the case of turtle excluder devices, see W. Gullett "Enforcing bycatch reduction in trawl fisheries: Legislating for the use of turtle exclusion devices" (2003) 20 *Environmental and Planning Law Journal* 195–210.

24 1996 Protocol to the Convention on the Prevention of Marine Pollution by Dumping of Wastes and other Matter (1997) 26 *International Legal Materials* 7, to the 1972 Convention on the Prevention of Marine Pollution by Dumping of Wastes and other Matter, as amended 1046 UNTS 120.

25 This Convention is not yet in force.

26 See, for example, M. Haward "Fisheries and oceans governance in Australia and Canada: From sectoral management to integration?" (2004) 26(1) *Dalhousie Law Journal* (in press).

27 R. Harding and L. Fischer "The Precautionary Principle in Australia" in T. O'Riordan and J. Cameron (eds) *Interpreting the Precautionary Principle* (Earthscan Publications, London: 1994) 252; A. Devill and R. Harding *Applying the Precautionary Principle* (Federation Press, Sydney: 1997).

28 Tasmania *Information Guide: Resource Management and Planning System* 2nd edition (Department of Environment and Land Management, Hobart: 1996). For more information on the RMPS, see www.rpdc.tas.gov.au/planning/pln_docs/rmps_guide.htm (accessed 1 December 2003).

29 Environmental Defenders Office *The Environmental Law Handbook* (Environmental Defenders Office, Hobart: 1999).

30 *Tasmanian State Coastal Policy* (Department of Environment and Land Management, Hobart: 1996).

31 Barton, note 8 at 509, argues "Acceptance of the principle by Australian courts supports the conclusion that the precautionary principle is evolving into a tenant of Australian common law."

32 (1993) 81 LGERA 270 (NSW Land and Environment Court).

33 (1992) 31 *International Legal Materials* 818.

34 See in particular the decisions in *Nicholls* v. *Director-General of National Parks and Wildlife* (1994) 84 LGERA 397; *Simpson* v. *Ballina Shire Council* (1994) 82 LGERA 392; *Greenpeace Australia* v. *Redbank Power Co* (1995) 86 LGERA 143. See assessment of these decisions in Barton, note 8 at 535–43.

35 For example, in *Bridgetown/Greenbushes Friends of the Forest Inc* v. *Executive Director of Conservation and Land Management* (1997) 18 WAR 102, Justice Wheeler concludes that the plaintiff had failed to demonstrate "a serious question to be tried concerning breach of the 'precautionary approach' in a case concerning an experimental logging program;" for comment on this decision see the Queensland cases of *Yamauchi* v. *Jondaryan Shire Council* (1998) QPLR 452, and *Yulara Pty Ltd* v. *Rockhampton City Council* (1999) QPELR 296.

36 (2000) SASC 238.
37 *Conservation Council of South Australia Inc* v. *Development Assessment Commission* (1999) SAERDC 86.
38 (2000) SASC 238 at 30 per Doyle CJ.
39 (1997) FCA 55.
40 See the decisions in *Dietrich* v. *R* (1992) 177 CLR 292; *Newcrest Mining* v. *Commonwealth* (1997) 147 ALR 42.
41 See Commonwealth of Australia *Aquaculture Industry Action Agenda: Discussion Paper* (National Aquaculture Development Committee, Agriculture, Fisheries and Forestry Australia, Canberra: 2001).
42 Commonwealth of Australia *Marine Matters: Atlas of Marine Activities and Coastal Communities in Australia's South-east Marine Region* (Bureau of Rural Science, Agriculture, Fisheries and Forestry Australia, Canberra: 2002). Production has increased from a total production value of around AUD$236 million in 1990–91 to over AUD$600 million in 2000–01: *Aquaculture Industry Action Agenda* note 41; Australian Bureau of Statistics *Yearbook Australia* (Australian Bureau of Statistics, Canberra: 2001); ABARE *Australian Fisheries Statistics* (ABARE, Canberra: 1998). This figure is estimated by the Commonwealth government to be more than 25 percent of the total national production value of all fisheries. Australia *Aquaculture* (Bureau of Rural Science, Agriculture, Fisheries and Forestry Australia, Canberra: 2002).
43 *Aquaculture Industry Action Agenda*, note 41, at 11.
44 *Aquaculture*, note 42.
45 National Oceans Office *Uses* (National Oceans Office, Hobart: 2002).
46 *Aquaculture*, note 42.
47 *Marine Matters*, note 42.
48 993 UNTS 243.
49 *Marine Matters*, note 42.
50 For discussion on these issues see *Tuna Boat Owners Association of SA* v. *Development Assessment Commission* (2000) SASC 238.
51 Commonwealth of Australia *National Strategy on Aquaculture in Australia* (Department of Primary Industry and Energy, Canberra: 1993) at 15.
52 Commonwealth of Australia *National Strategy on Aquaculture in Australia: Implementation Review* (Department of Primary Industries and Energy, Australian Government Printing Service, Canberra: 1997).
53 Adapted from ibid.
54 Ibid., at 5.
55 See *Australia's Oceans Policy: Specific Sectoral Measures*, note 9 at 12:

> The Government will ensure ecologically sustainable aquaculture practices by developing and implementing policies that promote ecologically sustainable practices and technologies that minimize waste and environmental damage.

56 Commonwealth of Australia *A Code of Conduct for Australian Aquaculture* (Australian Aquaculture Forum, Canberra: 1998).
57 Australian Seafood Industry Council *A Code of Conduct for a Responsible Seafood Industry* (Australian Seafood Industry Council, Canberra: 1998).
58 Commonwealth of Australia *AQUAPLAN: Australia National Strategy Plan for Aquatic Animal Health 1998–2003* (Agriculture, Fisheries and Forestry – Australia, Canberra: 1999).
59 M. Dadswell, Deputy Manager, Aquaculture Fisheries and Aquaculture, Department of Agriculture, Fisheries and Forestry – Australia, Canberra (Personal communication, 2002).
60 *Aquaculture Industry Action Agenda*, note 41 at 11.

61 For example, the Queensland Government and aquaculture industry have developed a *Code of Practise for Aquaculture in Queensland* that has been accredited by the Environment Protection Authority.
62 *Aquaculture Industry Action Agenda*, note 41.
63 Ibid., at 19.
64 M. Dadswell, note 59.
65 *Marine Matters*, note 42.
66 National Oceans Office *Resources, Using the Ocean: The South-east Regional Marine Plan Assessment Reports* (National Oceans Office, Hobart: 2002).
67 K. Anutha and D.O'Sullivan (eds) *Aquaculture and Coastal Zone Management in Australia and New Zealand: A Framework for Resource Allocation* (Turtle Press, Hobart: 1994).
68 *Marine Matters*, note 42 at 158.
69 *Tasmanian State Coastal Policy*, note 30.
70 Sustainable Development Advisory Council *State of the Environment Tasmania Volume 2: Recommendations* (State of the Environment Unit, Land Information Services, Department of Environment and Land Management, Hobart: 1997).
71 Department of Primary Industry and Fisheries *Policy and Background Information to the Draft Marine Farm Development Plan for the Pipe Clay Lagoon and Georges Bay* (Department of Primary Industry and Fisheries, Hobart: 1998) at 14.
72 N. Coleman *A Review of Introductions of the Pacific Oyster* (Cassostrea gigas) *Around the World and a Discussion of Possible Consequences of Introducing the Species in Victoria* Australia Marine Science Laboratories Technical Report Number 56 (Fisheries Wildlife Service, Victorian Department of Conservation Forests and Lands, Queenscliff: 1986); I.A. Mitchell, A. Jones and C. Crawford *Distribution of Feral Pacific Oysters and Environmental Conditions* (Tasmanian Aquaculture and Fisheries Institute, Hobart: 2000).
73 National Oceans Office *Impacts, Identifying Disturbances: The South-east Regional Marine Plan* (National Oceans Office, Hobart: 2002).
74 See L.K. Kriwoken "Australian biodiversity and marine protected areas" (1996) 31 *Ocean and Coastal Management* 113–32.
75 *Great Barrier Reef Marine Park Act 1975* (Cth), s. 5.(1).
76 National Oceans Office *South-east Regional Marine Plan: Implementing Australia's Oceans Policy in the South-east Marine Region* (National Oceans Office, Canberra: 2004) at viii.
77 Ibid.
78 This was the Green Island Marine Park off the coast of northern Queensland in the vicinity of Cairns. See P. Bridgewater and A. Ivanovici "Achieving a Representative System of Marine and Estuarine Protected Areas for Australia" in A. Ivanovici, D. Tarte and M. Olsen (eds) *Protection of Marine and Estuarine Areas – A Challenge for Australians* (World Conservation Union, Sydney: 1993) at 23.
79 Australia and New Zealand Environment and Conservation Council (ANZECC) *Strategic Plan of Action for the National Representative System of Marine Protected Areas: A Guide for Action by Australian Governments* (ANZECC Task Force on Marine Protected Areas, Environment Australia, Canberra: 1999) at 1.
80 Ibid.
81 Australia and New Zealand Environment and Conservation Council (ANZECC) *Guidelines for Establishing the National Representative System of Marine Protected Areas* (ANZECC Task Force on Marine Protected Areas, Environment Australia, Canberra: 1998) at 6.
82 IMCRA Technical Group *Interim Marine and Coastal Regionalisation for Australia: An Ecosystem-based Classification for Marine and Coastal Environments* Version 3.3

(Environment Australia for the Australian and New Zealand Environment and Conservation Council, Canberra: 1998).

83 Tasmania *Joint Policy for the Establishment and Management of Marine Reserves in Tasmania* (Tasmanian Government, Hobart: 1990).

84 L.K. Kriwoken and M. Haward "Marine and estuarine protected areas in Tasmania, Australia: The complexities of policy development" (1991) 15 *Ocean and Shoreline Management* 143–63.

85 Tasmania *Draft Tasmanian Marine Protected Area Strategy* (Marine and Marine Industries Council, Department of Primary Industries, Water and Environment, Hobart: 2001).

86 *Resource Planning and Development Commission Inquiry into the Establishment of Marine Protected Areas within the Davey and Twofold Shelf Bioregions: Final Recommendations Report* (Resource Planning and Development Commission, Hobart: 2003).

87 National Oceans Office, note 66 at 41.

88 See L. Zann *Our Sea, Our Future: Major Findings of the State of the Marine Environment Report for Australia* (Department of Environment, Sport and Territories, Canberra, 1995) and comment in C. Williams "Combatting marine pollution from land-based activities: Australian initiatives" (1996) 33 *Ocean & Coastal Management* 87–112.

89 National Oceans Office, note 76 at viii.

Part IV

Ecosystem-based management

8 Marine ecosystem management

Is the whole greater than the sum of the parts?

Bruce G. Hatcher and Roger H. Bradbury

Introduction

The emergence over the last 20 years of a paradigm for ecosystem management parallels recognition of the necessity for holistic approaches to controlling human use of natural resources. Like most paradigm shifts, the move towards an ecosystem management regime is driven by necessity born of the failings of existing regimes, and runs the risk of uncritical acceptance at the expense of established paradigms.[1] Theory and practice have not always been well linked, and at present the conceptual development exceeds implementation in most arenas. If we accept a simple definition of ecosystem management as "... the manipulation of the ecosystem by man,"[2] it is clear that we have been managing terrestrial ecosystems at least since the use of fire by Aboriginal peoples to direct forest productivity, and continue to do so in restricted ecosystems through the practices of agriculture and aquaculture. Only recently, however, has the approach infiltrated the highly sectoral governance structures of the over-developed nations as a formal management process for the use of ocean space and marine resources.[3] The knowledge and tools required for ecosystem management in the marine environment are less available, and the concepts and applications less well developed than those on land. The opacity, remoteness, complexity and connectivity of marine ecosystems make integrated management a pragmatic necessity and a serious conceptual and logistic challenge. Agencies responsible for marine management in both Australia and Canada are adopting ecosystem-based approaches with varying degrees of commitment and success.

In this chapter we review the theory and practice of marine ecosystem management in the Canadian and Australian contexts by considering two questions: What are the essential precepts of marine ecosystem management? How are these being incorporated into the mandate and practices of marine resource management and ocean governance?

Essential precepts of marine ecosystem management

Ecosystem-based approaches to management have received considerable description and synthesis in the literatures of natural science and resource management.[4] There is no universally accepted definition of ecosystem management, and the variable use of the term has been justly criticized.[5] Three fundamental aspects of ecosystem management must, however, be specified in any operational definition:

1. the intended objectives or outcomes of management (i.e. preservation of marine biodiversity, conservation of exploited marine resources, or reservation of rights of access and usage),
2. the ecological entities in which management occurs (i.e. the ocean space, marine or coastal ecosystem), and
3. the governance structures and processes by which management is executed (i.e. legislation, regulation and compliance).

Of these, how the marine ecosystem is defined scientifically largely determines how management is prescribed to operate at this level of understanding in order to achieve stated outcomes: ecosystem management is predicated upon the understanding of an ecosystem. The actual management techniques employed, and the degree to which management actually achieves these outcomes, however, have less to do with the scientific basis of the management policy and regime than they do with the practical limitations on our ability to control ecosystem components (especially human behavior). For example, the boundaries and regulatory framework of many marine protected areas (MPAs), a popular ecosystem management tool, are well matched to our understanding of marine ecosystem structure and function.[6] Yet, the majority of MPAs fail to achieve their management objectives because of problems of implementation of those regimes.[7] Indeed, it is too early in the history of marine ecosystem management to assess rigorously the effectiveness of various models and techniques. All we can do at this stage is assess the degree to which management prescriptions match our current best models of ecosystem function and best practice of natural resource management.

The main thesis of this chapter is that the goals of ecosystem management are well-enough prescribed by our current scientific knowledge, but that this understanding, while necessary, is insufficient to prescribe the techniques of ecosystem management. Comparative analysis of various experiments in ecosystem management is the way ahead.

In search of holism – the justification for ecosystem approaches

The diversity and complexity of the natural world is structured by multiple, interacting processes and patterns that are difficult to see or comprehend (especially underwater). Natural scientists deal with this complexity by reduction to smaller units or integration to larger units. Reductionism dominates, perhaps because it is easier; historically because it is philosophically more acceptable (the world is a machine: to understand its working, break it down into its constituent parts and figure out how they work). Failure to predict adequately the effects of human activities on a wide range of ecological entities (from landfill rat populations to coral reef communities) using reductionistic models have stimulated applied ecologists to experiment with holistic methodologies that extend the theoretical underpinnings. Ecosystem management is a case in point.

Holism is rare and more tenuous (bordering on superstition at the limit, e.g. Gaia "theory"), but at its best distils the emergent properties from the system's structure or function into models for understanding, prediction or control.[8] Both approaches are essential to the advancement of our understanding of the natural world because they simplify the unmanageably complex. They are best used in a complementary way, rather than in opposition, to figure out how nature works. The ecosystem is ecology's unit of holism, and ecosystem approaches to marine management universally identify the ecosystem as the object of management, and demand holistic or integrative models and methods.[9]

The concept of the ecosystem

The ecosystem concept and the usage of the term "ecosystem" are hardly contentious in the field of ecology.[10] It is a flexible entity that has no universal definition, yet has near universal currency and meaning in the community of natural scientists. It is used widely and variably to refer to an interacting system of living and non-living components in a scale-dependent context. Thus one scientist measures the flux of nitrogen from bacteria to interstitial water in the surficial sediment ecosystem (100 mm scale), and another measures the flux of nitrogen from mixed layer plankton communities to slope depicentres in the continental shelf ecosystem (100 km scale). The universality of the concept approaches that of the biological species (which is a remarkably liquid entity, commonly defined as what a competent taxonomist says it is!), but the rigor of the supporting theoretical and empirical evidence is nowhere nearly as well developed.[11] The different maturity of these two scientific concepts contrasts curiously with the extent and strength of policy and regulation based on them (e.g. numerous national and international endangered species acts compared with a few based on the ecosystem). Nonetheless, the ecosystem word increasingly finds its way into soft and hard law, with or without explicit definition. For example, by international convention the ecosystem is "a dynamic complex of plant, animal and microorganism communities and their non-living environment interacting as functional unit,"[12] while in Canada's most recent national legislation for marine management the term is used 17 times without formal definition.[13] In Australia, the planning processes for the Great Barrier Reef Marine Park Authority and the more recent regional marine plans of the National Oceans Office are both explicitly ecosystem based.

At least 40 definitions of "ecosystem" appear in the large scientific literature on the ecosystem concept and its application to natural resource management. Rather than list them, we summarize the key elements of the definitions, and their frequency of occurrence in the literature (Table 8.1).

Of these, two essential attributes distinguish marine ecosystems from other categories of ecological organization (e.g. the species population, the community or the habitat). First, marine ecosystems explicitly include the interactions among organisms and the viscous, energetic environment they

Table 8.1 Definitional attributes of marine ecosystems (derived from 40 published definitions)

Ecosystem attributes *The definition of ecosystem:*	*Frequency of use (%) (n = 40)*	*Representative reference*
Includes all living organisms (biodiversity, species, human beings)	100	van Dyne 1969
Includes the abiotic environment (habitat, substratum, fluid medium)	100	Golley 1993
Includes biotic–abiotic interactions (trophic structure, competition, succession)	80	Odum 1994
Includes bio-physical processes (connectivity, bio-geochemical cycling)	75	Lie 1985
Is dimensionally explicit (spatially fixed boundaries, temporal stability)	70	Likens 1992
Exhibits system integrity (homeorhesis, resilience, closure)	63	Hatcher 1997
Is hierarchically organized (emergent properties, vertical symmetry)	25	O'Neill *et al.* 1986
Includes adjacent terrestrial systems (watersheds, intertidal zone, coastal development)	20	Sherman 1994

Sources: George M. van Dyne (ed.) *The Ecosystem Concept in Natural Resource Management* (Academic Press, New York: 1969); Frank B. Golley *A History of the Ecosystem Concept in Ecology* (Yale University Press, New Haven: 1993); Howard T. Odum *Ecological and General Systems: An Introduction to Systems Ecology* (University Press of Colorado, Denver: 1994); Ulriche Lie "Marine Ecosystems: Research and Management" in J. Richardson (ed.) *Managing the Oceans: Resources, Research, Law* (Lomond Publ. Inc., Mt. Airy, MD: 1985); Gene. E. Likens "The Ecosystem Approach: Its Use and Abuse" in O. Kinne (ed.) *Excellence in Ecology, Volume 3* (Ecology Institute, Oldendorf/Luhe: 1992); Bruce G. Hatcher "Coral reef ecosystems: How much greater is the whole than the sum of the parts?" (1997) 16 *Coral Reefs* 77–91; Robert V. O'Neill, Donald L. DeAngelis, John B. Wade and Thomas F.H. Allen *A Hierarchical Concept of Ecosystems* (Princeton University Press, Princeton: 1986); Kenneth Sherman "Sustainability, biomass yields and health of coastal ecosystems: An ecological perspective" (1994) 112 *Marine Ecology Progress Series 277–301.*

inhabit (i.e. the processes at the bio-physical interface). The fluid medium of the ocean connects marine populations, communities, habitats and pools of biogeochemicals far more intimately than their terrestrial counterparts.[14] Second, marine ecosystems implicitly exhibit some form of integrity reflected in emergent properties (i.e. the whole is greater than the sum of the parts, Figure 8.1), and organizational or thermodynamic closure (i.e. internal transformations exceed transboundary fluxes).[15] Without these attributes, it is not an ecosystem in the functional sense, although it may have the apparent structure of one.

A potential problem for the translation of the scientific concept of the ecosystem to the practice of ecosystem management is that the functional attributes are necessary but not sufficient criteria upon which to define any particular ecosystem. Indeed, the term "ecosystem" is used so widely and variably as to be essentially useless in governance structures that require a spatially and temporally defined entity as the object of management focus. The very definitional flexibility that allows ecosystems to be defined in terms of thermodynamics (e.g. dissipative structures) or in terms of a dominant species (e.g. kelp beds) is acceptable for science, but jeopardizes the incorporation of the concept into civil policy and legal instruments (Figure 8.1). It is one thing to identify a spatially discrete ecosystem like a lake and its watershed as a management unit. It is another, more difficult challenge to identify an ocean production system, such as an upwelling region, as an object of management, much less to predict the outcomes of various management interventions on ecosystem function. Yet many ecologists and some marine managers believe that the ecosystem is the most appropriate level of biophysical organization at which to attempt regulatory control.[16] For scientists, prediction is primarily used for the goal of understanding. For managers, however, prediction is part of the more difficult goal of sustaining the delivery of marine ecosystem goods and services to humans (Figure 8.1).

Hierarchical organizations

A fundamental feature of ecosystems is hierarchical organization.[17] Nesting is everywhere evident in the natural world: organelles within cells within organisms within populations within communities, etc. It is a truism that no matter how circumscribed the unit of investigation, there will be a smaller one still embedded within, and a larger one (an environment) surrounding it. And in biology at least, each level is likely to have as much apparent complexity, and as many counter-intuitive surprises, as those above and below.

Two fairly distinct hierarchies have emerged in ecology: one based on structural organization (word-sketched above); the other on functional organization (flows of energy and materials). Ecosystems exist at the apex of both hierarchies, but the structural and functional attributes do not map

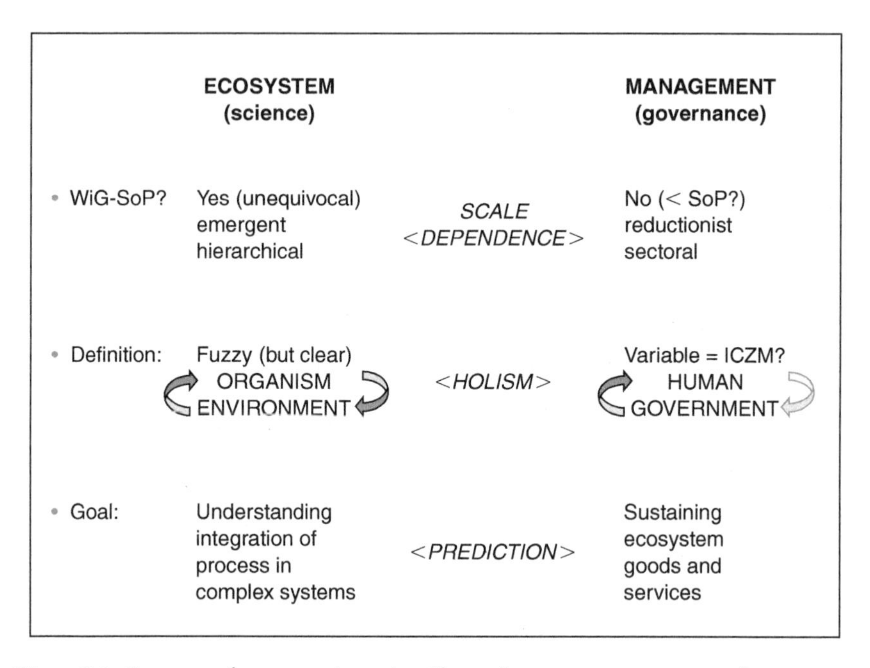

Figure 8.1 Contrasts between the scientific and governance aspects of ecosystem management. Emergent properties of ecological hierarchies (e.g. resilience) are well-known attributes of ecosystems that cannot be predicted from reduction to their parts. High-level management structures, by contrast, are rarely more, and sometimes even less effective than the aggregation of often overlapping management units (e.g. federal vs. state governments). Ecosystems are well-defined scientifically, but poorly defined as objects of management. Interactions among organisms and their environment are strong and reciprocal, while those among people and their governments are decidedly stronger from top-down than bottom-up. The unachievable but worthwhile goal of predictive ecosystem science is complete understanding, while the essential goal of ecosystem-based management is ecologically and economically sustainable human development. WiG-SoP = Whole is Greater than Sum of Parts; ICZM = Integrated Coastal Zone Management.

cleanly onto one another.[18] An emerging body of theory strives to derive the rules of assembly within levels of a hierarchy and rules of transference across levels.[19] The latter are virtually always non-linear, such that predictions of marine ecosystem trajectories cannot be derived by simply summing the trajectories of component populations (even if this were possible). While great strides have been made in the field over the last decade (e.g. in the case of the ecosystem effects of fishing),[20] it is not pessimistic to say that models capable of predicting quantitatively how management actions will propagate through the hierarchy of marine ecosystems are still decades away.

Hierarchies are not unique to the organization of ecosystems: most anthropogenic social, political, legal and economic structures are hierarchical (Figure 8.2). A potentially powerful, but little explored approach to reconciling ecosystem and management structures is to attempt to match levels of management capacity to levels of ecological organization.[21] Thus, a residential human community might take responsibility for the management of an adjacent salt marsh while provincial agencies might effectively manage the estuary in which the marsh is embedded. While this strategy is not explicit in current marine management practice in Canada, the tendency is for the management of small, coastal ecosystems to be increasingly co-managed at the municipal level with the local communities of residents (e.g. Environment Canada's Atlantic Coastal Action Program).[22] The management responsibility for the larger, coastal seascape is held primarily by the

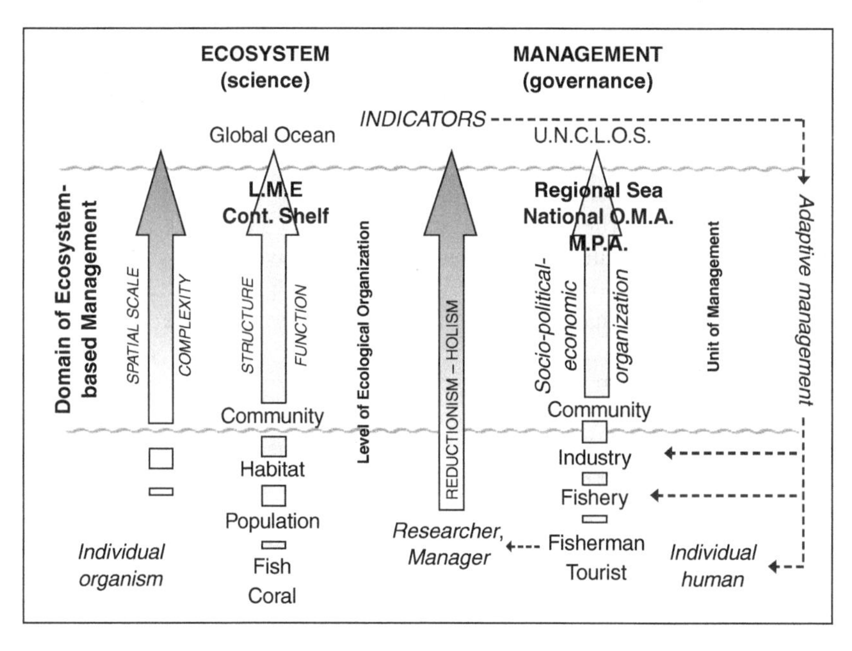

Figure 8.2 Differential scaling of the scientific and governance aspects of ecosystem management in the ocean realm. The dual hierarchies of ecosystem and management structures are portrayed from the smallest, least complex levels (bottom) to the global system (top). The domain of ecosystem-based management extends from the level of the ecological and human community to the large marine ecosystem (LME), as attempted, for example, through the Regional Seas Programme of the United Nations Environment Programme. The two hierarchies are not fully coincident. Scientists and managers increasingly produce indicators of objectives at the higher, ecosystem levels of organization, but the objects of management intervention are concentrated at the lower levels of management units. MPA = Marine Protected Area. OMA = Ocean Management Area. UNCLOS = United Nations Convention on the Law of the Sea.

provincial level of government, while the control of massive, offshore ecosystems in which these coastal zones are embedded falls in the federal domain (e.g. the large Ocean Management Area of the Scotian Shelf).[23] Similarly in Australia, the Draft South-east Regional Marine Plan (the first plan to be developed by the National Oceans Office) suggests a hierarchical mapping of management responsibility with ecosystem level.[24] Unfortunately, the separation of powers and competition between these levels of government in both countries often means that, in practice, they will function in parallel rather than hierarchically.

Spatial-temporal scaling

Intimately linked to the hierarchical organization of marine ecosystems is the relationship between the spatial and temporal scales at which ecological structures evolve and persist.[25] In this realm the biological and physical components of marine ecosystems diverge. Physical processes such as mixing, advection and exchange turnover, and the oceanographic structures they engender such as eddys, gyres and coherent water masses tend to scale linearly in space–time (i.e. smaller is faster, larger is slower in mathematically tractable fashion). Biological processes such as photosynthesis, growth and predation, and the ecological structures they engender such as algal blooms, fish stocks and coral reefs are disjunctive and patchy in their space–time distribution. The beautiful allometry that characterizes the physiology of individual organisms (e.g. respiration scales exponentially to body mass) starts to break down at the level of the population, and is mathematically intractable at the level of the community.[26] Thus, while there is a rough scaling of higher levels of organization to larger–longer domains, it is not a relationship readily extrapolated to ocean management units (e.g. estuaries will not always be larger or older than the populations of salt marsh organisms that inhabit them).

As a key aspect of ecological holism and ecosystem management is the "up-scaling" from the small sections of ecosystems that can be experimentally manipulated, monitored and patrolled to the entire ecosystem: the scaling relationships of process and structure are very important. Scale mismatches between measured and predicted systems comprise the single most serious source of error in ecology. It is in part for this reason that many ecologists conceptualize ecosystems as processors of materials rather than ensembles of organisms, and then invoke physical models and controls that can be up-scaled more confidently than biological interactions.[27]

A creative tension exists between those who perceive ecosystems as primarily controlled by physical processes and those who see biological interactions as the dominant structuring forces. As is invariably the case in ecology, it is always a mix of both types of factors that yield a particular ecosystem function or configuration at any particular place and time. A parsimonious approach to ecosystem delineation is to determine the variability explained by the physical processes, and if that is not sufficient for prediction, then turn to the biological processes. This is particularly defensible in the context of decision support for management because of the relative ease of defining the geographical and temporal domains of the physical entities and processes. Thus, for example, the zoning of benthic ecosystems for biodiversity preservation on both the western and eastern shelves of Canada is predicated primarily on the distribution of topography, substratum and sediment grain size, itself a reflection of hydrodynamic and geological processes.[28] There has been a similar, if perhaps less rigorously developed approach in Australia, where geomorphic and oceanographic processes are being used to define the large-scale management entities, within which more biologically-based management entities are nested.[29]

Relationships between ecosystem structure and function

Perhaps the most important development in marine ecology this decade has been the rapprochement of the fish counters and carbon fluxers under the combined banners of biodiversity conservation and ecosystem goods and services. Growing maturity in both streams of the young discipline of ecology allowed new collaborations focused on determining how the structure of ecosystems (e.g. the relative proportions of different feeding types in a community) influence key functions of the ecosystem (such as the remineralization of organic inputs). Important concepts and metrics such as keystone species and functional redundancy were rejuvenated or developed to explain how, for example, two ecosystems with very different species compositions could produce very similar harvestable yields or create bioherms at identical rates. The implications for the selection and protection of conservation management or resource exploitation targets are obvious, but are virtually unexplored. This is a particularly rich vein for research.

Although most of the arguments above draw on ecosystem theory and data in the generic sense, there exist profound differences between marine and terrestrial ecosystems, which limit the relevance of many of the large majority of theoretical constructs, case studies and models stemming from research in terrestrial ecosystems. In brief, marine ecosystems are more open, more variable, more biologically diverse, and more highly interconnected than their terrestrial counterparts. These attributes translate into more extensive geographical areas and more diffuse boundaries associated with marine ecosystems. As oceans management moves inevitably towards the owning and zoning of marine territory (the dominant terrestrial management strategy), these large and fuzzy lines will be hard to transfer onto legal deeds, leases and zoning plans for human usage.[30] This is a problem that will not be solved by more or better ecological research. It must be resolved by a change in the methodology of zoning as it is adapted to the marine environment. There is a need for legal, institutional and governance structures to adapt to the reality of marine ecosystems, not the reverse. This adaptation will be challenged by the mismatches between the scaling of marine ecosystems and of ocean governance systems (Figure 8.2). One of the most serious challenges to their reconciliation is the fact that even the most advanced and adaptive ecosystem management is ultimately effected at the level of the individual human or sectoral peer group (Figure 8.2). We do not yet have ecosystem-scale agents of management. For this reason, ecosystem management is correctly termed "ecosystem-based management", as our management actions do not operate directly on the ecosystem itself in the vast majority of ocean applications.

Another key ecological concept relating ecosystem structure to function is that of the source–sink relationship. These may be fluxes of genetic information, fish larvae, migrating whales or atmospheric carbon among ecosystems and ecosystem components: the concept remains applicable throughout the hierarchy of ecosystems. Identifying sources and sinks is an ecological research challenge that implies a creative variety of tools from genetic micro-satellites to earth-orbiting space satellites. In the end, it results in a biogeographical product that is directly transferable to zoning schemes and upstream regulatory protocols. We see this as the most promising of all current ecological endeavors in the context of ecosystem management.

Dealing with variability and uncertainty

Natural systems are inherently variable and complex (i.e. middle number systems), and shallow marine ecosystems comprise an end member of the continuum because of their strong connections to the adjacent atmospheric and terrestrial systems. Heteroscedasticity, non-linear dynamics and chaotic behavior are the norm, homogeneity, linearity and predictability the exceptions. No one can know the number of fish in the sea: they can only state a range of statistical probability.[31] Vehicular traffic patterns and macroeconomic time series are paragons of determinism by comparison. This is an area where ecologists seeking more money and time to "understand the ecosystem" should be treated with skepticism by the manager with urgent decisions to make. The management sciences and their tools (e.g. risk analysis, multiple criteria analysis) are better equipped to deal with the intractability than the natural sciences when it comes to ecosystem-based management. The challenge is to incorporate formally the best ecological statistical approaches and our growing understanding of ecosystems as complex systems[32] into adaptive, ecosystem-based, decision support systems.[33]

Incorporation of ecosystem science into marine management (with a Canadian perspective)

Operationalizing the ecosystem concept in oceans governance and marine resource management is a top agenda item in environmental and development agencies worldwide.[34] Its top-down implementation in Canada has been cautious and sporadic. The federal and provincial agencies responsible for the conservation of wildlife and pristine habitat (e.g. Environment Canada and Parks Canada), supported by academia and non-governmental organizations, led the way by identifying and sequestering parks and protected areas on the basis of representivity and species at risk.[35] The main focus of this work so far has been the classification of Canada's marine ecosystems with the explicit or implicit goal of designating a network of marine protected areas that preserve the nation's marine and coastal biodiversity (especially birds and mammals). Lately, the inclusion of resident communities of humans as co-managers of these areas has gained momentum.

In parallel, the apparent degradation of Canada's most intensely used "seas" (the Great Lakes) fostered the development of holistic approaches to environmental quality assessment for management decision support through the monitoring of new metrics of ecosystem integrity.[36] The key inputs here were from academia, with some excellent implementation by governmental (including municipal) agencies. The Great Lakes clean up is one of the few, well-documented success stories of ecosystem-based management.

Not surprisingly, the federal agency responsible for the management of exploited living marine resources, the Department of Fisheries and Oceans (DFO), was the last to join the ecosystem management bandwagon. A 25-year investment in systems ecology[37] was never translated into decision support for resource management, and the focus of this agency was predominantly on the management of the great fisheries using single species stock assessment and population modeling with gear and quota-based interventions. The crises of the collapsing ground and anadromous fisheries in the 1990s, followed by the visionary *Oceans Act* of 1996, have caused many of the agency's fisheries biologists to question their assumptions and to reach out to more holistic approaches. Simultaneously, the oceans branch (as distinct from the science and fisheries management branches) was created and invigorated with a cadre of new or reformed staff who brought an ecosystem approach *a priori*. In short order, three major initiatives have been made in marine ecosystem management: the delineation of large ocean management areas for zoning-based management (e.g. the Eastern Scotian Shelf Integrated Ocean Management Plan),[38] the identification of candidate marine protected areas,[39] and the incorporation of ecosystem objectives and metrics into fisheries management planning.[40]

On the latter issue, there is healthy resistance to ecosystem-based management within the agency, based on scientific uncertainty and enforcement practicality. For many fisheries scientists and managers in DFO, suitable indicators and reference points for ecosystem management consist of the additive combination of several single species stock assessments. The focus on maintaining yields and verifiable management outcomes is encapsulated in a statement of the objective of ecosystem management circulated in the agency: "The prevention of reductions in the productivity of resources, or modification of the function of ecosystems in ways that are difficult or impossible to reverse, as a result of human activities."[41] The tension between approaches to marine ecosystem management based on management of individual components informed by the whole, and management of the whole without attempting to micro-manage its components (Figure 8.3), reflects the theoretical and practical uncertainties facing an agency confronting the challenge of ocean management with the tools of exploited resource management.

Observing and working with this large, science-based agency as it grapples with the major challenges of implementing the ecosystem management mandate of Canada's *Oceans Act* is an education and a privilege. Here we consider the main scientific issues of marine ecosystem management as it is developing in Canada.

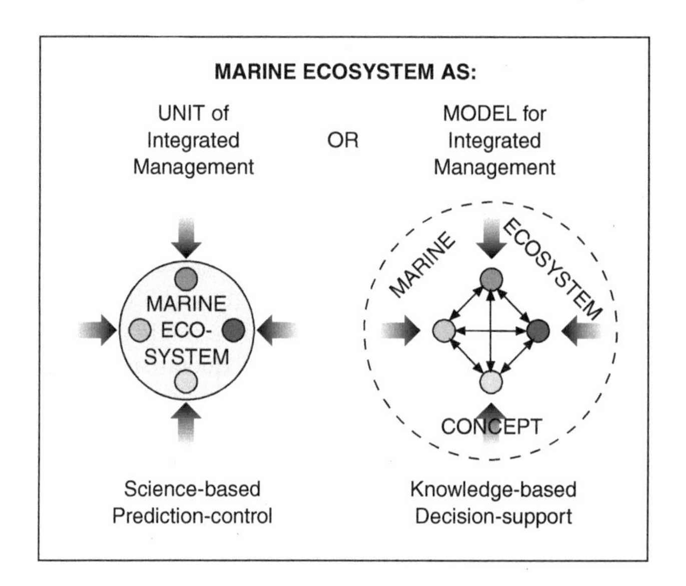

Figure 8.3 Two operational approaches to marine ecosystem-based management. Small, patterned circles represent different components or sub-systems (e.g. coral reef, fishery) of a large marine ecosystem designated for integrated management. The arrows represent management actions (e.g. protection of biodiversity, prevention of land-based sources of pollution). In the first approach (left), management actions are directed at a well-defined ecosystem (closed circle), and are designed to contribute to management objectives for all contained ecosystem components. In the second approach (right), the ecosystem is poorly defined (dashed circle), and management actions are directed at better-known components of the ecosystem in the context of their interactions within it. This is the *status quo* for ecosystem-based management, but the first approach is increasingly the desired target.

Models and currencies of ecosystem management

The dual hierarchies of ecosystem organization reflect a duality of ecological models and metrics by which ecosystem management outcomes may be assessed. Those ecologists who count things have dominated the field in numbers, theory development and empirical data collection. Their ecosystems are defined by the distribution and abundance of communities of plants and animals, and their theory and models are predicated on the dynamics of the species population. If the ecosystem is incorporated into this knowledge, it is as a suite of abiotic and biotic factors (i.e. externalities) that influence the trajectory of the population. This is the realm of wildlife and fisheries biology, which has dominated marine resource management in Canada since Confederation. Apparent failings of resource management based on population biology[42] have been a major incentive for the development of more holistic (ecosystemic) approaches to oceans governance in Canada.[43] In essence, this represents a paradigm shift, a transition from a focus on the parts to a focus on the whole as the object of management (Figure 8.3).

Those ecologists who measure ecosystems in terms of fluxes are a vocal minority. Their ecosystems are defined by the pools and exchange networks of materials, energy and information. Their theory and models are predicated on the physical-chemical principles of thermodynamics and conservation of mass, and on the mathematical analysis of pattern and information flows.[44] The incorporation of the ecosystem is explicit, as these models and metrics are constituted at the higher levels of organization (e.g. nutrient sequestering by an estuary, tidal subsidy to the productivity of the Bay of Fundy). This is the realm of biogeochemical cycling, landscape ecology and ecosystem modeling.

During the 1960s and 1970s, a large investment was made in ecological modeling based on trophodynamic and thermodynamic theory and principles.[45] The desired goals were ambitious and exceeded the capacity of the poorly developed theory and analytical techniques. A vast amount of modeling was done; very little of this stood up to rigorous verification and testing, and the good stuff[46] largely got buried in the piles of disillusioned rubbish. After a decade's hiatus, some of this material is being recycled, reinvented and extended to ecosystem-based management. The theory and tools are more robust, and the modelers more humble. The new hybrid, coupled physical–biological numerical models, are particularly promising for predicting source–sink relationships relevant to the design of networks of marine protected areas (MPAs).[47] The massive data demands of empirical description of even the simplest marine ecosystem mean that modeling will always be a primary tool of ecosystem-based management.

The contributions of ecosystem modeling to ecosystem-based management in Canada have been best developed in the Great Lakes,[48] while major initiatives in marine applications during the 1960s and 1970s (e.g. the

Marine Ecology Laboratory at the Bedford Institute of Oceanography),[49] were abandoned in favor of species-based approaches during the 1980s. There is, at present, little apparent effort on the part of Canadian agencies charged with the funding of scientific research or the management of marine ecosystems to re-invest in these approaches.

Both types of natural science-based models and their currencies have strengths and weaknesses in their capacity to explain and predict ecosystem structure and function in the context of anthropogenic effects. Both should be used, according to their availability and relevance to particular cases, to provide qualitative and quantitative inputs to formal decision support systems for marine ecosystem-based management. A significant research challenge is to contrast rigorously the different models in the same ecosystem management scenario, and to derive guidelines for managers to select the less risky and most cost-effective currencies for assessing management effectiveness. Towards this end, DFO has identified three main conservation objectives of marine ecosystem-based management: to maintain biodiversity, productivity and physical-chemical properties of the oceans.[50] The models and techniques for achieving these objectives are yet to be finalized, but the tendency is to apply population-based models and currencies to the biodiversity objective, biogeochemical models to the environmental quality objective, and a combination of both to the productivity objective.

Measures of ecosystem integrity and health

The greatest promise of the ecosystem approach to marine management is the potential to identify single metrics of the integrated response of an ecosystem to an anthropogenic perturbation (be it a "push" or a "press").[51] An example is the ratio of new to recycled primary production (the "f-ratio") in photic zone water masses at bay to shelf scales.[52] This single measure, relatively easily obtained, encapsulates the trophic status of the water column, the dependence of the biotic community on allochthonous inputs of nutrients (e.g. land-based sources), and the capacity of that community to export biomass (e.g. potential fish yields). Similar indices have been developed for coral reef ecosystems.[53] In general, however, these measures disappoint when it comes to management decision support, usually because the sensitivity is low and the inferred chain of causality cannot produce the "smoking gun" that unequivocally identifies a target of management action. It is difficult to envision an expert witness ecologist successfully defending such a measure in a court of law. Yet, if ecosystem-based management is to flourish, robust measures of the integrity and health of marine ecosystems must be developed so that its effectiveness can be quantified. This is arguably the most important area of research and development for marine ecosystem-based management,[54] because managers are generally involved by stating clearly what they need, rather than waiting for ecologists to tell them what they can have.

Universal indicators of marine ecosystem integrity or health should not be a priority because of the variety of ecosystems and management objectives. Three common attributes may, however, be identified from the existing experiences with ecological indicators of the effectiveness of ecosystem management. A good indicator should:

1 Capture in a few metrics a substantial portion of the structural and functional complexity inherent to the ecosystem at lower levels of organization (i.e. use emergent properties).
2 Exhibit a measurable response to planned management actions that provides a clear guide for decision making (i.e. be statistically robust).
3 Be practically and financially feasible for measurement at appropriate spatial and temporal scales (i.e. involve monitoring).

Such indicators are currently best developed for terrestrial and freshwater ecosystems in Canada in the management contexts of protected areas and water quality maintenance.[55] Most commonly, mappable changes in habitat characteristics are used to indicate change in biodiversity based on empirical relationships between habitats and their biota. Specification of indicators for Canada's large "ocean management areas" is a recent initiative.[56] The approach involves the identification of ecosystem objectives (e.g. maintain trophic structure) and specific characteristics (e.g. predator–prey relationships) within each of the three, overarching management goals (conserving biodiversity, productivity and environmental quality). Then, one to five indicators are suggested for each characteristic (e.g. a measure of food-web complexity) and reference points are specified (e.g. proportion of prey to predator species). Finally, an achievable, operational objective of management is detailed (e.g. retain historical patterns of abundance of predators and their prey). The resulting matrix includes no less than 76 indicators of ecosystem status or function, begging the question of whether meeting the third criterion (above) is possible.

Including humans in ecosystems

Our focus to this point has been almost exclusively on the application of natural science products to marine ecosystem management. This bias contradicts a tenet of ecosystem management as "... a complex and interdisciplinary type of management in which human beings are the fundamental elements...."[57] It reflects the history of marine resource management, however, that targets the non-human components of ecosystems (e.g. fish), while depending almost exclusively on management methods directed at the control of human activities within arbitrary portions of marine ecosystems (Figure 8.2). Until recently, most marine ecologists preferred to conduct their research in ecosystems with minimal human influence (preferably none, other than themselves of course!). Fisheries biologists are an obvious exception, where the assumption is that fishing mortality is the dominant control on fish populations. Even when the human component is large and evident, measurements of ecosystem processes typically treat the people as externalities rather than components whose behavior is to be studied as part of the whole. This is a significant problem for ecosystem management that can only be redressed through the marriage of the natural and social sciences in bioeconomic and governance models.[58]

The reasons for explicitly including the human organism in any program of research in support of marine ecosystem-based management are obvious and compelling. First, they are often the dominant organism in the system: top predator, major cause of disturbance or main source of biogeochemical loading. It follows that neglecting to measure the human role jeopardizes the entire research. Even if the human component is negligible at present, it will not be in the future. Second, the primary methods used to exploit, protect or restore marine ecosystems involve the regulation of human behavior and activity through education, legislation and enforcement. Building the distribution, abundance and behavior of humans into ecosystem models from the outset means that the research products will be more likely to provide sound support for management decisions.

A potentially effective, but little-used way to include humans in research and monitoring for ecosystem management is to use the exceptionally well understood physiology and psychology of the organism to provide metrics of ecosystem integrity and health. For example, rates of mortality, morbidity, malnutrition and migration, as well as various epidemiological and economic indices of human health and livelihoods in coastal fishing communities, can provide directly relevant measures of the health of the adjacent marine ecosystems upon which these people depend.[59]

While explicitly stating the goal of sustaining the human usage of marine ecosystem goods and services,[60] the strategies and tools for marine ecosystem management in Canada make scant use as yet of social science models or research products. However, several experimental initiatives in community-based co-management of marine ecosystems are underway,[61] and the recently formed Ocean Management Research Network[62] is an interdisciplinary academic group with a strong social science orientation.

Setting the boundary conditions

Major simplifying assumptions are necessary to deal with the problem of the complexity of natural ecosystems and human societies in ocean management: the domains of the management units and issues must be sharply delineated. Setting the boundary conditions on the ecosystem and the parameters of control is the first, difficult part of ecosystem management. Even the apparently straightforward business of establishing the spatial boundaries of the management unit is problematic because of the difficulty in matching ecosystem domains to the spatial patterns of enforceable management units.

The edges of marine ecosystems are often hard to locate and their positions vary through time. Similarly, the habitat components of ecosystems form shifting mosaics at multiple spatial scales.[63] In the case of the management goal of preserving benthic biodiversity on the shelf ecosystems of Canada, for example, the characteristic spatial scales of distinct varieties of seabed habitat (and their associated assemblages of organisms) range from decimeters to kilometers, and the positions of the boundaries of these units may change by similar distances inter-annually.[64]

As the dominant operational tool for marine management of ocean space is multiple use zoning of various human activities, the spatial scales at which marine ecosystems are classified become a prime determinant of the location and sizes of the various management zones. Marine ecosystem classification is a topic of intense research globally,[65] and is steadily leading to the mapping of ocean habitats and communities at ever-finer scales.[66] Such detailed delineation of marine ecosystems may exceed the practical requirements of management. For example, the smallest management unit that can be reliably monitored and enforced on the Scotian Shelf is approximately 10 km by 10 km with not more than six way points, which sets an operational limit on the scale at which benthic ecosystems need be mapped for management purposes.[67]

Given the present state of ecosystem theory and practice, it seems to us that we should look first to the limitations of the management tools to set the boundaries of the ecosystems we attempt to manage. This suggestion does not in any way preclude good natural science. Managers should set the research agenda in terms of verifiable indicators of effective management, and boundary conditions of the monitoring science should be matched to the spatial resolution at which practical control of human activity is possible.

Putting ecosystem-based management into practice (with an Australian perspective)

The real advances in marine ecosystem-based management to date are more a matter of expedient necessity than of the precautionary application of theoretical understanding. Australia leads the way in actually developing marine management regimes focused on spatially explicit ecosystems and their components. The small human population and low ratios of marine scientists to coastline and ocean space have encouraged Australians to seek alternatives to reductionism, resource-by-resource and sector-by-sector management structures. The scientific traditions of the nation have not generally been conducive to holism. The British-model universities are strongly disciplinary (still separating botany and zoology on most campuses). The dominant government research agency, the Commonwealth Scientific and Industrial Research Organization (CSIRO), has an industry-by-industry heritage. The American (Odum) school of ecosystem ecology, focused on trophodynamics and biogeochemical cycling, had little impact on the development of Australian marine ecological sciences, which focused strongly on population dynamics and community structure and function. The few, early, research initiatives focused at the level of entire marine ecosystems (e.g. CSIRO's Port Hacking Estuary project of the late 1970s[68] and the Western Australian Coastal Ecology Program of the 1980s)[69] had little impact on management, even at the level of the main commercially exploited species.

By contrast, Australia's government management authorities at both state and federal levels, and the emerging power of "green" politics from the mid-1970s on produced a series of marine ecosystem management initiatives that stimulated and demanded science for management at this level. Most notable of these was the creation of the Great Barrier Reef Marine Park (GBRMP), which led to the massive expansion of the Australian Institute of Marine Science (AIMS) and an explosion of academic research under a unifying theme of the nation's flagship ecosystem. Perceived, ecosystem-wide threats to the integrity of the Great Barrier Reef, such as the Crown of Thorns starfish and nutrification from coastal development further served to rally scientists and managers to a common cause that sometimes transcended their disciplinary and institutional separations.[70] National advisory boards (e.g. the Australian Marine Science and Technology Advisory Council) and academic research funding agencies such as the traditional Australian Research Grants Council, and the more specific Marine Sciences and Technologies Grants Scheme dedicated attention and resources to support research on marine ecosystems by university and government scientists. The innovative program of establishing well-funded Collaborative Research Centres (CRC) focuses public and private sector partners on particular ecosystems judged to require integrated management, such as the CRC for the Sustainable Management of the Great Barrier Reef in Townsville and the CRC for Antarctic Research in Hobart. Australian policy-makers and the Australian public understand what marine ecosystems are and accept the necessity and advantage of management at that level.

The division of responsibility for the management of marine resources between state and federal governments in Australia (so much more equitable than in Canada) has allowed the federal agencies to invest heavily in research and experimentation for alternative, ecosystem-based management regimes in a few key areas. The Northwest Shelf of Australia falls within federal jurisdiction and produces large amounts of both renewable and non-renewable resources, contributing substantially to the nations' economy. CSIRO, in collaboration with Western Australian institutions and AIMS is investing heavily in the development of science-based decision support tools for ecosystem-based management, including coupled bio-physical models for conserving fisheries and preserving marine biodiversity in the face of intense offshore gas exploitation in partnership with Indonesia.[71] Another ecosystem focus has been Antarctica, in which Australia claims a major stake. Through its participation in the Convention on the Conservation of Antarctic Marine

Living Resources (CCAMLR), Australia has made significant contributions to the ecosystem-based management of the southern ocean krill fishery in a model of shared international responsibility for ocean governance.[72]

The National Oceans Office (NOO) represents a multi-ministry, national policy body that has supported and facilitated a biogeographical approach to the implementation of ecosystem-based coastal and marine management in Australia by providing a scientific basis for the identification of bioregions, while remaining flexible on the locale-specific modes of implementation.[73] The approach is explicitly spatial, and the boundary conditions for the management units are clearly to be set on the basis of the best available science. (But the lack of complete scientific knowledge or understanding is not permitted to delay the process.) The focus of implementation is on a consensual, multi-stakeholder process involving multiple levels of government and nongovernmental agencies, according to their jurisdictions and capacities (which vary greatly from region to region around the nation's coasts). All of the bioregions and many of the lead management agencies have been identified, but NOO is taking a step-by-step approach, starting with one of the better known, but most societally and managerially complex regions, the southeast coast.[74] It is too early to judge the efficacy or success of this approach in implementing ecosystem-based marine management in Australia. Progress is obvious on several fronts; it appears to have captured the enthusiasm of a good proportion of the nation's marine scientists, and the interest at least of many sectors in the country's intensely used coastal and marine areas. As with the more embryonic experiments in the ecosystem-based management of large ocean areas in Canada, the degree to which lessons learned in one region can be generalized to others remains to be tested. Certainly, some of the lessons from the GBRMP planning process have been transferred.

Finally, although one cannot dismisses regional differences in Australia given the strong assigned and residual powers of the states, the geopolitical isolation of the nation and the restricted scale of politics in Australia mean that rapid and definitive decisions can be made on marine management issues in which the entire nation is engaged. Designation and protection of islands and seamounts in the nation's southern ocean are a case in point.[75]

By explicitly identifying marine ecosystems as objects of management (Figure 8.3), and allocating resources for research to support decision making in that context, Australia revolutionized its marine science cadre in the space of two decades. Note that the politicians and managers drove the transition from reductionistic to holistic approaches, but there was never any question that marine science was the key requirement for ecosystem-based management. In short, while Canadians have talked about and legislated for marine ecosystem-based management, the Australians have started implementing it as a matter of necessity with the policy and legal tools available, even in the absence initially of a formal, national policy and dedicated harmonized legislation. These are now being developed, and will be the better for the extensive experience gained. This is adaptive management at the highest level.

MPA implementation as a metric of ecosystem-based management

The marine protected area (MPA),[76] with its explicit geographical boundaries and multiple management objectives, is the most developed manifestation of marine ecosystem-based management. While the natural and social science underpinning the design and function of MPAs is largely in place (i.e. oceanography, fisheries and conservation biology, ecological economics, public administration), empirical tests and demonstrations of their effectiveness in achieving desired goals have not yet accumulated to the point that these interventions may be considered science-based or proven.[77] This has not hindered the establishment of hundreds of MPAs worldwide, but rarely are they located, configured and managed according to results and predictions of marine ecosystem science (e.g. nutrient-production budgets, food web structures, source–sink relationships, meta-population dynamics). Rather, MPAs are selected primarily in response to levels and patterns of human use and perceptions of value. This is not to say that MPAs designated on social grounds (e.g. aesthetic value) will not enhance ecosystem integrity (e.g. maintain biodiversity), or that MPAs managed for societal benefit (e.g. maintenance of tourism profits) cannot be evaluated with ecosystem-based metrics of management effectiveness (e.g. stability of species–area relationships).

Regardless of the role of ecosystem science in their implementation, MPAs are currently the management tool of choice for governments and communities seeking to maintain the goods and services (and non-use values) of marine ecosystems. A crude measure of a society's progress towards ecosystem-based management of marine resources is the number or area of MPAs relative to the total area of ocean space. Of course, not all MPAs are of equal value in meeting marine management objectives, nor do they share the same objectives. In both Australia and Canada, MPAs are designated by multiple agencies for different purposes, operating at two levels of government in Australia, and only one in Canada. Despite the fact that Canada has chosen to legislate for MPAs by statute law (the *Oceans Act*) and Australia through policy-driven applications of existing legislation, the many similarities between the two countries (governments, economies, population distributions, cultures, etc.) suggest that a comparison of the status of MPA establishment may be interpreted as an indication of their relative progress in the implementation of marine ecosystem-based management. Both nations have similar-sized EEZ areas, but federal and state agencies in Australia have designated at least 178 MPAs covering a total of 646,155 km^2 (5.38 percent of the EEZ).[78] Even correcting for the fact that only about 33 percent of the world's largest nominal MPA (the Great

Barrier Reef Marine Park) is actually protected from fishing, the numbers are substantial. In contrast, three federal agencies in Canada have established only four true MPAs (i.e. "no-take" zones) covering approximately 3,700 km^2 (0.06 percent of Canada's EEZ), with a further 12 proposed to cover another 7,800 km^2. In addition to these unequivocal MPAs, there are some 56 parks, sanctuaries and fishery closures that include marine space or species, and afford some degree of protection from certain human activities. The majority of these are marine bird sanctuaries. At the most generous estimate, partially protected marine space would cover a total of about 116,000 km^2 (barely 1.8 percent of Canada's EEZ).[79] The Australian MPAs are more likely to be designed to serve multiple objectives and to be embedded within multiple-use management areas. In Canada, MPAs often stand in isolation and are focused on the preservation of a single component of a marine ecosystem (e.g. a migratory bird population, a whale observation site, a localized coral community). The few attempts to establish MPAs that protect entire environments and conform to obvious boundaries of marine ecosystems have been slow to develop in Canada.[80]

It would appear on the basis of this criterion that Australia has made considerably more progress on the implementation of marine ecosystem-based management than has Canada. The reasons for this are several and complex, but we suggest that the devolution of power over ocean resources to levels of government closer to resident communities of users (i.e. state governments), and the healthy separation of the research and management agencies,[81] coupled with effective programs of education, produces an environment of greater public trust in the benefits ecosystem-based approaches bring to the management of ocean space and marine resources in Australia.

Conclusion

Marine ecosystem management is an oxymoron. It will never happen in the way that we manage agricultural or aquacultural ecosystems, for example, because it is operationally limited to the human components of poorly known and uncontrollable ecosystems. Rather, it is a short-hand term for "... a strategy for the integrated management of land, water and living resources that promotes conservation and sustainable use in an equitable way ... based on the application of appropriate scientific methodologies focused on levels of biological organization that encompass the essential processes, functions and interactions among organisms and their environment."[82] To be effective, ecosystem management should be "... goal-driven ... based on a collaboratively developed vision of desired future conditions that integrates ecological, economic, and social factors ... applied within a geographic framework defined primarily by ecological boundaries."[83] In short, ecosystem management is part and parcel of what we also call integrated coastal zone management and good oceans governance.

It is hard to conceive of a more demanding and important challenge than translating ecosystem theory into management policy and practice with the goal of reducing, halting and reversing the anthropogenic degradation of the ocean ecosystems. Besides the support of those who use the ocean, it will require fully interdisciplinary collaboration among specialists in the natural, social and management sciences, as well as economics and law. Perhaps more significantly, it requires the emergence of transdisciplinary generalists who build their ideas in the interstices among the specialties (collectively, the developing field of marine affairs). Organizations such as the ACORN foster this emergence.

Notes

1 D.S. Wilcove and R.B. Blair "The ecosystem management bandwagon" (1995) 10 *Trends in Ecology and Evolution* 345.
2 S.H. Spur "The Natural Resource Ecosystem" in G.M. van Dyne (ed.) *The Ecosystem Concept in Natural Resource Management* (Academic Press, New York: 1969).
3 Government of Canada Bill C-26 *An Act Respecting the Oceans of Canada* S.C. 1996, c. 31; Commonwealth of Australia *Australia's Oceans Policy* (Environment Australia, Canberra: 1998).
4 M.H. Belsky "Developing an Ecosystem Management Regime for Large Marine Ecosystems" in K. Sherman and L.M. Alexander (eds) *Biomass Yields of Large Marine Ecosystems* (American Association for the Advancement of Science Symposium 111, Washington, DC: 1989) 443–60; R.H. Bradbury "Strategies for the Analysis of Large Marine Ecosystems" in L.M. Chou and K. Boto (eds) *Living Coastal Resources: Data Management and Analysis* (National University of Singapore, Singapore: 1988) 6–15; G. Likens *An Ecosystem Approach: Its Use and Abuse* (Ecology Institute, Oldendorf: 1992); R. Everett "An Overview of Ecosystem Management Principles" in M.E. Jensen and P.S. Bourgeron (eds) *Ecosystem Management: Principals and Applications* (Pacific Northwest Research Station General Technical Report No. 318, Portland, OR: 1994) 5; J. Lubchenco "The Scientific Basis of Ecosystem Management: Framing the Context, Language, and Goals" in *Ecosystem Management – Status and Potential* 103rd Congress, 2nd Session (U.S. Government Printing Office, Washington DC: 1994) 33–9; R.E. Grumbine "Reflections on 'What is ecosystem management?'" (1997) 11 *Conservation Biology* 41–7; N.L. Christensen, A.N. Bartuska, J.H. Brown, S. Carpenter, C. D'Antonio, R. Francis, J.F. Franklin, J.A. MacMahon, R.F. Noss, D.J. Parsons, C.H. Peterson, M.G. Turner and R.G. Woodmansee "The report of the Ecological Society of America Committee on the scientific basis for ecosystem management" (1996) 6 *Ecological Applications* 665–91; H.J. Cortner, M.A. Shannon, M.G. Wallace and S.M. Burke *Institutional Barriers and Incentives for Ecosystem Management* (Pacific Northwest Research Station General Technical Report No. 354, Portland, OR: 1996); K.L. Jope and J.C. Dunstan "Ecosystem-Based Management: Natural Processes and Systems Theory" in G.R.W. Wright (ed.) *Natural Parks and Protected Areas* (Blackwell Science Publishing, Cambridge, MA: 1996); S.L. Yafee "Three faces of ecosystem management" (1999) 13 *The Journal of the Society for Conservation Biology* 713–25; Global Environment Facility (GEF) *Integrated Ecosystem Management* (Global Environment Facility Operational Paper No. 12, GEF, New York: 2000).
5 R. Schlaepfer *Ecosystem-Based Management of Natural Resources: A Step Towards Sustainable Development* (IUFRO Occasional Paper No. 6, International Union of Forest Research Organizations, Washington, DC: 1997).

6 T.S. Agardi *Marine Protected Areas and Ocean Conservation* (RG Landes Company, Austin: 1997); P.K. Dayton, E. Sala, M.J. Tegner and S. Thrush "Marine reserves: Parks, baselines, and fishery enhancement" (2000) 66 *Bulletin of Marine Science* 617–4; E. Sala, O. Aburto-Oropeza, G. Paredes, I. Parra, J.C. Barrera and P.K. Dayton "A general model for designing networks of marine reserves" (2002) 298(5600) *Science* 1991–3.
7 G. Kelleher, S. Wells and C. Bleakley *A Global Representative System of Marine Protected Areas* (The World Bank, Washington, DC: 1995); J. Alder "Have tropical marine protected areas worked?: An initial analysis of their success" (1996) 24 *Coastal Management* 97–114.
8 C.S. Holling (ed.) *Adaptive Environmental Assessment and Management* (John Wiley & Sons, London: 1978); R.H. Bradbury, R.E. Reichelt and D.G. Green "Explanation, Prediction and Control in Coral Reef Ecosystems III. Models for Control" in J.T. Baker, R.M. Carter, P.W. Sammarco and K.P. Stark (eds) *Proceedings of the Inaugural Great Barrier Reef Conference, Townsville, August 28–September 2, 1983* (James Cook University Press, Townsville: 1983) 165–9.
9 See N.L. Christensen *et al.*, note 4; H.R. Skjoldal, C. van Dam, E. Degré, P. Funegård, L. Hedlund, B. Marchant, A. Schippers and W. Zevenboom *Workshop on the Ecosystem Approach to Management and Protection of the North Sea* (Nordic Council of Ministers, Oslo: 1998); GEF, note 4.
10 A.G. Tansley "The use and abuse of vegetational concepts and terms" (1935) 16 *Ecology* 284–307; H.T. Odum *Systems Ecology* (Wiley, New York: 1993); F.B. Golley *A History of the Ecosystem Concept in Ecology: More Than the Sum of the Parts* (Yale University Press, New Haven: 1993); K.H. Mann and J.R.N. Lazier *Dynamics of Marine Ecosystems: Biological-Physical Interactions in the Oceans* (Blackwell Science, London: 1996) 257; R.H. Bradbury, J. van der Laan and D.G. Green "The idea of complexity in ecology" (1996) 27 *Senckenbergiana maritima* 89–96; M.D. Bertness, S.D. Gaines and M.E. Hay *Marine Community Ecology* (Sinauer Associates, Sunderland, MA: 2001) 550.
11 J. Major "Historical Development of the Ecosystem Concept" in G.M. van Dyne (ed.) *The Ecosystem Concept in Natural Resource Management* (Academic Press, New York: 1969); R.V. O'Neill, D.L. DeAngelis, J.B. Waide and T.F.H. Allen *A Hierarchical Concept of Ecosystems* (Princeton University Press, Princeton: 1986) 253.
12 United Nations Environment Programme (UNEP) *From Policy to Implementation: Decisions from the Fifth Meeting of Parties to the Convention on Biodiversity* (UNEP, Nairobi: 2000) 36.
13 See Government of Canada, note 3.
14 J. Steele "The ocean 'landscape'" (1989) 3 *Landscape Ecology* 185–92.
15 S. Woodley, J. Kay and G. Francis (eds) *Ecological Integrity and the Management of Ecosystems* (University of Waterloo Press, Waterloo: 1993); B.G. Hatcher "Coral reef ecosystems: How much greater is the whole than the sum of the parts?" (1997) 16 *Coral Reefs* 77–91.
16 L.M. Alexander "Large marine ecosystems: A new focus for marine resource management" (1993) 17 *Marine Policy* 186–98; Exploratory Steering Committee (ESC) *Millennium Ecosystem Assessment: Strengthening the Capacity to Manage Ecosystems for Human Development – Project Overview* (Exploratory Steering Committee, Seattle, WA: 2000) 20; GEF, note 4.
17 See O'Neill *et al.*, note 11.
18 H.A. Mooney, J.H. Cushman, E. Medina, O.E. Sala and E-D. Schulze (eds) *Functional Roles of Biodiversity: A Global Perspective* (Scientific Committee on Problems in the Environment, John Wiley and Sons, New York: 1996) 393–429.

19 R.V. O'Neill "Perspectives in Hierarchy and Scale" in J. Roughgarden, R.M. May and S.A. Levin (eds) *Perspectives in Ecological Theory* (Princeton University Press, Princeton: 1989) 140–56; T.F.H. Allen and T.W. Hoekstra *Towards a Unified Ecology* (Columbia University Press, New York: 1993); S. Hubbell *Unified Theory of Biodiversity and Biogeography* (Princeton University Press, Princeton: 2001).
20 S.J. Hall *The Effects of Fishing on Marine Ecosystems and Communities* (Blackwell Science Publishing, London: 1999); S. Jennings *Ecosystem Effects of Fishing* (Chapman & Hall, New York: 2001).
21 F. Berkes and C. Folke (eds) *Linking Social and Ecological Systems. Management Practices and Social Mechanisms for Building Resilience* (Cambridge University Press, Cambridge, UK: 1998).
22 See "What is the Atlantic Coastal Action Program (ACAP)?" Online at: www.ns.ec.gc.ca/community/acap/index_e.html (accessed 19 January 2004).
23 R. O'Boyle (ed.) *Proceedings of a Workshop on the Ecosystem Considerations for the Eastern Scotian Shelf Integrated Management (ESSIM) Area* (Canadian Stock Assessment Secretariat Proceeding Series 2000/14, Fisheries and Oceans Canada, Dartmouth, NS: 2000) 92.
24 National Oceans Office (NOO) *Draft South-east Regional Marine Plan: Implementing Australia's Ocean Policy in the South-east Marine Region* (NOO, Hobart: 2003). The Plan was adopted in May 2004. Online at: www.oceans.gov.au/se_implementation_plan.jsp (accessed 28 April 2005).
25 T.M. Powell "Physical and Biological Scales of Variability in Lakes, Estuaries, and the Coastal Ocean" in J. Roughgarden, R.M. May and S.A. Levin (eds) *Perspectives in Ecological Theory* (Princeton University Press, Princeton: 1989); S.A. Levin "The problem of pattern and scale in ecology" (1992) 73 *Ecology* 1943–67.
26 J.H. Brown *Scaling in Biology* (Oxford University Press, New York: 2000); R.H. Gardner *Scaling Relations in Experimental Ecology* (Columbia University Press, New York: 2001).
27 B.G. Hatcher, J. Imberger and S.V. Smith "Scaling analysis of coral reef systems: An approach to problems of scale" (1987) 5 *Coral Reefs* 171–81.
28 M.A. Zacharias, D.E. Howes, J.R. Harper and P. Wainwright "The British Columbia marine ecosystem classification: Rationale, development, and verification" (1998) 26 *Coastal Management* 105–24; J.C. Roff and M.E. Taylor "National frameworks for marine conservation – A hierarchical geophysical approach" (2000) 10 *Aquatic Conservation: Marine and Freshwater Ecosystems* 209–23; B.G. Hatcher "Literature Review" and "Workshop Outputs" in V.E. Kostylev (ed.) *Maintenance of the Diversity of Ecosystem Types* (Canadian Science Advisory Secretariat Proceedings Series 2002/023, Fisheries and Oceans Canada, Dartmouth, NS: 2002) 12–92.
29 See note 24.
30 Penny Doherty (ed.) *Ocean Zoning: Can it Work in the Northwest Atlantic? Workshop Proceedings* (Marine Issues Committee Special Publ. No. 14, Ecology Action Centre, Halifax: 2005).
31 D.R. Ludwig, R. Hilborn and C. Walters "Uncertainty, resource exploitation and conservation: Lessons from history" (1993) 260 *Science* 1736.
32 R.H. Bradbury "Futures, predictions and other foolishness" in M.A. Janssen (ed.) *Complexity and Ecosystem Management: The Theory and Practice of Multi-agent Systems* (Edward Elgar, Cheltenham: 2002) 48–62.
33 C. Walters *Adaptive Management of Renewable Resources* (MacMillan Publishing Company, New York: 1986).
34 See Christensen *et al.*, note 4; Anonymous *Report of the Workshop on the Ecosystem Approach* (UNEP/CBD/COP/4/ Inf.9, Lilongwe, Malawi: 1998); Skjoldal, note 9; GEF, note 4.
35 J.R. Harper, G.A. Robilliard and J. Lathrop *Marine Regions of Canada: Frame-*

work for Canada's System of National Marine Parks (Parks Canada, Ottawa: 1983) 49; L. Harding, H.E. Hirvonen and J. Landucci *Marine Ecological Classification System for Canada* (Marine Environmental Quality Advisory Group, Environment Canada, Ottawa: 1994) 21; H.E. Hirvonen, L. Harding and J. Landucci "A National Marine Ecological Framework for Ecosystem Monitoring and State of the Environment Reporting" in N.L. Skackell and J.H.M. Willison (eds) *Marine Protected Areas and Sustainable Fisheries* (Acadia University Press, Wolfville, NS: 1995) 117–29; J. Day and J.C. Roff *Planning for Representative Marine Protected Areas: A Framework for Canada's Oceans* (World Wildlife Fund Canada, Toronto: 2000) 147.

36 D.J. Rapport "What constitutes ecosystem health?" (1989) 1 *Perspectives in Biology and Medicine* 121–32; C.J. Edwards and H.A. Reiger *An Ecosystem Approach to the Integrity of the Great Lakes in Turbulent Times* (Special Publication 90-4, Great Lakes Fishery Commission, Ann Arbor: 1989); J.J. Kay "A non-equilibrium, thermodynamic framework for discussing ecosystem integrity" (1991) 15 *Environmental Management* 483–95; R.E. Munn "Monitoring for Ecosystem Integrity" in S. Woodley, J.J. Kay and G. Francis (eds) *Ecological Integrity and the Management of Ecosystems* (University Waterloo Press, Waterloo, ON: 1993) 105.

37 The Marine Ecology Laboratory was founded by the Fisheries Research Board in 1962 at the St. Andrew's Biological Station, and was dismantled by the Department of Fisheries and Oceans in 1987 at the Bedford Institute of Oceanography. Kenneth H. Mann, personal communication.

38 See O'Boyle, note 23. ESSIM Planning Office, *Eastern Scotian Shelf Integrated Ocean Management Plan (2006–2011) Draft for discussion* (Oceans & Coastal Management Division, Oceans and Habitat Branch, Maritime Region, Fisheries and Oceans Canada, Dartmouth, NS: 2005).

39 W.G. Harrison and D.G. Fenton (eds) *The Gully: A Scientific Review of Its Environment and Ecosystem* (Canadian Stock Assessment Secretariat Research Document 98/93, The Secretariat, Ottawa: 1998) 282; G.S. Jamieson and J. Lessard *Marine Protected Areas and Fishery Closures in British Columbia* (Canadian Special Publications of Fisheries and Aquatic Sciences 131, NRC Press, Ottawa: 2000).

40 M. Sinclair, R. O'Boyle, L. Burke and S. D'Entremont *Incorporating Ecosystem Objectives Within Fisheries Management Plans in the Maritimes Region of Atlantic Canada* (ICES CM 1999/Z:03, International Council for the Exploration of the Sea, Copenhagen: 1999) 20; G. Jamieson, R. O'Boyle, J. Arbour, D. Cobb, S. Courtnay, R. Gregory, C. Levings, J. Munro, I. Perry and H. Vandermeulen *Proceedings of the National Workshop on Objectives and Indicators for Ecosystem-Based Management* (Canadian Science Advisory Secretariat Proceedings Series 2001/09, DFO, Nanaimo, BC: 2001) 140.

41 DFO *Canada's Ocean Strategy* (Department of Fisheries and Oceans, Ottawa: 2002); *Canada's Oceans Action Plan – Phase 1*. Online at: www-comm.pac.dfo-mpo.gc.ca/pages/release/bckgrnd/2005/bg002_e.htm (accessed 5 March 2005).

42 J.A. Hutchings and R.A. Myers "What can be learned from the collapse of a renewable resource? Atlantic cod, *Gadus morhua*, of Newfoundland and Labrador" (1994) 51 *Canadian Journal of Fisheries and Aquatic Science* 2126–46.

43 DFO *Strategic Plan: Moving Ahead With Confidence and Credibility* (Communications Directorate, Department of Fisheries and Oceans, Ottawa: 2000).

44 R.E. Ulanowitz and T. Platt (eds) "Ecosystem theory for biological oceanography" (1985) 213 *Canadian Bulletin of Fisheries and Aquatic Science* 1–260; Kay, note 36; Odum, note 10.

45 B.C. Patten (ed.) *Systems Analysis and Simulation in Ecology Volumes 1-4* (Academic Press, New York: 1976).

46 E.P. Odum "The emergence of ecology as a new integrative discipline" (1977) 195 *Science* 1289–93.
47 E. Wolanski (ed.) *Oceanographic Processes of Coral Reefs: Physical and Biological Links in the Great Barrier Reef* (CRC Press, Baton Rouge: 2000); R.K. Cowen, K.M.M. Lwiza, S. Spnaugle, C.B. Paris and D.B. Olsen "Connectivity of marine populations: Open or closed?" (2000) 287 *Science* 857–9.
48 See Edwards and Reiger, note 36.
49 See note 37.
50 See Jamieson and Lessard, note 39.
51 E.P. Odum "Trends expected in stressed ecosystems" (1985) 35 *BioScience* 419–21.
52 T. Platt, P. Jauhari and S. Sathyendranath "The Importance and Measurement of New Production" in P.G. Falkowski and A.D. Woodhead (eds) *Primary Productivity and Biogeochemical Cycles in the Sea* (Plenun Press, New York: 1992) 273–84.
53 T.J. Done and R.E. Reichelt "Integrated coastal zone and fisheries ecosystem management: Generic goals and performance indices" (1998) 8 *Ecological Applications* S110–S118.
54 See Rapport, note 36; B.D. Haskell, B. Norton and R. Costanza "What is Ecosystem Health and Why Should We Worry About It?" in *Ecosystem Health: New Goals for Environmental Management* (Island Press, Washington, DC: 1992); Munn, note 36; S.-H. Liang and B.W. Menzel "A new method to establish scoring criteria of the index of biotic integrity" (1997) 36 *Zoological Studies* 240–50; D.J. Rapport, R. Costanza and A.J. McMichael "Assessing ecosystem health" (1998) 13 *Trends in Ecology and Evolution* 397–402; B.D. Smiley, D. Thomas, W. Duval and A. Eade *Selecting Indicators of Marine Ecosystem Health: A Conceptual Framework and an Operational Procedure* (State of the Environment Report Occasional Paper Series No. 9, Environment Canada, Ottawa: 1998).
55 See R.J. Steedman "Modification and assessment of an index of biotic integrity to quantify stream quality in Southern Ontario" (1988) 45 *Canadian Journal of Fisheries and Aquatic Sciences* 492–501; Edwards and Reiger, note 36; Environment Canada *Compendium of Ecosystem Health, Goals, Objectives, and Indicators: Community Initiatives Quick Reference Guide* (Environment Canada, Ottawa: 1997).
56 See O'Boyle, note 23; Jamieson and Lessard, note 39.
57 See Schlaepfer, note 5.
58 See Berkes and Folke, note 21; A.T. Charles *Sustainable Fishery Systems* (Blackwell Science Publishing, Oxford, UK: 2001).
59 B.G. Hatcher and G.H. Hatcher "A question of mutual security: Exploring interactions between the health of coral reef ecosystems and coastal communities" (2004) 1 *EcoHealth* 229–35.
60 See note 3; Environment Canada, note 53; Jamieson and Lessard, note 39.
61 See ACAP, note 22; Bay of Fundy Ecosystem Partnership (BoFEP). Online at:www.bofep.org/ (accessed 19 January 2004).
62 See Ocean Management Research Network (OMRN). Online at: www.omrn.ca/eng_home.html (accessed 19 January 2004).
63 See Levin, note 25.
64 See D.S. Davis, P.L. Stewart, R.H. Loucks and S. Browne *Development of a Biophysical Classification of Offshore Regions for the Nova Scotia Continental Shelf* (Proceedings of the Coastal Zone Canada Conference, Halifax, NS: 1994) 2149–57; Day and Roff, note 34; P. Elsner *The Scotian Shelf Seascape: Using GIS for Quantitative Resource Evaluation and Management – A Case Study* (M.M.M. Thesis, Marine Affairs Program, Dalhousie University, Halifax, NS: 1999) 94; V.E. Kostylev, B.J. Todd, G.B.J. Fader, R.C. Courtney, G.D.M. Cameron and R.A. Pickrill "Benthic habitat mapping on the Scotian Shelf based on multibeam

bathymetry, surficial geology and sea floor photographs" (2002) 219 *Marine Ecology Progress Series* 121–37; Zacharias *et al.*, note 28.

65 See L.M. Cowardin, V. Carter, F.C. Golet and E.T. LaRoe *Classification of Wetlands and Deepwater Habitats of the United States* (U.S. Department of the Interior, Fish and Wildlife Service, FWS/OBS-79/31, Washington, DC: 1979) 103; M.N. Deither *A Marine and Estuarine Classification System for Washington State* (Department of Natural Resources, Corvallis, WA: 1990) 56; J.R. Harper, J. Christian, W.E. Cross, H.R. Frith, G. Searing and D. Thompson *Classification of the Marine Regions of Canada* (Environment Canada, Ottawa, ON: 1993) 77; Hirvonen *et al.*, note 35; K. Hiscock (ed.) *Classification of Benthic Marine Biotopes of the North-East Atlantic* (Proceedings of a BioMar Workshop, Peterborough Joint Nature Conservation Committee, Cambridge, UK: 1995) 105; C.L.K. Robinson and C.D. Levings "An overview of habitat classification systems, ecological models, and geographic information systems applied to shallow foreshore and marine habitats" (1995) *Canadian Manuscript Report of Fisheries and Aquatic Sciences* 2322; Jane Watson *A Review of Ecosystem Classification: Delineating the Strait of Georgia* (DFO Science Branch, Pacific Region, Vancouver, BC: 1997) 81; H.G. Greene, M.M. Yoklavich, R.M. Starr, V.M. O'Connell, W.W. Wakefield, D. Sullivan, J.E. McRae and G.M. Cailliet "A classification scheme for deep seafloor habitats" (1999) 22 *Oceanologica Acta* 663–78.

66 See H.R. Frith, G. Searing and P. Wainwright *Methodology for a B.C. Shoreline Biotic Mapping System* (Report to B.C. Environment, Lands & Parks, LGL Ltd., Sydney, BC: 1994) 38; Ronald I. Miller (ed.) *Mapping the Diversity of Nature* (Chapman & Hall, Melbourne: 1994) 341; P.J. Mumby and A. Harborne "Development of a systematic classification scheme of marine habitats to facilitate management and mapping of Caribbean coral reefs" (2000) 88 *Biological Conservation* 155–63; Kostylev *et al.*, note 64.

67 B.G. Hatcher "Scaling Management Units to Ecosystem Components in Ocean Management Areas" paper presented at the *First Ocean Management Research Network Annual Meeting*, Ottawa, 2002.

68 E. Parker (ed.) *The Port Hacking Estuary Research Program* (CSIRO, Cronulla, NSW: 1977) at 47.

69 W.J. Wiebe, C.J. Crossland and R.E. Johannes *Coastal Ecology Program, Annual Report* (CSIRO, Marmion, WA: 1985) at 53.

70 J. Sapp *What is Natural? The Crown-of-Thorns Starfish Threat to the Great Barrier Reef* (MacMillan Press, Toronto: 1996) at 284.

71 K. Sainsbury, personal communication (August, 2000).

72 For more information, see CCAMLR website. Online at: www.ccamlr.org/ (accessed 20 January 2004).

73 National Oceans Office *Australia's Oceans Policy* (NOO, Canberra: 2001) 21; E.N. Eadie, "Evaluation of Australia's Oceans Policy as an Example of Public Policy Making in Australia" (2001) 120 *Australian Centre for Maritime Studies* 1–13.

74 See note 24.

75 A. Kostlow, personal communication (October, 2000).

76 The World Conservation Union (IUCN) defines marine protected areas (MPAs) as: "Any area of intertidal or sub-tidal terrain, together with its overlying waters and associated flora, fauna and historical and cultural features, which has been reserved by legislation or other effective means to manage and protect part or all of the enclosed environment." See IUCN *Guidelines for Protected Area Management Categories* (IUCN, Gland: 1994).

77 See J. Alder, note 7.

78 A complete, categorized and dated listing of all 17 types of marine protected areas in Australia may be obtained from the Australian Commonwealth

Department of Environment and Heritage. Online at: www.deh.gov.au/parks/nrs/capad/2002/index.html (accessed 25 April 2005). The Commonwealth (federal) government manages the large majority of the protected marine area (616,635 km^2) in only 14 MPAs, of which the Great Barrier Reef Marine Park, at almost 350,000 km^2, is the largest.

79 There is no one source of information about Canada's MPAs. Those created by Fisheries and Oceans Canada under the *Fisheries Act* and *Oceans Act* are described at www.dfo-mpo.gc.ca/canwaters-eauxcan/oceans/mpa-zpm/dmpa_e.asp (accessed 25 April 2005). Those created by Parks Canada under the *National Parks Act* and the *National Marine Conservation Areas Act* are described at www.pc.gc.ca/progs/amnc-nmca/plan/prog_e.asp (accessed 25 April 2005). Those created by Environment Canada under the *Canada Wildlife Act* and the *Migratory Bird Convention Act* are described at www.cws-scf.ec.gc.ca/habitat/protected_e.cfm (accessed 25 April 2005).

80 For example, the Sable Island Gully, a shelf-edge canyon ecosystem off Nova Scotia, has been identified as a priority site for protection since the early 1990s, but was only designated under the *Oceans Act* in May 2004. See Canadian Wildlife Service *Notes from the Workshop – Evaluation of the Natural Values of "The Gully" and Sable Island Offshore Areas* (Bedford Institute of Oceanography, Dartmouth, NS: 1994); Online at: www.mar.dfo-mpo.gc.ca/oceans/e/essim/gully/essim-gully-e.html (accessed 25 April 2005). The Endeavour Hydrothermal Vents Area off British Columbia, the first MPA designated under the *Oceans Act* in 2003, also encompasses a distinct marine ecosystem that is small, remote and unexploited. See Online at: www.pac.dfo-mpo.gc.ca/oceans/mpa/Endeavour_e.htm (accessed 25 April 2005).

81 The Australian Institute of Marine Science, the Commonwealth Scientific and Industrial Research Organization, and the universities that conduct the research relevant to MPA selection and assessment in Australia are separate from those agencies that have the legislative authority to create and manage MPAs (i.e. state and federal government departments such as the Department of Conservation and Land Management in Western Australia and the Great Barrier Reef Marine Park Authority).

82 See UNEP, note 12.

83 See Schlaepfer, note 5.

9 Ecosystem bill of rights

Richard J. Beamish and Chyrs-Ellen M. Neville

Introduction

New information about the factors that affect the dynamics of fish populations and new policies that expand the responsibilities of managers beyond single species are changing our attitudes of how we should be stewards of ecosystems. For example, there is accumulating evidence that climate, climate change and the ocean environment affect the abundance trends of commercially important fishes[1] in a manner that we previously believed was exclusive to fishing impacts. This new information indicates that fish populations are affected by atmospheric impacts as well as by human interventions into their ocean habitats.

We also are learning that there are natural trends in abundance of fish that persist for periods, then may shift to new states.[2] Some investigators suspect that the frequency and extent of these shifts will change as the impacts of greenhouse gas accumulations are realized.[3] Thus, there is a new appreciation of the relevance of the impacts of the ocean habitat of a species when assessing the levels of catch that safely ensure that a population is not prevented from replenishing itself. In Canada, it is not only an appreciation, it is policy, entrenched in new legislation.

In Canada, the *Oceans Act*[4] and the *Species at Risk Act* (SARA)[5] are new policies that identify a requirement to consider ecosystem-based management approaches and provide protection for any species that is at a critically low abundance. It is clear from the *Oceans Act* that Canada intends to move beyond a single-species-based management approach (Table 9.1). It is true that developing an ecosystem-based management approach is difficult because fisheries are often regulated at the single species level. It is also true that it is difficult for science to determine fishing quotas using multi-species assessments because the required models are poorly developed. Perhaps our own organizations perpetuate a single-species orientation by assigning the management and research of single species to individual investigators rather than creating multi-species task teams. SARA treats all species equally and legislates protection and recovery for any species that is determined to be in need of protection (Table 9.2). In fisheries management there has been a preference for some species because of their taste or charisma and an indifference or even animosity and fear for others. A legislative requirement to protect species and stocks at risk will eventually require that there is an assessment of the impacts of fishing on associated species. It will be in the best interest of the fishing industry to ensure that by-catch, as well as targeted catch, are not overfished. Irradication fisheries, such as existed for spiny dogfish (*Squalus acanthias*) in British Columbia in the late 1950s and early 1960s[6] obviously will no longer be tolerated.

Table 9.1 Highlights from Canada's Ocean Act that relate to ecosystem-based management

1	WHEREAS Canada promotes the understanding of oceans, ocean processes, marine resources and marine ecosystems to foster the sustainable development of the oceans and their resources.
2	WHEREAS Canada holds that conservation, based on an ecosystem approach, is of fundamental importance to maintaining biological diversity and productivity in the marine environment.
3	WHEREAS Canada promotes the wide application of the precautionary approach to the conservation, management and exploitation of marine resources in order to protect these resources and preserve the marine environment.
4	WHEREAS the Minister of Fisheries and Oceans, in collaboration with other ministers, boards and agencies of the Government of Canada, with provincial and territorial governments and with affected aboriginal organizations, coastal communities and other persons and bodies, including those bodies established under land claims agreements, is encouraging the development and implementation of a national strategy for the management of estuarine, coastal and marine ecosystems.

Table 9.2 Highlights from Canada's Species at Risk Act that relate to the legislative requirement to protect species and stocks at risk

1	Create a legislative base for the scientific body that assesses the status of species at risk in Canada.
2	Prohibit the killing of extirpated, endangered or threatened species and the destruction of their residences.
3	Provide authority to prohibit the destruction of the critical habitat of a listed wildlife species anywhere in Canada.
4	Lead to automatic recovery planning and action plans through the listing of species at risk.
5	Provide emergency authority to protect species in imminent danger, including emergency authority to prohibit the destruction of the critical habitat of such species.
6	Wildlife has international value and that providing legal protection for species at risk would in part meet Canada's obligations under the United Nations Convention on Biological Diversity.

Ecosystem rights

We suggest that a logical next step is to agree that ecosystems have rights. A statement of rights is recognition that our own health and quality of life is related to the other species that share our habitat. Most humans expect that our marine ecosystem will continue to function in a manner that will sustain our fisheries and will not threaten our own well-being. All species compete for space and humans are no exception to this generalization. A statement of the rights of ecosystems does not inhibit our intervention into the habitat of other species, but it does make a commitment that we will attempt to understand the consequences of our interventions and be responsible stewards of the species that share our ecosystem. In Canada, we place great value on our Charter of Rights and Freedoms. We believe that people in other countries deserve similar rights. Canadians and others have defended these rights, sometimes with horrific consequences. An ecosystem bill of rights would be an extension of the values we hold for ourselves and a reminder that we cannot remain indifferent to the health of our environment.

We illustrate the need for responsible changes in the stewardship of marine ecosystems using two areas off Canada's west coast. One area is Bowie Seamount, located 180 kilometers west of the Queen Charlotte Islands and the other is the Strait of Georgia. Approximately 74 percent of the population of British Columbia lives close to the shores of the Strait of Georgia.[7] The Strait of Georgia is the most important rearing area for juvenile Pacific salmon and habitat for a variety of other species of commercial and charismatic importance.

Bowie Seamount is one of the shallowest seamounts in the Northeast Pacific. It rises from 3,100 meters to 25 meters below the surface where the area above 1,000 meters is approximately 120 km^2 (Figure 9.1). It is a discrete ecosystem that has a fish fauna similar to coastal ecosystems but simpler. Surprisingly, the top trophic levels are well represented, and the mid-trophic levels, such as small pelagics, appear to be reduced in diversity. Pacific halibut (*Hippoglossus stenolipis*), sablefish (*Anoplopoma fimbria*) and rougheye rockfish (*Sebastes aleutiaus*) appear to be abundant. There are 19 other species of rockfish and 29 other species of fish.[8] Fishing is permitted on the seamount, but there are no stock assessments made for any of these species.[9] There have not been any evaluations about the impacts of the permitted fishing activity on the species being fished or on any of the associated species. There is no understanding of recruitment processes for the species that are fished that extends beyond some creative speculations. There are no restrictions on halibut fishing. Any licensed fisherman can fish on the seamount and theoretically remove halibut down to the last fish. Licensed sablefish fishermen give themselves one trip per year to the seamount. There is no apparent limit to this catch, and the catch is not counted either in the individual's quota or the total annual quota. Apparently, it is believed that these sablefish are surplus to the population that is being managed. There

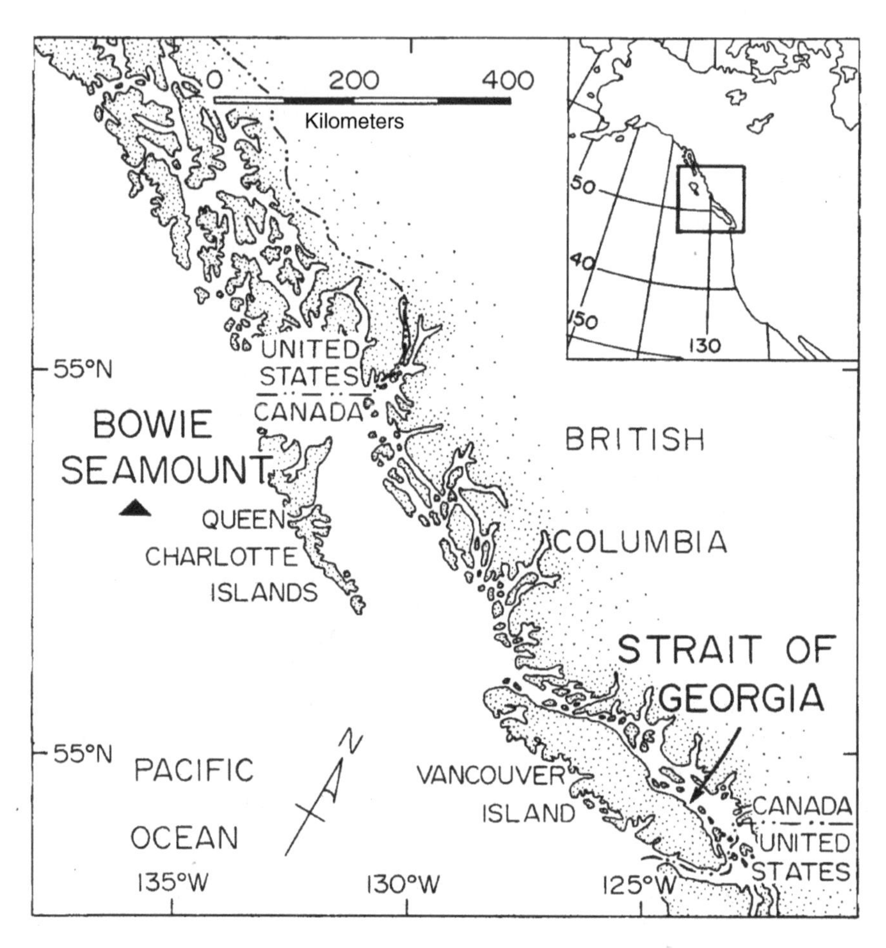

Figure 9.1 The west coast of Canada showing the location of Bowie Seamount (180 km southwest of the Queen Charlotte Islands) and the Strait of Georgia (between Vancouver Island and the mainland).

has been some research on sablefish from Bowie Seamount and it indicates that there is some continuing movement of individuals between the coastal population and the Bowie Seamount population[10] indicating that the fish on the Seamount are part of the coastal population.

Rockfish catch on Bowie Seamount is also additional to annual quotas determined for the commercial fishery. There has been some research on the rockfish stocks on the Seamount, although there are no separate assessments and the impact of the fishing on the other species has not been assessed. Fishing was regulated by special permit, sometimes referred to as an "experimental" permit. In some years the catch was large relative to the annual quota. In 1999, the catch of rougheye rockfish was about 21 percent of the total allowable catch.[11] Since 2000, rockfish have only been allowed to be caught incidentally in other fisheries.

We suggest that management of fish stocks on Bowie Seamount is an example of single species management taken to the extreme. Clearly, the aggregate of fish have been considered to be outside of the structure of the "coastal" population that must be managed in a responsible manner. Fish on the seamount have been treated as some fortuitous discovery of wealth. It has only been the economic cost of acquiring and transporting the fish that has slowed the exploitation. The individual species all live in an area less than 13 percent of the size of the city of Vancouver, but the agencies associated with the management of the particular species have never considered the impact of their decisions on the dynamics of the other species, and only recently on the ecosystem in general. The reason goes beyond the "out of sight, out of mind" nature of human behavior. Rather, the current management structure has never needed to consider Bowie Seamount as a whole ecosystem. There was no policy, no directive and apparently no interest in the impacts of the removal of animals from the top of the food chain in this small and discrete ecosystem.

Beamish and Neville (2002) constructed a model of the Bowie Seamount ecosystem using the trophic accounting model Ecopath.[12] The relationships (Figure 9.2) are based on relatively poor information and thus speculative as there has been limited study of the relationships among species. The biomass estimates and diets for key functional groups can only be considered

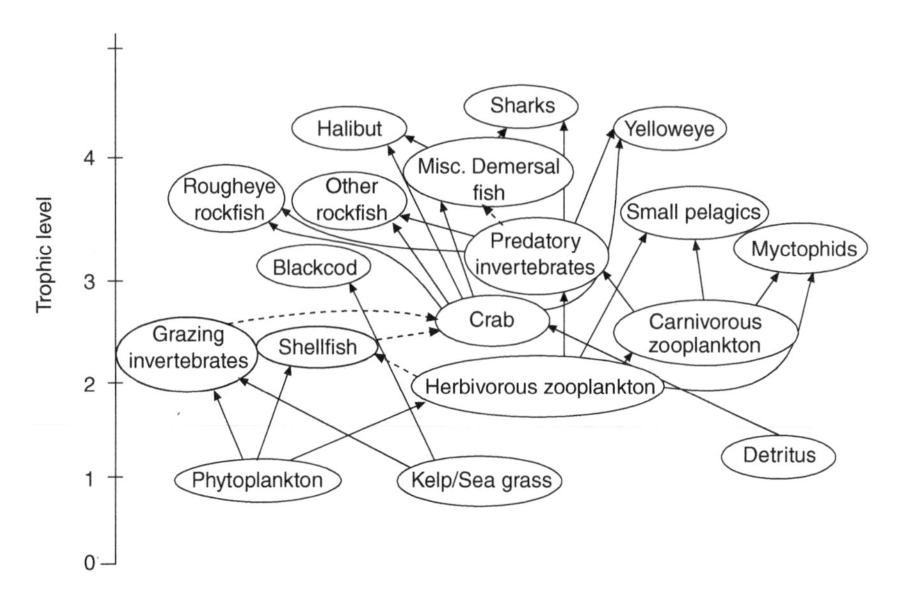

Figure 9.2 Ecopath model of major functional groups at Bowie Seamount (Beamish and Neville 2002).

"best guesses." However, there is a requirement within the ecosystem model to ensure that there is a balance between the food produced or entering the system and the consumption or export from the system. Therefore, the model provides a good starting point to examine the species relationships within the Bowie Seamount ecosystem.

It is interesting that there is an apparent lack of small pelagics in the Bowie Seamount community.[13] Beamish and Neville (2002) also identified the possible impacts of large removals of the top predators (i.e. Pacific halibut, sablefish, rougheye rockfish) in the ecosystem. Major reductions of these species, and the reduced diversity of the small pelagic community, put increased pressure on the lower trophic levels and caused imbalances in the system that could potentially last for decades due to the long life span of these predators.

It is apparent that additional studies at Bowie Seamount are essential. The most critical information includes basic diet information, biomass estimates of key species and estimates of age of the populations resident at the seamount. As this information becomes available, the ecosystem model can be modified and the understanding of the dynamics of the ecosystem at Bowie Seamount will improve.[14]

Our second marine ecosystem example is virtually the opposite of Bowie Seamount. The Strait of Georgia is located between the British Columbia mainland and Vancouver Island (Figure 9.1). It is a semi-enclosed sea[15] that supports a diverse fish fauna, including a large number of juvenile Pacific salmon during their early marine residence. Two of the major fish predators in the Strait of Georgia are lingcod (*Ophiodon elongatus*) and spiny dogfish (*Squalus acanthius*). Lingcod are overfished.[16] Oral reports of their historic abundance would indicate that they were the dominant large predator, growing to sizes of 14 kilograms in about 15 years.[17] We propose that the large reduction in lingcod abundance could have resulted in an increase in the abundance of fish in the small pelagic community. It is only speculation, but the current large abundance of seals (*Phoca vitulina*)[18] may result from the abundant prey at the small pelagic level. The seals and sea lions also feed on the resident Pacific salmon, which annoys those trying to increase their abundance through management and those trying to land a sport-caught salmon before it is eaten by a seal. There have been recent unpublished proposals by scientists and fishers to correct a perceived "trophic imbalance" by eradicating seals. We suggest that a better-managed lingcod fishery may have restricted the growth rate of the seal population by limiting prey, thus avoiding the current issue that agonizes managers, fishers and the admirers of small marine mammals. Spiny dogfish, however, have few admirers. In the late 1950s they were identified as being a nuisance.[19] Apparently this meant that they would eat the bait designed to catch more charismatic species. People were paid to kill spiny dogfish, and 11 separate eradication programs were supported by government agencies who did not know better and scientists who should have known better.

We explored the possible impact of a much larger population of lingcod on the abundance of associated species using an ecosystem model for the Strait of Georgia in 1998.[20] There are obvious difficulties with this approach, but our intent was to show that lingcod were a major predator and a major influence in the abundance of prey species.

The largest recorded commercial catch of lingcod in the Strait of Georgia occurred in 1944.[21] This catch was approximately 130 times larger than commercial hook and line catch in 1996–98.[22] This increase does not directly measure biomass, however it does indicate a large-scale change in abundance. We estimated that the biomass of lingcod in the 1940s might be at least 100 times the current biomass in our 1998 ecosystem model (Figure 9.3). We increased the biomass of lingcod 100 times and examined the impact on other species. The impacts are not dynamic as the resulting trophic level changes were not modeled. However, it is possible to show that the prey of lingcod would be dramatically affected. Obviously, the other species that consume these prey would also be affected. In the Strait of Georgia, the current dominant fish biomasses are Pacific herring, Pacific hake and aggregates of species in the small pelagic and miscellaneous demersal fishes categories. The modeled increased abundance of lingcod resulted in a significant imbalance of the model as there was no longer enough annual production from these major categories for the various predators including lingcod (Figure 9.4). This impact illustrates the important associations between lingcod and the abundance of Pacific hake and other small pelagic species. Thus, it is possible that the collapse of the lingcod population was linked to the increase in the Pacific hake population. In fact, the

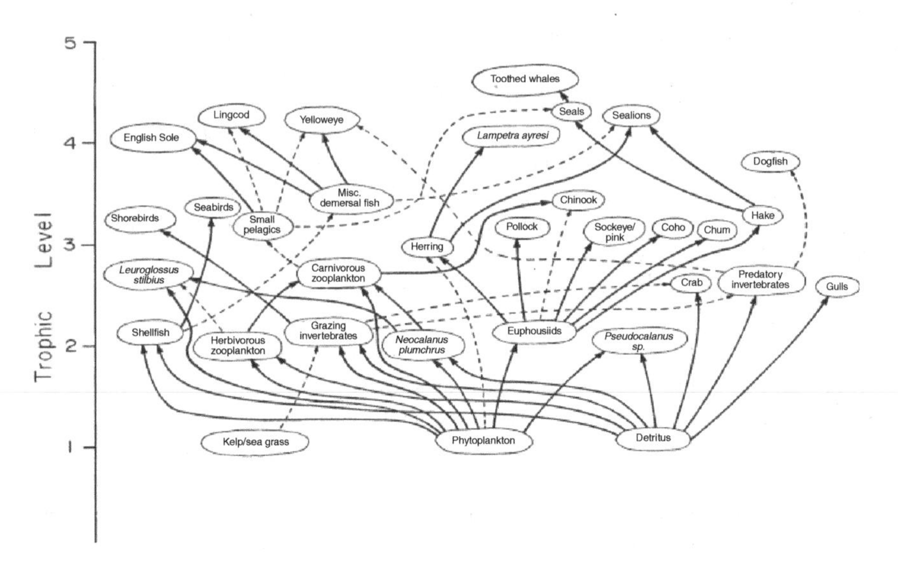

Figure 9.3 Ecopath model of major functional groups in the Strait of Georgia (Beamish *et al.* 2001).

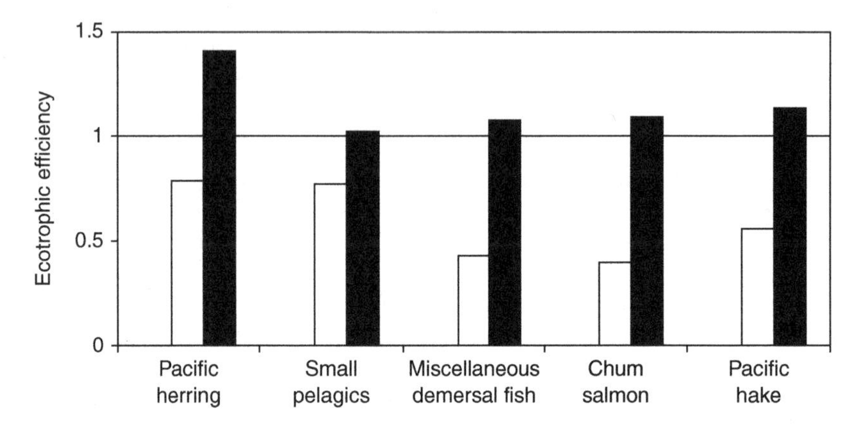

Figure 9.4 Changes in the ecotrophic efficiency of the major species or groups of species in the Strait of Georgia when lingcod abundance is increased 100 times. Ecotrophic efficiency is the proportion of annual production that is consumed or exported from the ecosystem. A value of 1 means that all of the production is lost. Values greater than 1 are theoretically not possible in the model, but for our hypothetical exercise, values greater than 1 would indicate that the population would decline.

large abundance of Pacific hake was first detected in the early 1970s, possibly indicating that abundance in the 1940s was low. Our point is that an ecosystem-based management approach would not have allowed the severe overfishing of lingcod, which increased the abundance of small pelagic fishes. According to this logic, we created the "seal problem" when we overfished lingcod.

These are only two examples of interventions into the Strait of Georgia ecosystem that were undertaken without a commitment to understand the consequences to the whole ecosystem. There are many more. Our point is that the Strait of Georgia is in the center of the human community in British Columbia. Fortunately this community values the health of their environment. However, it is difficult for people to assess the health of the Strait of Georgia ecosystem, as there are few standard procedures that can be used as evaluation criteria that are meaningful to most British Columbians. We have measures of our own health that are useful indicators of the proper relationship among the cells and systems that keep our human ecosystem functioning. Proper care of our own systems usually allows us to live out a natural life span. Identifying the criteria that promote a healthy Strait of Georgia ecosystem or a Bowie Seamount ecosystem will begin the development of ecosystem-based thinking in the general population. The requirement to do this is no longer "a nice thing to do," it is legislated. We actually have no choice but to change. Having established the *Oceans Act* and SARA, it is no longer possible to ignore the changes in the fish community that result from fishing, or other interventions into our marine ecosystems.

An ecosystem bill of rights

We propose that the following five principles, the basis for an ecosystem bill of rights, are statements that honest people can support. Honesty means that people have taken the time to evaluate the benefits and difficulties of committing to managing ecosystems. The principles are statements that recognize that change is needed. The change is in attitude, and this is the first step. We suggest that the second, third and subsequent steps will be much easier.

Principle 1: *Interventions into the dynamics of marine ecosystems occur naturally, intentionally and unintentionally. Ecosystem management must improve our understanding of these interventions and communicate the knowledge to the public.*
This first principle recognizes that ecosystems will change. It is not necessary to know in advance how specific events such as fishing will change ecosystems, but there is a commitment continually to improve our ability to understand how ecosystems respond to changes. This principle ensures that the new information is communicated to the general public. Clients must be kept informed about what we know, what we do not know, and how interpretations differ. In many cases, client information is not typical of daily news, thus innovative methods of communication need to be developed.

Principle 2: *All natural species in an ecosystem are recognized as being important to the health of the ecosystem.*
This principle is similar to the intent of SARA, which treats all species equally. Recognizing the importance of all species would not mean that we must assess the dynamics of all species. However, it would require, for example, that by-catch be managed. It would mean that we would never support eradication programs as was done in the past. Principle 2 would encourage us to begin to assess the impacts of fishing on other species. What happens when top predators such as halibut are fished? We would also want to understand how global warming impacts affect the relationship among species. Principle 2 could be viewed as a minority rights statement. We need to be stewards of spiny dogfish as well as seals and killer whales. Some may view Principle 2 as potentially restricting economic development if the principle is used to block human expansion. However, we consider that Principle 2 is a statement that the impact of our intervention needs to be evaluated; evaluation should not prevent all interventions. There is a reality that humans have babies and expand their habitat. Principle 2 requires that there is an honest evaluation of impacts when we intervene in marine ecosystems.

Principle 3: *Surplus production of some species may be available for human consumption, but estimates of surplus production must include consideration of the impact on associated species.*
This principle reflects basic fishing theory that states that natural populations produce a surplus yield that can be harvested. Some scientists[23] have

proposed that by removing production that is surplus to the production required to replenish the population, it is possible to stimulate more growth in the total population. The concept is best visualized by relating fish production to forest production. Very old trees in a forest do not grow much each year, but the amount of wood in the forest is large. Removing the old trees encourages young trees to grow and the total amount of wood produced each year becomes greater than that produced by the old trees. The removal of the old trees begins a new cycle of succession among plants. Our Principle 3 is a commitment to begin to understand how the ecosystem changes when preferred species are fished and sometimes overfished. For example, using our two marine ecosystems examples, what are the consequences of overfishing lingcod in the Strait of Georgia, or removing large numbers of the very slow-growing and long-lived rougheye rockfish from Bowie Seamount?

Principle 4: *Ecosystems must be able to re-organize naturally which may result in declines of charismatic species.*
There are natural variations in the trends of fish abundance that are large and occur quickly, and may be more important for management than fishing effects in well-managed fisheries. There is a tendency for some to believe that fish abundance will remain constant in a well-managed fishery. The belief is founded on an incorrect interpretation of early scientific theory that reported that the population dynamics of a species could be fitted to an equation that would allow calculations to be made of maximum sustainable catch. This was an exciting idea because it proposed that the correct level of catch could be determined and the stock would become more productive. Fisheries science no longer supports the idea of maximum sustainable yield, although variations of the idea are still in use. Unfortunately, there is still a mythology that good management means that the abundance of preferred species will be high. Levels lower than expectations are sometimes seen as management failures that can be corrected through large commitments of new money. Professionals willing to accept these funds are not hard to find. Principle 4 recognizes that responsible stewardship will let ecosystems change and let key species respond naturally. Principle 4 informs managers and the general public to "let it be."

Principle 5: *Humans are part of the ecosystem and will introduce change, but because of our trophic level we must be stewards of our changes.*
This is an essential principle because it acknowledges that humans compete for space like all animals and plants. However, we credit ourselves with a level of intelligence that obligates us to understand and manage our changes. As our population expands, we will displace other species. Fishing removes species and involves by-catch and possibly habitat damage. Principle 5 recognizes these impacts, but requires that the changes minimize the impact on the ecosystem. Fishing needs to be managed to minimize impact

on both the target species and associated species. Fishing must not restrict the natural ability of a stock to replenish itself. However, it is important to recognize that the term overfishing is difficult to quantify once long-term natural variability is included in the definition. If large, natural fluctuations occur over long periods of time,[24] is it possible to identify a "virgin" biomass that can be used to scale the current impacts of fishing? Implicit in this principle is an understanding that excessive fishing pressures will occur despite the best scientific support. It is logical to assume that when science has not been able to prevent excessive fishing, stewardship should be applied to promote recovery of stocks. This means that traditional knowledge that is wise is an acceptable basis for decision making. Traditional knowledge in this case is wisdom that comes from the old, the young, fishermen, Aboriginal people and local communities.

Conclusion

This ecosystem bill of rights is a first step in changing our thinking. The five principles are public commitments to understand the impacts of our society on the species that occupy our marine ecosystems. Such statements are necessary to draw the attention of humans who do not think about the large majority of fish that are not as charismatic as killer whales or Chinook salmon. The five principles are commitments to change how we think. We need to move away from single species management and we need a new reward system for professionals who learn how to work cooperatively to achieve ecosystem-based management. An essential component of our proposal is public communication, which is not a trivial task. Informing the public is a skill that recognizes that knowledge must be seen to be entertainment as well as education. In Canada, there is new legislation requiring that we begin to change, but perhaps the time has come to modernize Canada's *Fisheries Act*[25] formally to embrace ecosystem principles. This ecosystem bill of rights is a step in changing our attitude about other species in our habitat.

Notes

1 Richard J. Beamish and Daniel R. Bouillon "Pacific salmon production trends in relation to climate" (1993) 50 *Canadian Journal of Fisheries and Aquatic Sciences* 1002–26; Steven R. Hare and Robert C. Francis "Climate Change and Salmon Production in the Northeast Pacific Ocean" in Richard J. Beamish (ed.) *Climate Change and Northern Fish Populations* (1995) 121 *Canadian Special Publication of Fisheries and Aquatic Sciences* 357–72; Jeffrey J. Polovina "Decadal variation in the trans-Pacific migration of northern bluefin tuna (*Thunnus thynnus*) coherent with climate-induced change in prey abundance" (1996) 5 *Fisheries Oceanography* 114–19; Richard J. Beamish, Chrys-Ellen M. Neville and Alan J. Cass "Production of Fraser River sockeye salmon (*Oncorhynchus Nerka*) in relation to decadal-scale changes in the climate and the ocean" (1997a) 54(1) *Canadian Journal of Fisheries and Aquatic Sciences* 543–54; R.J. Beamish, C. Mahnken and C.M.

Neville "Hatchery and wild production of Pacific salmon in relation to large scale, natural shifts in the productivity of the marine environment" (1997b) 54 *ICES Journal of Marine Science* 1200–15; Nathan J. Mantua, Steven R. Hare, Yuan Zhang, John M. Wallace and Robert C. Francis "A Pacific interdecadal climate oscillation with impacts on salmon production" (1997) 78 *Bulletin of the American Meteorological Society* 1069–79; D.J. Noakes, R.J. Beamish, L. Klyashtorin and G.A. McFarlane "On the coherence of salmon abundance trends and environmental factors" (1998) *North Pacific Anadromous Fish Commission Bulletin No. 1* 454–63; R.J. Beamish, D.J. Noakes, G.A. McFarlane, L. Klyashtorin, V.V. Ivanov and V. Kurashov "The regime concept and natural trends in the production of Pacific salmon" (1999) 56(1) *Canadian Journal of Fisheries Aquatic Sciences* 516–26; Richard J. Beamish, Gordon A. McFarlane and J.R. King "Fisheries Climatology: Understanding Decadal Scale Processes that Naturally Regulate British Columbia Fish Populations" in Paul J. Harrison and Timothy R. Parsons (eds) *Fisheries Oceanography: An Integrative Approach to Fisheries Ecology and Management* (Blackwell Science, Oxford: 2000a) 94–145; R.J. Beamish, D. Noakes, G. McFarlane, W. Pinnix, R. Sweeting and J. King "Trends in coho marine survival in relation to the regime concept" (2000b) 9 *Fisheries Oceanography* 114–19.

2 Beamish *et al.* 2000b, note 1; Gordon A. McFarlane, Jacquelynne R. King and Richard J. Beamish "Have there been recent changes in climate? Ask the fish" (2000) 47 *Progress in Oceanography* 147–69.

3 Intergovernmental Panel on Climate Change (IPCC) "Climate Change – Impacts, Adaptation, and Vulnerability" in J.J. McCarthy, O.F. Canziani, N.A. Leary, D.J. Dokken and K.S. White (eds) *Contribution of Working Group II to the Third Assessment Report of the Intergovernmental Panel on Climate Change* (Cambridge University Press, Cambridge, UK: 2001).

4 S.C. 1996, c. 31 (Canada). Online at: www.laws.justice.gc.ca/en/o-2.4/87839.html (accessed 20 January 2004).

5 S.C. 2002, c. 29 (Canada). Online at: www.laws.justice.gc.ca/en/s-15.3/99968.html (accessed 20 January 2004).

6 K.S. Ketchen "The spiny dogfish (*Squalus acanthias*) in the Northeast Pacific and a history of its utilization" (1986) 88 *Canadian Special Publication of Fisheries and Aquatic Sciences.*

7 Government of British Columbia "The 2001 census of population and housing" (2001). Online at: www.bcstats.gov.bc.ca/data/cen01/c2001pop.htm (accessed 20 January 2004).

8 John F. Dower and Frances J. Fee *The Bowie Seamount Area – Pilot Marine Protected Area in Canada's Pacific Ocean* (Oceans Background Report, Fisheries and Oceans Canada, Sidney, BC: 1999).

9 Richard J. Beamish and Chrys-Ellen M. Neville "The Importance of Establishing Bowie Seamount as an Experimental Research Area" in J. Beumer, A. Grant and D. Smith (eds) *Aquatic Protected Areas – What Works Best and How Do We Know? Congress on Aquatic Protected Areas, Cairns, Australia, August 2002* (Australian Society for Fish Biology, North Beach, Western Australia, 2003).

10 Ibid.

11 Ibid.

12 V. Christensen and D. Pauly "ECOPATH II – A software for balancing steady-state ecosystem models and calculating network characteristics" (1992) 61 *Ecological Modelling* 169–85.

13 Beamish and Neville, note 9.

14 Following the designation of the Bowie Seamount as a potential marine protected area (MPA) in 1998, there has been increased scientific research in the

waters surrounding the seamount and adjacent seamounts, as well as examination of the requirements for a management regime for the Bowie Seamount MPA. In August 2003, a five-member dive team conducted a comprehensive biological survey of the Bowie Seamount. See N. McDaniel, D. Swanston, R. Haight, D. Reid and G. Grant *Biological Observations at Bowie Seamount, August 3–5, 2003* (Preliminary Report prepared for Department of Fisheries and Oceans, October 2003). Online at: www.pac.dfo-mpo.gc.ca/oceans/mpa/bowsupport_e.htm (accessed 19 January 2004). See also, AXYS Environmental Consulting Ltd. *Management Direction for the Bowie Seamount MPA: Links between Conservation, Research, and Fishing* (Prepared for the World Wildlife Fund Canada, Pacific Region, Prince Rupert, BC: June 2003). Online at: www.wwf.ca/NewsAndFacts/Supplemental/BowieSeamountReport2003.pdf (accessed 19 January 2004); R.R. Canessa, K.W. Conley and B.D. Smiley *Bowie Seamount Pilot Marine Protected Area: An Ecosystem Overview Report* (Canadian Technical Report of Fisheries and Aquatic Sciences 2461, Science Branch, Fisheries and Oceans Canada, Sidney, BC: 2003).

15 Richard E. Thomson "Oceanography of the British Columbia coast" (1981) 56 *Canadian Special Publication of Fisheries and Aquatic Sciences* 291.

16 Jacquelynne R. King *Assessment of Lingcod in the Strait of Georgia* (Canadian Science Advisory Secretariat Research Document 2001/132, The Secretariat, Ottawa: 2001).

17 Alan J. Cass, Richard J. Beamish and Gordon A. McFarlane *Lingcod (Ophiodon elongatus)* (Canadian Special Publication of Fisheries and Aquatic Sciences 109, Department of Fisheries and Oceans, Ottawa: 1999).

18 Peter Olesiuk *An Assessment of the Status of Harbour Seals* (Phoca vitulina) *in British Columbia* (Canadian Stock Assessment Secretariat Research Document 99/33, The Secretariat, Ottawa: 1990).

19 Ketchen, note 6.

20 R.J. Beamish, G.A. McFarlane, C.M. Neville and I. Pearsall "Changes in the Strait of Georgia ECOPATH Model Needed to Balance the Abrupt Increases in Productivity that Occurred in 2000" in G.A. McFarlane, B.A. Megrey, B.A. Taft and W.T. Peterson (eds) *PICES-GLOBEC International Program on Climate Change and Carrying Capacity: Report of the 2000 BASS, MODEL, MONITOR and REX Workshops, and the 2001 BASS/MODEL Workshop* (PICES Scientific Report No. 17, North Pacific Marine Science Organization (PICES), Sydney, BC: 2001) 5–9.

21 Cass *et al.*, note 17.

22 King, note 16.

23 William E. Ricker "Stock and recruitment" (1954) 11 *Journal of Fisheries Research Board Canada* 559–623.

24 Bruce P. Finney, Irene Gregory-Eaves, Jon J. Sweetman, Marianne S. V. Douglas and John P. Smol "Impacts of climatic change and fishing on Pacific salmon abundance over the past 300 years" (2000) 290 *Science* 795–99.

25 R.S.C. 1985, c. F-14.

Part V

Community-based management

10 Community involvement in marine and coastal management in Australia and Canada

Marian Binkley, Alison Gill, Phillip Saunders and Geoff Wescott

Introduction

The continuing review and development of ocean and coastal policy in Australia and Canada in recent years, coupled with periodic efforts at rethinking and restructuring the institutions and processes of ocean governance, have stimulated a renewed interest in the role of the community in marine and coastal affairs. This chapter examines the evolving role of the community in ocean and coastal management in Australia and Canada, with reference to case studies from each country.

Community-based management and co-management

The increasing interest in the role of the community in ocean and coastal management is reflective of a broader effort that has been devoted to developing a range of options for the devolution of management control over fisheries and other marine resources, or over coastal areas in general. Emerging in part from the literature on common property resources,[1] this trend is reflected in several different terms, including territorial use rights in fisheries,[2] local level management, community-based resource management and co-management.[3] Perhaps the most expansive, and potentially inclusive, term in common use is community-based management (CBM), and this is the descriptor adopted for much of this chapter. However, the related notion of co-management, which can be thought of as a sub-set of CBM, is also of particular relevance with respect to fisheries.

These general concepts have received some sanction at the international level, particularly in the agreements and other instruments connected with the 1992 United Nations Conference on Environment and Development (UNCED). For example, Principle 22 of the Rio Declaration on Environment and Development explicitly addressed this issue in the following terms:

> Indigenous people and their communities, and other local communities, have a vital role in environmental management and development because of their knowledge and traditional practices. States should

> recognize and duly support their identity, culture and interests and enable their effective participation in the achievement of sustainable development.[4]

Similarly, Agenda 21 called for the increased involvement of communities and resource users in the planning and implementation of environmental and resource management measures. Chapter 17 specifically called on coastal states to

> [i]ntegrate small-scale artisanal fisheries development in marine and coastal planning, taking into account the interests, and where appropriate, encouraging representation of fishermen, small-scale fishery workers, women, local communities and indigenous people.[5]

It is beyond the scope of this chapter to provide a comprehensive global review of CBM and co-management, or to settle the terminological questions referred to above. It is useful, however, to consider some recent commentaries in order to provide a general working definition of CBM for the case studies discussed below.

In 1997 Hildebrand[6] reviewed trends in "top-down" and "bottom-up" approaches to the implementation of integrated coastal management (ICM), and described community-based coastal management with reference to the following characteristics:

> [W]herein the people who live and work in coastal areas and depend on their resources, are enabled to take an active and responsible role, and increasingly share planning and decision-making responsibilities with government.[7]

It is clear that there is no single, cohesive approach to this issue, and even the use of terminology varies considerably. The mere use of terms such as "co-management" or "community-based management" provides little information about the substance of a program or policy, and it is possible to overstate the impact of "formal" exercises in devolution.[8] Hildebrand noted the need for tighter definition of terms such as "public involvement," "co-management" and "empowerment," but emphasized that the common direction is towards partnerships between governments and community-based organizations or "in essence," to use his words, *power sharing*.[9]

Pursuing the call for tighter definitions of various terms, including community-based management, Harvey *et al.*[10] quote the following description of CBM from Ferrer and Nozawa,[11] which incorporates some of the rationale for increasing the community's planning and decision-making responsibilities. In this construction, CBM is

> people-centred, community-oriented and resource-based. It starts from the basic premise that people have the innate capacity to understand and act on their own problems. It begins where the people are, i.e. What the

people already know, and build (*sic*) on this knowledge to develop further their knowledge and create new consciousness. It strives for a more active people's participation in the planning, implementation and evaluation of coastal resource management programs.

Harvey *et al.*[12] have produced a very useful diagram (Figure 10.1), building on similar attempts by Hale *et al.*[13] and Ellsworth *et al.*,[14] that sets out a continuum of public involvement in coastal management, which will assist in consideration of the case studies that follow.

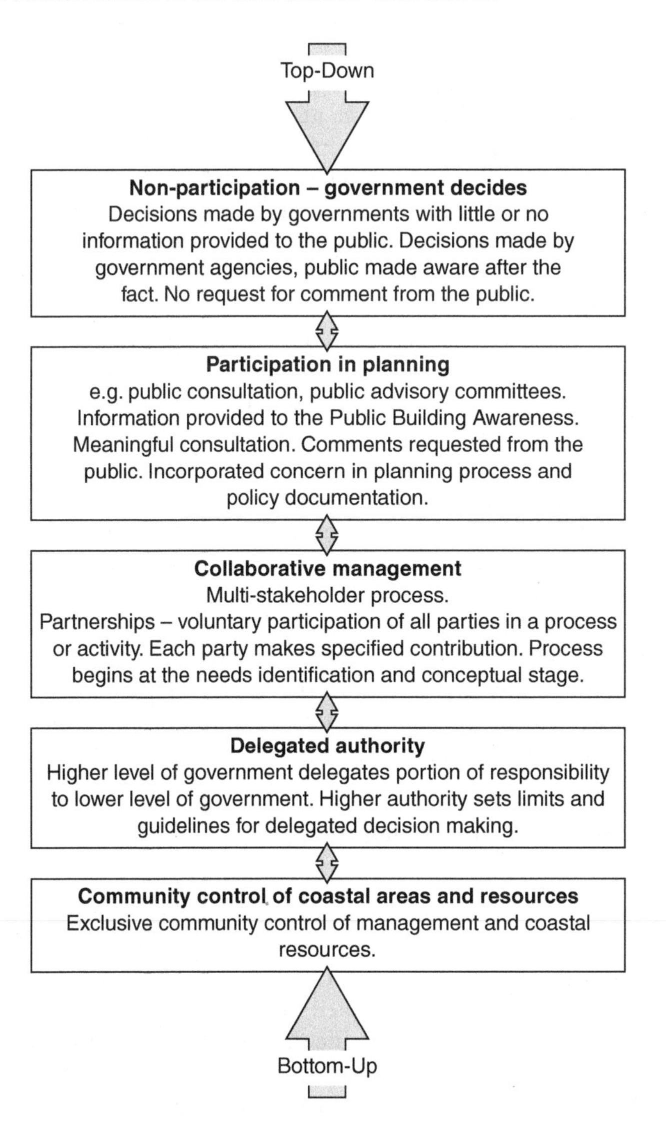

Figure 10.1 Continuum of community involvement in coastal management.

Implicit in all of the studies and policy initiatives relating to CBM is a fundamental contention that devolution of authority is in general a good thing, leading not only to enhanced management of resources, but to an improved quality of life for the people who depend upon those resources.[15] This understanding of the value of community involvement, which is certainly central to the relevant Australian and Canadian policy developments, is supported by a number of underlying premises. These include a belief in the value of encouraging a sense of "stewardship"[16] in the community, which can lead to a greater sense of collective responsibility or "ownership." This in turn is seen as leading to more sustainable use and better conservation outcomes in marine and coastal environments.

As is noted in the definition of CBM suggested by Ferrer and Nozawa (above), movements towards CBM are also justified by the argument that communities have a close association with, and knowledge of, a specific area and resource that may be qualitatively different from that which is possible within a government agency. More broadly, local communities are also more likely to be intimately aware of underlying socio-economic problems confronting their area and to have distinctive ideas on how to address, if not rectify, those problems.

There is an additional practical impetus that could be seen both as an opportunity, and as a cautionary note for overextended community institutions. It seems clear that in both Australia and Canada, as resources for government agencies have decreased in recent years, there has been a corresponding increase in the desirability of replacing agency management efforts with management by community groups.[17]

All of these important justifications for community involvement in marine and coastal affairs, as well as the explanatory continuum in Figure 10.1, provide the frame of reference for the case studies that follow. These case studies, it should be noted, are not intended to provide an exhaustive survey of coastal and marine CBM activities in Australia and Canada. Rather, they are meant to give a sampling of the kinds of policies and initiatives which come within the general rubric of community-based management or co-management in both countries.

Canadian East Coast fisheries – "legalizing" co-management

The past two decades have been a period of major change and readjustment in the fisheries of the east coast of Canada, as fish stocks have declined dramatically throughout the North Atlantic. The latest crisis in the Atlantic Canada region began in the 1980s and peaked in the early 1990s;[18] by 1993 the cod fishery had experienced a catastrophic decline, which resulted in the federal government issuing a moratorium on almost all cod fishing in Atlantic Canada. In those areas exempt from this moratorium (including the waters off the shores of southern Nova Scotia, from Halifax to Yarmouth,

Georges Bank and the Bay of Fundy), the cod fishery and other groundfisheries such as pollock, halibut and haddock were severely curtailed. Although scientists speculated on a variety of ecological and environmental reasons for the decline, the chief cause was overfishing – too many boats chasing too few fish.

Throughout the region, governments responded to the crisis by putting in place management policies that increased the economic efficiency of the fishing industry while promoting sustainability of the marine resources. Ensuing discussions of the fisheries crisis revolved around the concerns of managing the fish stocks, methods for harvesting and marketing the resources, and plans for the economic development of areas hardest hit by the crisis. Discussions involving government, marine scientists, economists and fishers covered a wide range of issues, including co-management, quota systems and other marine resources management practices,[19] and local knowledge systems and their applicability to marine resources management.[20]

The moratorium on cod fishing led to a drastic restructuring of the fishing industry and of the communities that support it and were supported by it. Fishing companies downsized, dramatically reducing the size of their fleets and closing processing plants. Small-scale fishers diversified their catches, replaced hired help with family members, and went further from shore in search of the few remaining fish. As of 1993, the northern cod closures had eliminated over 40,000 fishery jobs.[21] Susan Williams[22] estimated that, in Atlantic Canada, 50,000 people working in the fishing industry and another 47,000 people working in fishery-dependent sectors saw their employment modified by the fisheries' crisis. In 1995, the federal government responded to the crisis through the Atlantic Groundfish Strategy (TAGS), in conjunction with early retirement packages for plant workers and the retirement of groundfish licenses by fishers. The "package" compensated plant workers and fishers for lost earnings and offered them retraining programs. In the summer of 1997, the federal government stopped the retraining component of TAGS (one year early) and the remainder of the program ended in the summer of 1998.

Around the world, many coastal communities, previously dependent on marine resources, have dramatically downsized their fisheries and have turned to other industries, most notably tourism. Like all communities undergoing massive restructuring, coastal communities were altered culturally,[23] economically[24] and politically.[25] The processes of globalization and restructuring reduced the subsistence production of small-scale fishers as large multinational companies controlled access to more and more of the marine resources. As more fishers and their families were deprived of their access to these resources, they increasingly depended on wage labor and on government welfare benefits,[26] yet these same governments decreased social benefits. The few small-scale fishers who managed to retain their access to the marine resources were economically squeezed and became dependent on the large multinationals for the sale and processing of their products in the global market. Fishing-dependent communities also paid the price for this economic remedy through higher unemployment rates, increased emigration, higher costs of living, and increased de-skilling of labor.[27] International responses to the world fisheries' crisis have been similar, including rapid development of aquaculture, technology transfer, importation of scientific management regimes, development of the tourist industry, and the liberalization of trade and of direct foreign investment.[28]

Against this backdrop, resort to some degree of co-management or community-based management has been a recurrent theme in proposals to modernize the governance of the industry that survives in Atlantic Canada (including newly significant fisheries). The industrial fishery is, however, a highly regulated sector in which the possibility for change is always limited by what is permissable in law. One aspect of co-management which has received limited attention has been the implementation of such activities in *legal* structures, a task in which the terminological imprecision referred to above is likely to raise particular problems. There is some consensus that the lack of appropriate legal instruments may be an easier problem to identify than it is to solve, as reflected in the following conclusion from a FAO-sponsored consultation on community-based management:

> With regard to legal aspects, it was noted that the devolution of management authority to the local level would require, in many countries, a major or even drastic revision of fisheries laws and possibly other related legislation. . . . For . . . cases where complex political and socio-economic conditions prevail, the required legal changes may be difficult to accomplish.[29]

There is currently little or no explicit legislative basis for the general development of community-based fisheries management approaches in Atlantic Canada, apart from the usual wide-ranging ministerial discretion that applies to much of the fisheries management process. The *Oceans Act*[30] does provide for the Minister of Fisheries and Oceans to involve "coastal communities" in the development and implementation of both a national oceans strategy and integrated management plans, but this is very generally stated and does not appear to be directly applicable to fisheries.

Despite this gap, which was the subject of one abortive attempt at legislative amendment to provide for legally binding "partnering" agreements under the *Fisheries Act*,[31] the broad powers of the Minister have been employed to put in place a number of arrangements which fall under the general umbrella of co-management or community-based management. As the Department of Fisheries and Oceans (DFO) has acknowledged, and consistent with the confusion mentioned above, no one description or definition adequately addresses the variety of activities which might be lumped together under the rubric of "co-management."[32] Broadly considered, this could include most of the Department's efforts at consultation and involvement of so-called "stakeholders." More concrete versions of co-management have extended to participation in compliance and enforcement activities and the scientific work of DFO.[33]

The more formalized efforts at co-management of East Coast fisheries seem to date from a 1996 consultation with the Atlantic commercial fishing industry, which resulted in a framework for co-management that was in place by 1997.[34] This framework envisaged at least four levels of co-management:

> The first and most basic level that the Department considers to be co-management has user groups providing input to the Integrated Fisheries Management Plans. The second level has user groups, through their legally constituted, representative organizations, entering into agreements that reflect a greater involvement in the management of their specific fishery. At the third level, fishers would enter into formal partnering through legally binding arrangements that transfer greater responsibility to industry. This level would require changes to the existing *Fisheries Act*. The fourth level is co-management legislated under land claims settlements.[35]

The "legislated" option has not been a factor on the East Coast, and the potential for legally binding Fisheries Management Agreements (FMAs) was eliminated with the failure to enact the amendments to the *Fisheries Act* (as noted above). The minimal option put forth by DFO as a type of co-management refers to the participation of resource users and others in the development of Integrated Fisheries Management Plans (IFMPs),[36] which set out conservation requirements and management plans for a given fishery. In the view of DFO, the provision of input during the process of development of IFMPs "may be considered the basic form of co-management."[37] Within the framework established by IFMPs, however, more extensive approaches to co-management can be found, both in the Joint Project Agreement (JPA) structure and in the possibility of "community quotas."

The JPA is a means by which resource users may enter into a voluntary agreement with DFO covering "specific activities such as data collection, data analysis, science and/or fishery management activities."[38] The possibility of a JPA may be provided for as part of the IFMP for a fishery, as in the eastern Nova Scotia snow crab IFMP. In that IFMP, some idea of the details of what might be included in a JPA were provided in a draft or model included as an appendix, based on an agreement between the Minister and a fishing association.[39] The general purpose of this sample JPA is to "enhance the management" of the fishery, and specific project activities are set out in the areas of resource management and scientific research. For the latter, quite specific practical activities are listed, involving annual surveys and analytical tasks. With respect to other management requirements, however, the model focuses on a somewhat more general objective:

> The objective, from a resource management perspective, is to have the industry actively participate in the management of their industry. Through this Agreement, consultations will be held with the Association and its members to discuss and jointly decide upon development and implementation of policies, management plans, and quota and fishery monitoring requirements.[40]

This statement makes it clear that under this model, DFO, consistent with the applicable legislation, must continue to manage, and that the Minister's discretion on matters such as allocation is (at least formally) not fettered by the existence of a JPA.

DFO has not in practice been entirely constrained by the original options or "levels" of co-management set out in 1997. In a 1999 summary of "co-management projects" in the Atlantic and Gulf of St. Lawrence regions, seven JPAs were listed, of which one did not have the supposedly mandatory accompanying IFMP. In addition, there were eight "informal agreements" or "informal arrangements" of unspecified effect.[41]

DFO has in some cases also moved, within the general framework set out by IFMPs, to establish "community quotas" over which community management boards exert some degree of management control. The community quota approach was introduced in Atlantic Canada on an experimental basis in 1996, where it was applied to inshore fixed gear license holders who were divided into seven geographical groups based on counties or similar areas.[42] Within quotas established under the IFMPs for certain groundfisheries,[43] the community management boards give "industry associations the opportunity to develop conservation harvesting plans that address seasonal fishing patterns," and generally "develop, implement and monitor community fishing plans," including internal allocations.[44]

It is useful to note a few general characteristics of the approaches to "co-management" as they have developed in Atlantic fisheries.[45] First, it must be remembered that DFO has been careful to make it clear that management authority over key decisions such as allocation and access still rests with the Minister, as it must in the absence of amendments to the *Fisheries Act*. In this sense, JPAs might be considered to be a very limited form of co-management, especially given the usual requirement that there be an IFMP in place as a precondition to moving on to a JPA. That is, the central management decisions are, at least formally, made outside the context of the actual co-management arrangement, and the plan that dictates the management approach should be in place first. The same is true of community quota management, which must of necessity remain a subsidiary level, dependent upon both a quota system and the overall allocation of the quota.[46]

Second, the co-management options under JPAs and community quotas have been utilized mainly by sectors of the commercial fishing industry, and not by "communities" as such. With respect to the JPA, this is encouraged by the very structure of the arrangement, which requires that the cooperating party be a legal entity, whether as a commercial venture or an industry association.[47] Communities might form such an entity, but the thrust of the program has clearly been towards the inclusion of the organized, commercial fishery. Similarly, community quotas are in fact structured around gear sectors, and are based on historical involvement in the relevant fishery. They do not appear to open up the possibility for serious management involvement by the community at large, or even fishers who have not already obtained entry to the designated group and sector.

Third, these arrangements are built around temporary, time-limited agreements[48] and not the recognition of permanent claims to specific areas of the ocean.[49] As such, they are not particularly amenable to quasi-territorial claims of *communities* to ocean space, but rather are designed around accommodation of industrial sectors and their need for a degree of certainty of access and management measures over a sufficient period of time to justify an investment. Despite the repeated assurances of DFO that co-management in the form of JPAs and similar arrangements is not about privatization of fisheries management, or the limitation of access to those with the capability to enter into such agreements,[50] a significant degree of suspicion exists among what might be termed the small independent fishers. There is a widely shared concern that the partnership provisions brought in by DFO are in fact consistent with other initiatives, such as individual transferable quotas (ITQs) and the concentration on a "professional" core of fishers, that are generally intended to concentrate the fishery in the industrial sector and to reduce the numbers of smaller players.[51]

This question of the types of participants supported by current approaches to co-management, as noted above, raises a related problem. It appears that, with some limited exceptions, government policies have not been primarily concerned with *community-based* management, unless one includes associations, or even affiliations of a few participants in a fishery, as a "community." Even if this version of "community" is accepted, most arrangements concluded to date do not appear to have been based around the concept of community entitlements to particular *geographic areas* within which traditional rights might be recognized, an important subset of potential community-based approaches. Community quotas have a geographic element in the description of the relevant fleets, but not in defining the areas and resources to be managed.

If we accept the proposition that local community claims to some form of "marine tenure" based on long use and custom are well-established in parts of Canada, and that such claims could form a natural basis for cooperative management approaches, why have geographically defined community-based management schemes not emerged as an important factor?[52] This gap is not peculiar to the question of community-based management. If one considers the critical question of allocation of access to fisheries (again, setting aside the Aboriginal case), it seems clear that there has been limited room for allocation on the basis of community claims to some control or tenurial rights over marine areas. Other entities and collectives – whether "enterprises" or gear sectors – have been recognized as part of the allocation process, but local community claims to particular resources and spaces have not been included in the same way, although locality has been incorporated as one factor to be considered in some cases.[53]

It is, of course, possible to argue that this is not necessary to reflect marine property rights in law, in that they operate most effectively at a local level based on local custom, and will continue to do so within the legal framework of the formal system.[54] The most important reason given to explain the lack of recognition given to such rights, however, is that they do not have critical characteristics of property rights in a legal sense, and are therefore difficult to include in a system that bases its definition of rights on legal notions of property. Most significant, these "rights" are generally communal in nature, representing an intermediate form of private title, neither true "common property" nor individual ownership in a legal sense, and this is a form of property currently unrecognized in Canadian law (outside the Aboriginal context):

> Claims or ownership and control of property [are] centred in the community, and individual use-rights are derived from membership in the community. Although basically foreign to the system of owner/non-owner property relations dominant in capitalist-industrial societies, collectively based property claims and associated individual use rights represent a distinctive form of property relation.[55]

Future prospects

Existing approaches to fisheries co-management in Atlantic Canada, with some exceptions, do not really extend to what can properly be termed community-based management, if one considers the delineation of a "community" as requiring something more than narrow criteria rooted in fishing sectors. Indeed, the extent to which existing arrangements constitute co-management of the *oceans*, as opposed to enhanced participation for some in the normal *fisheries* management and allocation process, is doubtful. The caution with which this issue is being approached, and the slow pace of progress, is emphasized in the February 2001 DFO discussion document prepared as part of the Atlantic Fisheries Policy Review:[56]

> Government and industry have talked about co-management since the 1970s, but the concept has undergone an evolutionary maturation.
>
> In the initial stages, the emphasis was on effective consultation processes through advisory committees for individual fisheries and through ad hoc policy and planning conferences. In the 1990s, co-management took the more specific form of IFMPs and joint project agreements to share management responsibilities in particular fisheries . . .

> Today, the Department recognizes that efforts to share management responsibilities and costs with resources users on a fishery-by-fishery basis have to be supplemented by an expanded role for resource users in overall policy and planning. To this end, we offer the following definition: . . .
>
> *The sharing of authority and responsibility for fisheries management, and of accountability for results, between DFO and resource users.*[57]

It could be argued that the 30-year process of "maturation" seems to have led to something less than an expansive definition of "co-management," and none at all for community-based management. Furthermore, other references in the discussion document make it clear that co-management for Atlantic fisheries is still primarily viewed as a matter for cooperation between DFO and industry sectors, with "other" interested parties having input of a distinctly lower order. For example, after a description of the various means by which industry participants were drawn into the management process, there is the following description of the issues related to broader participation:

> [T]here is also the issue of public interest representation. Local community organizations, employers and workers in the fish and seafood processing sector, local and provincial governments, environmental and animal rights groups and other users of the marine environment *may sometimes look for opportunities to participate in fisheries management decision making.*[58]

As noted above, the discussion document in which this statement appears is part of an ongoing process of policy development for the Atlantic fisheries. In 2004, DFO released a document representing the next stage in this process: *A Policy Framework for the Management of Fisheries on Canada's Atlantic Coast*,[59] which provides a general policy framework as the first phase of a two-part process, the second of which will establish priorities and move towards implementation. There are numerous references to co-management in this policy, but it is evident that the concept is still being considered at a high level of generality. The tentative nature of the approach to co-management is reflected in the following definition, which emphasizes the future possibilities over the present reality, and acknowledges the lack of a legal underpinning:

> Co-management means the sharing of responsibility and accountability for results between Fisheries and Oceans Canada and resource users, and in time and with the required legislative amendments, the sharing of authority for fisheries management.[60]

It remains to be seen whether the next evolution of policy will result in an approach to fisheries co-management for the Atlantic coast that is closer in form to community-based management, as opposed to the more limited industrial co-management models of the past. There are signs that institutions involved in the management of community quotas, such as the Fundy Fixed Gear Council, may over time move to broader efforts at the sustainable management of fisheries, and engage as participants in more comprehensive community-based approaches to *ocean* management.[61]

Community-based management of fisheries on Canada's West Coast

There is a significant difference between the east and west coasts of Canada in terms of the nature of the fishing industry and the communities that depend on them. On the West Coast, wild salmon traditionally formed the basis of the fishery (as opposed to groundfish on the Atlantic coast), and there is a long history of its association with the culture and practices of Aboriginal peoples before white settlement. Fewer fishing-dependent communities developed on the British Columbia (BC) coast compared with Atlantic Canada. Gislason *et al.*[62] identify 50 BC communities of which only 11 are single-sector communities. These communities range from isolated First Nations villages to larger more diversified centers.

As on the East Coast, British Columbia's fisheries and their dependent communities have experienced severe decline over almost two decades. Beginning in the mid-1980s, salmon prices fell in response to global competition and declining access to adjacent resources. At the same time, salmon stocks began to decline due to factors such as overharvesting, poor ocean survival rates, habitat destruction and management cutbacks. Strict conservation measures were put in place to protect these stocks in the mid-1990s, including closures and reductions in fishing times.[63] This has resulted in more effort on groundfish, elasmobranchs and invertebrates. One of the first formal co-management approaches, introduced in 1992, was the Skeena Watershed Committee. During the 1990s, various other community-related initiatives were introduced, including the Coastal Communities Network (CCN). At conferences related to this network, and in various other public forums and workshops on fisheries held in BC coastal communities, there was a call for increased community participation in fisheries management in BC, and concern about the viability of fishing-dependent communities.[64]

There has been debate about the merits of community-based fisheries management in the province, particularly in the management of migratory salmon stocks.[65] Since the beginning of the 1990s, there have been a myriad of multi-stakeholder processes initiated in BC coastal areas that to some degree or another can be considered as co-management processes. These include watershed management initiatives, regional management associations, Aboriginal fisheries strategies and community-based organizations. The federal *Oceans Act 1997* has stimulated renewed attempts to pursue co-management approaches in the management of ocean resources.[66] A recent initiative is the West Coast of Vancouver Island (WCVI) Aquatic Management Board launched in February 2002. The three examples of co-management practices on the BC coast noted above – the Skeena Watershed Committee, the Coastal Communities Network and the WCVI Aquatic Management Board – are each discussed briefly below.

The first formal fisheries co-management arrangements made in BC were those associated with the Aboriginal Fisheries Strategy, which was funded by the federal Department of Fisheries and Oceans. In 1992–93, the program's first year, 80 agreements were made with Aboriginal groups on co-operative management projects and pilot projects for commercial salmon sales.[67] A widely cited example of fisheries co-management in BC is the Skeena Watershed Committee.[68] The Skeena River flows into the Pacific Ocean at Prince Rupert in northern BC, just south of the Alaska border. Two major allocation conflicts existed, one between commercial fishers near the mouth of the river and sport fishers upstream, and a second between commercial fishers and the Gitskan and Wet'suwet'en First Nations. The Committee comprised five "equal partners" including First Nations, commercial, recreational, federal and provincial interests. Its accomplishments included devising a three-year fishing plan for the Skeena River salmon fishery, public education programs and dialogue, selective harvest and sockeye enhancement initiatives, and increased data collection through a tagging program.[69] Despite this success, the conflict between sectors that motivated the initiative re-emerged in the difficult years of 1996–97 and the partners could not reach consensus over harvesting issues. The Committee was dissolved in March 1997 when the North Coast Advisory Board, representing commercial harvesters, withdrew its participation.[70]

The Coastal Community Network (CCN) promotes itself as the "big voice for small communities" in British Columbia. CCN's role is to link coastal communities, to develop common ground on resource and marine policy, and to articulate the needs of coastal communities to senior governments, industry, media and the general public. Their goals are to enhance the long-term viability of communities, increase their self-reliance, and ensure a balanced and fair approach to the development of public policy related to the marine environment. CCN grew out of the Coastal Communities Conference on Fisheries held in Port Alberni in April 1993. At this conference, representatives from community councils, native bands, tribal councils, fishers, processors, cooperatives and unions gathered to discuss the future economic development of all coastal regions of BC.

Organized and facilitated by Simon Fraser University, and in partnership with the CCN, a subsequent series of conferences were held in BC coastal communities in 1996–97. Three themes emerged from these meetings concerning the involvement of coastal residents in fisheries management.[71] First, coastal people have the right to share in benefits obtained from resources adjacent to them and to participate in related decisions. Second, local knowledge is important to ensuring sound fisheries management, and third, coastal communities have stewardship capabilities that cannot be cost-effectively matched by government in terms of data collection, enhancement, habitat protection and even enforcement.[72] The CCN is currently struggling with leadership and financial difficulties and its future is uncertain.

A recent example of co-management approaches is the WCVI Aquatic Management Board, which covers an area that corresponds with the traditional Nuu-chah-nulth territory on Vancouver Island. This is a new forum for individuals and organizations to participate in the integrated management of aquatic resources and represents the delivery of a new approach to governance of aquatic resources. The creation of the Board was influenced by growing pressure to consider different approaches to the management of aquatic resources. These pressures include: increased demand from coastal communities, the Province of British Columbia and various public interest groups for an enhanced role in decision making; the government's need to establish more extensive, localized and integrated consultation and advisory processes as outlined in the federal *Oceans Act*; and the First Nations' desire to redevelop management processes with an enhanced First Nations' jurisdictional role.[73]

The federal, provincial, Nuu-chah-nulth and local governments jointly established the Board as a three-year pilot project. This pilot will test the implementation of a community and area-based process that will allow local communities, in partnership with others, to provide input and have an influence over aquatic management issues affecting their area. The Board will address issues in areas related to fisheries and integrated oceans management, stewardship, aquaculture and community economic development. The Board's purpose is to lead, facilitate and participate in decision-making processes related to these issues. This will involve working closely with other regional advisory and management processes as well as other groups. Recommendations from the Board will be presented to the appropriate statutory authority for their consideration.

In conclusion, the success of various community-based fisheries management approaches has varied depending not only on the mix of participants and the availability of resources, but also on a host of external factors ranging from changing ocean conditions to federal and provincial resource policies that are beyond the capabilities of any local group to address.[74] Pinkerton[75] suggests that barriers to community conservation initiatives relate to two general areas: distrust and resistance of management agencies and lack of broadly organized political support. While Savioe *et al.*[76] concluded "there is no one model that can possibly fit all in co-managing the fishery," much experience has been gained through trial and error over the past decade. A planned evaluation of the WCVI Aquatic Management Board should provide future direction on how community voices can more effectively be incorporated into decision making in fisheries management.

The Central Coast Land and Coastal Resource Management Plan, British Columbia

Land and Resource Management Planning (LRMP) is a sub-regional integrated resource planning process for Crown land in British Columbia that began in 1993. Through a variety of participatory processes, alternative land and resource management scenarios are identified and evaluated by participants. The principles of LRMP include consideration of all resource values, public participation, inter-agency coordination and consensus-based decision making. As the planning guide suggests, "the intent is to develop a community forum for communication and understanding among residents and government agencies" and to serve as "a vehicle for education and promotion of long-term participation in resource management."[77] The LRMP forums are, however, only advisory to governments who are legally responsible for resource decision making.

The Central Coast Land and Coastal Resource Management Plan (CCLCRMP) began in 1996 as part of the LRMP process. CCLCRMP was distinguished from previous processes because land, water and coastal nearshore resources were considered simultaneously. This necessitated that federal and provincial governments work together with First Nations and the public in addressing long-term strategic and management directions. The planning area is large, covering 4.8 million hectares of marine foreshore and forested upland on the mainland west coast for British Columbia. Temperate rainforest, major watersheds, rugged shorelines and steep mountainous terrain characterize the area. The population of about 4,400 consists of predominantly First Nations people dependent on fishing and forestry for a living.[78]

Participation, which evolved during the process, includes about 60 stakeholders from all levels of government, industry, residents, environmental groups and outdoor recreation groups. Due to the size of the planning area, participants were divided into North and South Forums to allow better local representation. In addition, a special sub-committee, the Coastal and Marine Committee, was established to provide recommendations to the planning forums. The intent was to integrate provincial LRMP into the federal government's recently initiated integrated coastal zone management approach, for example, using the CCLCRMP process as the initial forum for recommending, by consensus, marine protected areas and marine study areas. At the strategic level only, the CCLCRMP addresses the management of marine and freshwater habitat and will provide recommendations on management objectives for aquaculture. A special First Nations Forum was also established to facilitate discussion among participating First Nations. An inter-agency planning team with representatives from federal and provincial agencies, local governments and First Nations provided technical analysis and mapping expertise.

The coastal component of the CCLCRMP began in 1997 with the *Coastal*

Zone Strategic Plan completed and approved by April 2001.[79] The key principle in the coastal zone plan is ecosystem-based management, which encompasses not only ecological integrity but also seeks to sustain social, economic and cultural activities.[80] However, the then newly elected provincial government reviewed the decision and decided it had been rushed and that while approved in principle, final discussions were needed with First Nations and Fisheries and Oceans Canada.[81] While the marine planning work has been completed, the terrestrial component adopted a more streamlined approach to facilitate completion and a final land use plan is expected by 30 June 2004.[82] In the meantime, the new (2001) provincial Ministry of Sustainable Resource Management introduced a coast sustainability strategy.[83] Within this strategy, the Central Coast plan will be coordinated with the plans from two other coastal LRMP processes.

The CDN$35 million Coast Sustainability Trust was also established to support the outcomes of the Central Coast plan. The Trust will ensure that the planning process continues and is designed to help workers, contractors, communities and companies whose interests are negatively affected by land-use decisions such as the establishment of protected areas and the implementation of ecosystem-based management practices.

Despite the fact that the coastal plan is not finally approved, work is going ahead on a key recommendation to undertake more specific integrated coastal planning in key areas in the southern part of the coastal area. The intent is to provide direction to government concerning tenure decisions for aquaculture, ecotourism and economic diversification opportunities. As part of the provincial government's coastal strategy, a Coast Information Team (CIT) was established. The CIT brings together the best available scientific traditional and local knowledge, environmental expertise and community experience to develop information and analyses to support the development and implementation of ecosystem-based management.[84]

The Central Coast LCRMP has faced several challenges, some unique to the planning area, and others inherent to many shared decision-making processes. First, the process has taken much longer than anticipated. In part this relates to the large size of the planning area, and to the fact that it includes a coastal component. This has required the federal and provincial governments to establish new working relationships, which was done through a special agreement. Further, First Nations participants have been concerned over their ability to meet the expectations of the planning process with respect to such issues as time, information sharing and the impact on treaty settlement.[85] Indeed, many participants expressed concern about their ability to dedicate the time and resources needed to complete the planning process. In similar processes elsewhere in the province, availability of time and burnout have also been identified as problems.[86]

Appropriate representation is also an important variable. Not all First Nations chose to be part of the process. However, due to the importance of having their support, the government has been required to engage in separate negotiations in some instances. However, the extensive involvement of many stakeholders from a comprehensive array of interests has lengthened the process. The new streamlined approach to complete the process employs a sectoral model with a collaborative approach to reaching decisions. As a result, there are fewer people at the table representing larger constituencies. This final "completion table" includes representatives from the environmental community, the forest industry, tourism, recreation, labor, small business forestry, local governments, the federal government and First Nations.[87]

To conclude, the Central Coast LCRMP, while a lengthy and as yet incomplete process, has already resulted in tangible results, especially in terms of identifying areas for protection. While the planning has been at a strategic level with maps at a scale of 1:250,000, local scale coastal planning (1:50,000) is already under way in some places and these integrated land use plans will provide important information for decision makers.[88] While there are many gaps in data and many policies that need reforming, the process has resulted in the beginnings of an integrated coastal management strategy for BC's coast that has engaged local residents in an active role in determining the future of their communities and livelihoods.

Coastcare in Australia

Coastcare is a community-based grants scheme run jointly by the Commonwealth government and state/territory governments. Coastcare is primarily funded from the Natural Heritage Trust fund that arose from the partial sale of the government-owned Telecommunications Corporation (Telstra). The parent program under the Natural Heritage Trust, the "Coasts and Clean Seas," had a total of AUD$141 million to distribute over the first phase of the program, which ended in June 2002. Of this amount, AUD$27.3 million went to Coastcare. The general purpose of the program is summarized as follows:

> Coastcare is a national program that encourages community involvement in the protection, management and rehabilitation of our coastal and marine environments. The program assists local communities to form partnerships with local Land managers to undertake projects that aim to improve and protect our coastal and marine habitats.[89]

The National Office of Coastcare is based in the headquarters of Environment Australia, with 30 regional facilitators spread across Australia. The program is jointly funded with state and territory governments, and local governments are vitally and actively involved. The stated aims of Coastcare include the following:[90]

- To engender in local communities, including local industries, a sense of stewardship for coastal and marine areas.

- To provide opportunities and resources for residents, volunteers, business and interest groups to participate in coastal management.
- To support community identification of natural and cultural resources.
- To facilitate interaction between the community and bodies with responsibility for managing coastal areas.

The focus of Coastcare was to assist on-ground work such as:

- protecting or rehabilitating dunes, estuaries and wetlands;
- rehabilitating coastal and marine habitats;
- removing threats to coastal environments;
- monitoring beach conditions, and coastal flora and fauna;
- helping to develop and implement local management plans;
- education and training activities that raise community awareness, knowledge or skills on coastal and marine conservation issues.

As can be seen from these objectives, the program is an example of "collaborative management" as set out in the continuum of public involvement in coastal management outlined earlier (see Figure 10.1).

Since the inception of Coastcare in 1995, over 2,000 projects around Australia have been funded,[91] with grants made ranging from AUD$1,000 to $30,000 per project. Community groups apply for funds to state-level state assessment panels (SAPs) which rank the applications, with the final decision resting with the federal minister. Applications are received in March each year and are assessed by regional and state assessment panels (both with strong community representation), prior to the announcement of grants in December.[92] According to Tarte, the combination of state, community and local government financial and "in-kind" contributions meant "each Commonwealth Government dollar contributed is matched at least two-fold by 'in kind' or direct financial support."[93]

The most comprehensive and independent evaluation of the Coastcare program has been that conducted by Harvey *et al.*[94] with respect to South Australia. This review, which included interviews with key stakeholders, provides a more detailed picture of program implementation at the state level. South Australia received AUD$150,000 in 1995/96 rising to $500,000 in 1998/99, with funds being matched by the state government. In addition to the direct grants, funds were used to employ a state co-coordinator and three regional coordinators whose role is to promote involvement of organizations, assist community groups in preparing applications and to facilitate local level participation in the Coastcare program.

Although total funds available increased over this time period, the number of grants decreased (from 81 to 64). This reflects the fact that most projects are modest (less than AUD$5,000) and the possibility that some groups continued to apply while others began to see the application procedure as too tedious for such small amounts of money.

Consistent with the national situation, the majority of the South Australia grants (80 percent) were for on-ground projects. In terms of the purpose of grants, protection and rehabilitation of sensitive areas (usually sand dunes) and enhancement of sustainable recreation and tourism (usually pathways and boardwalks) made up 66 percent of all projects. Community participation in development and implementation of management plans made up a disappointing 6.5 percent of projects, while community-based monitoring of the coast accounted for a mere 9.2 percent of projects. In general, the majority of projects are moderately funded, but directed to a decreasing number of volunteers (mainly conservation, resident and economic development groups). In terms of the degree of participation in coastal management (as reflected in Figure 10.1), this suggests a strong role in collaborative management but a very low participation in direct decision making. This has led observers to suggest that the community is being used as a substitute for work previously performed by government agencies.[95]

The authors of the South Australia study concluded that Coastcare has effectively harnessed community groups to work in partnership with local authorities and has led directly to capacity building for local communities.[96] In addition, they determined that the program had "kick started" community projects and had stimulated local interest and commitment based on a partnership with educational management agencies in which coastal expertise already existed. The challenge that remains is whether funds can be generated from new sources (other than government) to ensure that these positive benefits are sustainable in the medium to long term.[97] Furthermore, it remains to be seen whether governments are truly interested in moving along the participation spectrum and offering people a genuine involvement in decision making on the coast, rather than simply using the community as an unpaid workforce.

Harvey *et al.* point out that in the South Australia case, the lack of an overriding coastal policy has hampered the allocation of funds,[98] but that the factors that have worked for community-based management in Coastcare have been the role of the program in building community cohesion and capacity, which are less tangible benefits. They highlight the fact that the lack of evaluation of the projects' environmental value has also restricted how much could be learned about the potential of community-based management.

The experience in South Australia contrasts with the position in Victoria where the SAP uses the Victorian Coastal Strategy (the state's overall coastal policy) as a means of setting priorities in allocating Coastcare grants. As approximately half of the members of the SAP are members of the Victorian Coastal Council (the lead agency in Victoria), and the VCC has primary responsibility for preparation of the Victorian Coastal Strategy, the grants are consistent with integrated coastal management principles outlined in the Strategy. This allows for some continuity and coherence in the selection of projects (at least in theory). Despite this advantage, however, Wescott[99] nonetheless criticized the trend in Victoria to use Coastcare annual grant funding as a substitute for recurrent government funding, consistent with the critique of the South Australia program by Harvey *et al.*

In summary, Coastcare has been a valuable community-based program, despite its identified shortcomings. The federal government's political assessment of the program is reflected in its commitment to extend Coastcare until 2007 as part of a series of programs (including Landcare, Bushcare and Rivercare).[100] The overall prescription for change in Coastcare offered from Harvey *et al.* could be seen as a fair assessment of Coastcare as a community-based management program:

> There is a requirement for long term vision for community-based coastal management because without that vision there is the capacity for short-term funding rounds to influence the type of activity that local communities undertake.[101]

The Marine and Coastal Community Network in Australia

The Marine and Coastal Community Network (MCCN) grew out of the Ocean Rescue 2000 initiative of the Australian government, and stemmed from a recognition that progress in coastal and marine conservation could only come with a greater understanding (and consciousness) of the coastal, and in particular, marine environment amongst the general community.[102] The MCCN was a unique attempt to build an alliance of interested individual and groups (called "participants") to ensure the sustainability of the lessons learned by communities in coastal management, and to bring together like-minded people who shared a commitment to the coastal and marine environment (though they might well disagree on specific issues).

The first meeting of the Network was in Sydney in May 1993, in the form of the rather gloriously named National Implementation Committee. The Committee was formed by the Australian Marine Conservation Society, a non-government conservation organization that had been given the contract from Environment Australia to establish the Network. The Network's stated role was to raise consciousness of the marine and coastal environment and to promote ecologically sustainable use of the Australian coastal and marine environment. It was not to take sides in any debate on marine issues, nor become involved in partisan politics, but to draw together a wide spectrum of Australians interested in the marine and coastal environment. As such, it was seen as an "honest broker" that could act as an advocate for the marine and coastal environment.

The 1993 meeting put in place a mission statement and operating principles and set about its task. By October of that year, a Southern Regional Coordinator had been placed in Melbourne and a Northern Regional Coordinator in Darwin. The idea caught on rapidly and very soon afterwards regional coordinators were in place in Western Australia, New South Wales, South Australia and Tasmania. The Network was initially funded at approximately AUD$800,000 per year, and at $500,000 per year for the two years following June 2001. The budget remained at $500,000 for 2003–04. There is little likelihood of the budget being restored to its former level. Indeed, there is speculation that the only reason the budget was not cut further was because MCCN was awarded the "Gold Banksia" Environmental Award (Patron of the Prime Minister) as the most outstanding organization for environmental protection in 2003.

By December 2003, the Network was composed of over 9,000 registered participants from every possible interest group involved in coastal and marine matters, including government agencies, non-governmental organizations, indigenous organizations, conservation groups, coastal recreational users, fisheries associations, private industry, universities and other educational groups, and individuals. It is fair to say that all tiers of Australian society are represented in the Network.

Participants in the Network are kept informed of recent coastal and marine initiatives, including relevant conferences and workshops, through a national bi-monthly newsletter ("Waves") and state-level inserts ("Ripples"). MCCN has responded to the recent budget cuts by developing the Network's Internet resources.[103] Regional coordinators located in a "host" organization disseminate information and answer inquiries from the general community, business and media. The host organizations provide the physical infrastructure for the coordinator and are paid a contribution toward these costs. The types of organizations acting as hosts vary from state to state and include fisheries organizations, tertiary education institutions, conservation groups and local government agencies. The experience has been that the relationship with host organizations aids in expanding the diversity of the Network.

Each regional coordinator has a mentor who is a member of the National Reference Group (NRG). NRG members are drawn from a wide variety of backgrounds (including state and local government agencies, academia and conservation groups) and a range of disciplines. The NRG provides support and advice to the coordinators and develops the overall policies that guide the Network. The NRG and regional coordinators meet twice a year for 2–4 days.

Consistent with its original mandate, the Network is non-political and does not take "sides" in controversial issues. Rather it acts as honest broker through the dissemination of information and as a conduit for different opinions. Since its commencement in 1993, the Network has grown steadily, with full funding provided by Environment Australia (EA, the federal government environment department). The fact that the Network is seen as neutral and independent of government allows all groups in the community to use it as a vehicle for dissemination of information and as a mechanism for capacity building in coastal and marine affairs in Australia.

By way of illustration of the Network's community-based approach, it is useful to consider the role it played in the development of a constituency for *Australia's Ocean Policy*. The March 1997 discussion paper prepared by the Commonwealth government on the proposed 'comprehensive and integrated Oceans Policy' resulted, after a period for submissions of eight weeks, in only 63 submissions being received, predominantly from academics and non-governmental organizations. This was a rather disappointing response considering the difficulties an overarching policy was going to face from well-entrenched sectoral interests. The government department leading development of the policy (Environment Australia) and the Network set out to rectify this problem.

The agencies targeted a series of sectoral groups (e.g. conservation groups) who had not responded in the numbers expected. The Network had earlier surveyed its participants on the critical issues involved in the Policy, and now set up seminars and meetings across the country, cajoling the media and turning its concentrated efforts into increasing the response rate to the second set of documentation (the May 1998 *Issue Paper*). One of the key methods employed was to encourage people to find some specific issue or item in the Policy to comment on, despite the somewhat daunting breadth and complexity of the document. It was hoped that, once a party felt comfortable in dealing with the Policy, they would go on to comment on other issues. The strategy seemed to work, as the second round of public submissions drew 660 responses. This provided important input, and allowed the federal Minister to demonstrate wide-based support for the notion of an integrated, non-sectoral oceans policy.

The Network's regional coordinators have also worked closely with the regional facilitators of the Coastcare program to encourage community groups to become involved in Coastcare. Furthermore, the Network's participants database has been a valuable resource to publicize Coastcare grant rounds and to advertise Coastcare-related workshops and seminars.

The role of the Network in community-based management, therefore, has usually been indirect or facilitative, but as the work areas shown in Table 10.1 illustrate, it has been diverse. In addition, production of such items as the *Australian Marine Project Guide*, the *Blue Pages* (a directory of coastal and marine groups), and the management of projects such as "Dragon Search" (monitoring the numbers and conservation of seadragons) have led to more direct involvement.

The Network has regularly surveyed its participants. In a 2001 survey, 90 percent of respondents regarded the continuation of the Network as "highly important" with 70 percent rating the overall service provided as "excellent." The newsletters were regarded as "very useful" by 96 percent of respondents.[104] With reference to the continuum of public involvement set out in Figure 10.1, the Network is involved in some way in the middle three levels, but in reality its major input to community-based management does not fit well into this spectrum. Its primary role is stimulating the

Table 10.1 Major areas of work for the MCCN regional coordinators, mid-October 2001 to mid-February 2002

Category of work	*Vic*	*NT*	*Qld*	*NSW*	*Tas*	*SA*	*WA*	*NO*
Contract derived work								
Administration	×	××	××	××	×××	××	××	×××
Events	×××	–	×	××	–	××	×	–
Facilitating/organizing workshops	×××	–	–	×××	×××	×	×	–
Info collated for dissemination	××	××	×××	××	×××	×××	××	×
Networking/liaison	××	××	×	×××	×××	×	×××	××
Regional visits	×××	×	××	×	×××	×	×	××
Waves and Ripples	×××	××	×××	×××	××	××	××	×××
Web pages	××	×	××	×	×	×	×	××
Progress reports, etc.	×	××	×××	××	××	×	××	××
Oceans Policy	××	××	×	×××	×	×	××	××
Marine protected areas	×××	×	×	×××	××	×××	×××	×
Coast and Clean Seas	–	×	–	–	–	×	××	–
Participants' database	×	×	×	×	×	×	×	×××
National and Regional Reference Group and national meeting	–	–	××	×	×	×	×	×××
MCCN generated work								
Events	×	–	–	–	××	×	–	–
Projects	××	××	××	×××	×××	××	–	–
Short courses	–	×	–	–	–	–	–	–
Training for regional coordinators	–	–	–	–	–	–	×	×
Community radio	×	×	×	–	–	×	×××	–
New grant applications	××	–	–	–	–	–	–	–
MCCN info sheets/ dissemination including electronic info	–	–	–	××	–	×××	×	×
Externally generated work								
Assessment panels	–	–	–	–	×	–	–	–
Committees	×	–	×	××	××	××	××	××
Interview panels	–	–	–	–	–	–	–	–
Media	×××	×	×	××	××	×	×	–
Submissions	×	××	×××	×××	××	×××	×××	–
Workshop/conference attendance	××	–	×××	×××	–	××	×	××
Issues	×	–	××	×	×	×××	×××	×

community to become involved in community-based management, a role that may be difficult to track in concrete terms, but which is nonetheless essential.

Conclusion

The range of potential approaches for CBM and co-management, as reflected in the literature noted at the outset of this chapter, is borne out in practice in both Australia and Canada. It is clear that no one "package" of institutions and arrangements presents an ideal that will work in all circumstances; even the fundamental purpose for seeking community involvement can vary from participation in planning and conservation actions to some form of power-sharing in the regulatory context. The breadth of the experience in both countries would suggest that the development of CBM is, above all, a pragmatic and responsive process, with agencies and communities challenged to craft the particular variant that will best serve their interests. Given the strong demand for progress on the full participation of communities in coastal and ocean management, it seems certain that CBM, or co-management, will continue to be an important aspect of the development of new forms of governance. The cases examined here, however, suggest that a number of small cautions should be kept in mind.

First, while the quest for terminological precision should not be allowed to stifle creativity, and the names given to community-based activities are less important than the practical progress that can be made, proponents and analysts alike should still be clear on the degree of community involvement that is proposed. The overuse of terms like co-management to include, for example, routine consultation with no sharing of decision-making power is likely to lead to suspicion and cynicism on the part of communities. Second, and related to this problem, we must be aware of the danger in labeling a narrow interest group such as a fishery sector as the "community," when numerous other interests may be left as outsiders, with no option but active opposition to the new mechanisms. Finally, careful attention must be paid to the level of downloading and demands placed upon overstretched community institutions. Community involvement should be more than a means of shuffling off government responsibilities (and costs) to those with no institutional or financial base to support them.

Notes

1 For a survey of the early work in this area, see the collection of essays in B. McCay and J. Acheson (eds) *The Question of the Commons* (University of Arizona Press, Tucson: 1987).

2 See, for example, S. Siar, R. Agbayani and J. Valera "Acceptability of territorial use rights in fisheries: Towards community-based management of small-scale fisheries in the Philippines" 14 *Fisheries Research* (1992) 295–304.

3 See, for example, B. McCay and J. Acheson "Human Ecology of the Commons" in McCay and Acheson, note 1 at 31–4.
4 (17 June 1992) U.N. Doc A/CONF. 151/5 Rev. 1. (1992) 31 *International Legal Materials* 874.
5 Reprinted S.P. Johnson *The Earth Summit: The United Nations Conference on Environment and Development (UNCED)* (Graham & Trotman/Martinus Nijhoff, London: 1993) 123–508, para. 17.82 (a) at 321.
6 L. Hildebrand "Community-based coastal management" (1997) 36(1–3) *Ocean and Coastal Management* 1–9. This article appeared as an introduction to an entire issue of the journal devoted to "Community-based Coastal Management."
7 Ibid., at 2.
8 In the context of fisheries management, for example, despite the widespread retention of real management control in the hands of the state, it has been suggested that devolution efforts have led to a ". . . new paradigm in fisheries management. This is not a biological, model-driven approach with top down state managers recommending management interventions to state enforcers but a decentralized people/community-centred approach to resource management." See K. MacKay "Sustainable Management of Fisheries Resources: Common Property Issues" in *Philippines Coastal Resources Under Stress: Selected Papers from the Fourth Common Property Conference* (Marine Science Institute, University of the Philippines, Manila: 1995) at 2.
9 Hildebrand, note 6.
10 N. Harvey, B.D. Clarke and P. Carvalho "The role of the Australian Coastcare program in community-based management: A case study from South Australia"(2001) 44(3–4) *Ocean and Coastal Management* 163–4. In this work, Harvey *et al.* comment on the many examples of CBM assembled by Cicin-Sain and Knecht in their book, B. Cicin-Sain and R. Knecht *Integrated Ocean and Coastal Management – Concepts and Practices* (Island Press, Washington, DC: 1998).
11 E.M. Ferrer and C.M. Nozawa "Community-based coastal resources management in the Philippines – key concepts, methods and lessons learned," as cited in Harvey *et al.*, ibid. Online at: www.web.idrc.ca/en/ev-5994-201-1-DO_TOPIC.html (accessed 22 January 2004).
12 Harvey *et al.*, note at 10.
13 L.Z. Hale, M. Amaral, B.W. Mwandotto, M. Amaral and A. Issa "Catalysing coastal management in Kenya and Zanzibar: Building capacity and commitment" (2000) 28 *Coastal Management* 75–85.
14 J.P. Ellsworth, L.P. Hildebrand and E.A. Glover "Canada's Atlantic Coastal Action Program: A community-based approach to collective governance" (1997) 36(1–3) *Ocean and Coastal Management* 121–42.
15 Abregana, P. Gardiner-Barber, M. Maxino, P. Saunders and D. VanderZwaag *Legal Challenges for Local Management of Marine Resources: A Philippine Case Study* (CIDA/Environment and Resource Management Project, Halifax: 1996) 7–9.
16 G. Wescott "Partnerships for capacity building: Community, governments and universities working together" (2002) 45(9–10) *Ocean and Coastal Management* 549–71.
17 G. Wescott "Reforming coastal management to improve community participation and integration in Victoria, Australia" (1998) 26 *Coastal Management* 3–15.
18 R. Arnason and L. Felt *The North Atlantic Fisheries: Successes, Failures, and Challenges* (Institute of Island Studies, University of Prince Edward Island, Charlottetown: 1995).

19 See, for example, P. Copes "A critical review of the individual quota as a device in fisheries management" (1986) 62 *Land Economics* 278–91; B. McCay "Individual transferable quotas (ITQs) in Canadian and American fisheries" (1995) 28(1–3) *Ocean and Coastal Management* 85–116.
20 See, for example, B. McCay "User participation in fisheries management: Lessons drawn from international experiences" (1995) 19(3) *Marine Policy* 227–46; B. Neis, L. Felt, D. Schneider, R. Haedrich, J. Hutchings and J. Fischer *Northern Cod Stock Assessment: What Can be Learned from Interviewing Resource Users?* (Northwest Atlantic Fisheries Organization, Dartmouth: 1996).
21 "Atlantic Canada reels under fishery closures" (1993) *National Fisherman* (December) 14–15.
22 S. Williams *Our Lives Are at Stake: Women and the Fishery Crisis in Newfoundland and Labrador* (ISER Report No.11, Institute of Social and Economic Research, Memorial University of Newfoundland, St. John's: 1996) at 1.
23 Appadurai "Disjuncture and difference in the global economy" (1990) 7 *Theory, Culture and Society* 295–310.
24 M. Watts "Living under Contract: Work, Production, Politics and the Manufacturing of Discontent in a Peasant Society" in A. Pred and M. Watts *Reworking Modernity: Capitalism and Symbolic Discontents* (Rutgers University Press, New Brunswick: 1992) 65–105.
25 Harvey *The Condition of Postmodernity* (Basil Blackwell Press, Oxford: 1996).
26 J. Nash "Global integration and subsistence insecurity" (1994) 96(1) *American Anthropologist* 7–30.
27 R. Apostle *et al. Community, State and Market on the North Atlantic Rim: Challenges to Modernity in the Fisheries* (University of Toronto Press, Toronto: 1998); Watts, note 24.
28 S. Shrybman *A Citizen Guide to the World Trade Organisation* (Canadian Centre for Policy Alternatives, Ottawa: 1999); S. Stonich, J. Bort and L. Ovares "Globalization of shrimp mariculture: The impact of social justice and environmental quality in Central America" (1997) 10 *Society and Natural Resources* 161–79; A. Wilks "Prawns, profit and protein: Aquaculture and food production" (1995) 25(2–3) *The Ecologist* 120–5.
29 *Report of the FAO/Japan Expert Consultation on the Development of Community-Based Coastal Fishery Management Systems for Asia and the Pacific* (FAO Fisheries Report 474; FIDP/R474, Food and Agriculture Organization, Rome: 1993) at 7.
30 S.C. 1996, c. 31, Part II.
31 This was undertaken as part of a rewriting of the *Fisheries Act*, with *An Act Respecting Fisheries*, Bill C-62 (1996); the bill did not proceed past first reading. The degree of caution with which DFO approaches this is indicated by their distinction between "partnering" and "partnership" – with the latter seen as potentially referring to a legally binding agreement. See D. Savoie, G. Filteau and P. Gallaugher *Partnering the Fishery: Report of the Panel Studying Partnering* (Department of Fisheries and Oceans, Ottawa: 1998) at 4 (hereinafter *Savoie Report*).
32 See, for example, the discussion of "partnering" vs. "co-management" in the *Savoie Report*, ibid., which viewed co-management as a lesser form of cooperation that did *not* extend to any serious participation in actual management decisions.
33 It should be noted that some of the most advanced forms of co-management applied in Canada to date, with apparently mixed results, have been with respect to the Aboriginal Fishing Strategy and particularly legislated co-management under land claim settlements, both of which function under legal

authority not explicitly provided for other categories. See Marcus Haward, Rod Dobell, Anthony Charles, Elizabeth Foster and Tavis Potts "Fisheries and oceans governance in Australia and Canada: From sectoral management to integration?" (2004) 26(1) *Dalhousie Law Journal* (in press).

34 The current draft of the Framework dates from 1999 and removes references to the legally binding Fisheries Management Agreements that depended upon the revised *Fisheries Act*, which were dropped from the Guidelines pursuant to recommendations of the *Savoie Report*. See Department of Fisheries and Oceans *Draft Framework and Guidelines for Implementing the Co-management Approach* Volume 1 (DFO, Ottawa: January 1999) 10–11. Online at: www.dfo-mpo.gc.ca/Library/237529.pdf (accessed 19 January 2004) (hereinafter, *Draft Framework*).

35 Auditor-General of Canada *Managing Atlantic Shellfish In A Sustainable Manner* (Office of the Auditor-General, Ottawa: 1999) para 4.92. Online at: www.oag-bvg.gc.ca/domino/reports.nsf/html/9904cd.html#0.2.2Z141Z1.WEP9CA.KYSX9F.B7 (accessed 19 January 2004).

36 The Integrated Fisheries Management Policy was introduced in 1996, calling for the introduction of Integrated Fisheries Management Plans (IFMPs) for all fisheries, although this has not been achieved. The Policy foresaw plans for each fishery that would include "conservation, management and scientific requirements for a fishery and also spell out the process and implementation of resource management, conservation and protection measures." In addition, IFMPs were seen as providing the "basis upon which co-management and partnering can be developed." *Review of Integrated Fisheries Management Plans*, (DFO, Ottawa: 1997) (prepared by GTA Consultants and Review Directorate), section 1 and 2 (hereinafter *IFMP Review*). See also, the following examples of IFMPs: *Eastern Nova Scotia Snow Crab Integrated Fishery Management Plan* (DFO, Ottawa: 2000) (hereinafter, *Snow Crab IFMP*); *1998 Scotian Shelf Shrimp Management Plan* (DFO, Ottawa: 1998) (hereinafter, *Scotian Shelf Shrimp IFMP*).

37 *Draft Framework*, note 34 at 8. See also DFO Press Release NR-HQ-96-55E, 12 July 1996. This is a somewhat inflated assessment of the normal process by which the Department seeks input and consensus for what is still their plan, finalized under the discretion of the Minister and with no obligation to take account of any particular views.

38 *Draft Framework*, ibid., at 9.

39 *Snow Crab IFMP*, note 36, Appendix 7.

40 Ibid.

41 *Draft Framework*, note 34, Annex I.

42 See *Groundfish Integrated Fisheries Management Plan, Scotia-Fundy Fisheries: April 1, 2000–March 31, 2002* (DFO, Ottawa: 2000). Online at: www.mar.dfo-mpo.gc.ca/fisheries/res/imp/2000grndfish.htm (accessed 20 January 2004) (hereinafter, *2000 Groundfish IFMP*), sections 1.0 and 1.6.1. The seven areas were: eastern Nova Scotia, Halifax County (west of Halifax), Lunenburg/Queens Counties, Shelburne County (with two boards, based on differing interests), Yarmouth County, Digby and other counties on the Bay of Fundy in Nova Scotia, and southwestern New Brunswick.

43 The measures applied generally to cod, haddock and pollock fisheries, and there is also some involvement of the boards in harvesting plans for species not covered by community quotas. Ibid., at section 1.6.1. Community quotas have also been established for dogfish. See DFO News Release NR-MAR-02-07E, May 30, 2002.

44 *2000 Groundfish IFMP*, note 42, at section 1.6.1.

45 It should be noted that these comments may not be applicable to several ad hoc cases of cooperative approaches to fisheries, but the special cases do not

change the overall picture as it applies to the legal structure and the commercial fisheries.

46 See, for example, comments on co-management in the Atlantic shellfish industry at Office of the Auditor-General, note 35, paras 4.92–4.94.

47 See *Snow Crab IFMP*, note 36, Appendix 7; *Scotian Shelf Shrimp IFMP*, note 36, section 4.2.

48 Both one-year and multi-year plans have been utilized. The *Snow Crab IFMP*, ibid., used a one-year plan for some areas and a three-year plan for others.

49 This is distinct from defining an area within which the IFMP and/or any JPA might apply. The geographic limits are still accompanied by a restriction to a particular resource, so that there is no effective "marine" management in the broader sense.

50 See, for example, the 1996 DFO announcement of partnering initiatives in the revised *Fisheries Act* in which [then] Minister Mifflin offered reassurance that the process was "open to all sectors of the industry, be they rich or poor, large or small." DFO Press Release NR-HQ-96-88E, 6 November 1996.

51 These concerns, which were by no means shared by all participants, emerge from the record of consultations held in connection with the Atlantic Fisheries Policy Review in 2001. *Atlantic Fisheries Policy Review Public Consultations* (DFO, Ottawa: 2001). Online at: www.dfo-mpo.gc.ca/afpr-rppa/input_e.htm (accessed 19 January 2004). In this view, DFO's assurances that the existence of JPAs will not influence critical allocation decisions are simply formal statements made with a view to forestalling legal challenges to the Minister's exercise of discretion. That is, it seems unlikely that the parties to a JPA or similar arrangement would not be in a favored position with respect to allocation of limited quotas, particularly if they also fit the profile of the "preferred" participants in the industry. DFO has also indicated that co-management approaches are also at times considered as a means of reducing effort. See *2000 Groundfish IFMP*, note 42, section 5.3(c):

> The introduction of Community Boards to manage the activities of the fleets engaged in the competitive fixed gear fishery allows boards a certain flexibility in designing fishing plans that result in a better match of number of vessels to catch allocations . . . These measures have had the result of reducing the utilization of licenses in these fleets also.

52 Parts of this brief discussion appear as part of a broader review in P. Saunders "Marine Property Rights and the Development of Jurisdictional Regimes: Private Rights, Communal Tenure and State Control" in D. Vickers (ed.) *Marine Resources and Human Societies in the North Atlantic Since 1500* (ISER Conference Paper No. 5, Institute of Social and Economic Research, Memorial University of Newfoundland, St. John's: 1995) 245–74.

53 Prior to the community quota concept, there was a requirement for residency in an "area of fishing or home port" as an eligibility requirement of license issuance." Department of Fisheries and Oceans *Commercial Fisheries Licensing Policy for Eastern Canada* (Supply and Services Canada, Ottawa: 1992) at 23. This did not, however, extend to any real recognition of a *right* to fish in defined areas based on community, although community "adjacency" was a factor to be considered in allocation.

54 A related problem is the incompatibility of scientific and traditional management approaches, both in the knowledge sources upon which they rely and the management paradigms to which they apply. F. Berkes "Common Property Resource Management and Cree Indian Fisheries in Subarctic Canada" in McCay and Acheson, note 1 at 89.

55 Davis "Property Rights and Access Management in the Small Boat Fishery: A Case Study From Southwest Nova Scotia" in C. Lamson and A. Hanson (eds) *Atlantic Fisheries and Coastal Communities: Fisheries Decision-Making Case Studies* (Dalhousie Ocean Studies Programme, Halifax: 1984) 146.
56 Department of Fisheries and Oceans *The Management of Fisheries on Canada's Atlantic Coast: A Discussion Document on Policy Direction and Principles* (2001). Online at: www.dfo-mpo.gc.ca/afpr-rppa/Doc_Doc/discodoc_e.htm (accessed 19 January 2004).
57 Ibid., at section 4.6 (emphasis added).
58 Ibid., at section 4.6.1 (emphasis added). See also the statement of future intent which makes it clear that other interested parties can expect input, but not necessarily a management role: "Planning and decision making in the fisheries management system must provide opportunities for meaningful input from individuals who are not direct resource users but have an interest in the fishery or represent a broader public interest."
59 (DFO, Ottawa: 2004). Online at: www.dfo-mpo.gc.ca/afpr-rppa/Doc_Doc/policy_framework/Policy_Framework_e.pdf (accessed December 2004).
60 Ibid., at 33.
61 Significant progress in this regard has been made in the Bay of Fundy through the Bay of Fundy Fisheries Council and the Bay of Fundy Marine Resource Centre (Online at: www.bfmrc.ns.ca/ accessed 25 April 2005).
62 G. Gislason, E. Lam and M. Mohan *Fishing for Answers: Coastal Communities and the BC Salmon Fishery* (ARA Consulting Group Inc. for the BC Job Protection Commission, Victoria: 1996).
63 P. Gallaugher and K. Vodden "Tying it Together along the BC Coast" in D. Newell and R. Ommer (eds) *Fishing Places, Fishing People: Traditions and Issues in Canadian Small-Scale Fisheries* (University of Toronto Press, Toronto: 1999) 374.
64 Ibid.
65 *Savoie Report*, note 31 at 53; E. Pinkerton "Factors in overcoming barriers to implementing co-management" (1999) 3(2) *Conservation Ecology*. Online at: www.ecologyandsociety.org/vol3/iss2/art2/index.html (accessed 12 February 2004).
66 National Round Table on the Environment and the Economy *Sustainable Strategies for Oceans: A Co-Management Guide* (Renouf Publishing Co., Ottawa: 1998) 85.
67 T. McDaniels, M. Healey and R. Paisley "Cooperative fisheries management involving First Nations in British Columbia: An adaptive approach to strategy design" (1994) 51 *Canadian Journal of Aquatic Sciences*.
68 E. Pinkerton and M. Weinstein *Fisheries That Work: Sustainability Through Community-based Management*, A Report to the David Suzuki Foundation. (David Suzuki Foundation, Vancouver: 1995).
69 Ibid.
70 K. Vodden *Nanwakola: Co-management and Sustainable Community Economic Development in a BC Fishing Village*, Unpublished MA thesis (Department of Geography, Simon Fraser University, Burnaby, BC: 1999).
71 P. Gallaugher *Coastal Communities Taking Action: A Year of Dialogue along the British Columbia Coast* (Simon Fraser University, Burnaby: 1996) at 124.
72 Gallaugher and Vodden, note 63.
73 West Coast of Vancouver Island Aquatic Management Board. Online at: www.westcoastaquatic.ca (accessed 19 January 2004).
74 Vodden, note 70.
75 Pinkerton, note 65.

76 *Savoie Report*, note 31.
77 Integrated Resource Planning Committee *Land and Resource Management Planning: A Statement of Principles and Process* (Government of British Columbia, Victoria: 1993).
78 CCLCRMP. Online at: www.srmrpdwww.env.gov.bc.ca/lrmp/cencoast/news/bkgrnd111501.htm (accessed 23 January 2004).
79 The *Coastal Zone Strategic Plan* is a revised version of an earlier working document. Online at: www.srmwww.gov.bc.ca/cr/resource_mgmt/lrmp/cencoast/processcomp.htm (accessed 23 January 2004). The earlier working document is also available. Online at: www.srmwww.gov.bc.ca/cr/resource_mgmt/lrmp/cencoast/ prelim401.htm (accessed 23 January 2004).
80 Plan, Ibid., at 27.
81 These discussions have been deferred until after March 2004 when the Completion Table submits its final report on the terrestrial component of the CCLRMP. At that time, the marine plan will be reconciled with the land plan and the Trust (see below). John Bones, Director, Coast and Marine Planning Branch, Ministry of Sustainable Resource Management, Government of British Columbia (Personal communication, 23 January 2004).
82 "Central Coast Planning Table Reaches Consensus" Press Release 2003SRM0023-001103, 11 December 2003, Ministry of Sustainable Resource Management. Online at: www.srmwww.gov.bc.ca/cr/resource_mgmt/lrmp/cencoast/docs/cc_consenus001103.pdf (accessed 23 January 2004). The *Central Coast LRMP Completion Table Report of Consensus Recommendations to the Provincial Government and First Nations* was issued in May 2004. Online at: www.srmwww.gov.bc.ca/cr/resource_mgmt/lrmp/cencoast/table_rec.htm (accessed 25 April 2005).
83 See Coastal Sustainability Strategy website. Online at: www.srmwww.gov.bc.ca/rmd/coaststrategy/ (accessed 25 April 2005).
84 Online at: www.citbc.org (accessed 23 January 2004).
85 Central Coast LCRMP Newsletter, Fall 1999. Online at: www.srmwww.gov.bc.ca/cr/resource_mgmt/lrmp/cencoast/news/nl1299.htm (accessed 23 January 2004).
86 D. Duffy, L. Hallgren, Z. Parker, R. Penrose and M. Roseland *Improving the Shared Decision-Making Model: An Evaluation of Public Participation in Land and Resource Management Planning (LRMP) in British Columbia, Vol 1: Summary Report,* Report for Forest Renewal of British Columbia (Community Economic Development Centre, Simon Fraser University, Burnaby: 1998) 60.
87 Ibid. See also *Central Coast LRMP Completion Table Report*, note 82.
88 Online at: www.srmwww.gov.bc.ca/cr/resource_mgmt/lrmp/cencoast/mapfolio.htm (accessed 23 January 2004).
89 Commonwealth of Australia *Coastcare: Vision and Strategic Plan* (Canberra: 1997).
90 Ibid.
91 A mid-term review/audit of the program was conducted in 1999, but it concentrated on financial issues and did not comment on whether the program had met the objectives described above. Commonwealth of Australia *Improving Performance In Partnership: Mid Term Evaluation Of Coasts And Clean Seas 1997–1999* (Marine Group, Environment Australia, Canberra: 1999).
92 The author of this section is a member of the State Assessment Panel (SAP) for Victoria.
93 Ibid.
94 Harvey *et al.*, note 10.
95 Wescott, note 17.

96 Harvey et al., note 10.
97 Ibid.
98 Wescott, note 17.
99 Ibid.
100 See the Coastcare website. Online at: www.nht.gov.au/nht1/programs/coastcare/index.html#information (accessed 26 January 2004).
101 Harvey et al., note 10 at 179.
102 Wescott, note 17.
103 See the MCCN website. Online at: www.mccn.org.au (accessed 29 January 2004).
104 Ibid.

Part VI

Indigenous rights

11 Aboriginal title and oceans policy in Canada

Diana Ginn

Introduction

The *Oceans Act*[1] of Canada sets out a broad framework for the unified management of Canada's oceans based on an ecosystem approach. In particular, the *Oceans Act* calls on the Minister of Fisheries and Oceans to lead and facilitate the development of a national strategy to guide the management of Canada's estuarine, coastal and marine ecosystems.[2] The *Oceans Act* also reflects awareness that aboriginal rights may affect the development or implementation of policy surrounding oceans management. For example, s. 2(1) of the Act states that ". . . nothing in this Act shall be construed so as to abrogate or derogate from any existing aboriginal and treaty rights of the aboriginal peoples under section 35 of the Constitution Act, 1982."[3] The *Oceans Act* also provides for collaboration with aboriginal organizations[4] in the development and implementation of a national strategy and plans for integrated management of all activities affecting estuarine, coastal and marine waters, and provides for the possibility of aboriginal participation on certain advisory or management bodies,[5] thus creating an opportunity for aboriginal input into Canada's ocean policy in the future.

The extent and nature of the interaction between aboriginal rights and oceans policy may depend on a number of factors, both political and legal. Political factors include the extent to which successive federal governments perceive it to be feasible or necessary to incorporate the recognition of aboriginal interests into oceans policy in light of the collaboration envisaged by the Act. Legal considerations include the kinds of aboriginal rights that Canadian courts may be willing to recognize in relation to ocean areas and how courts interpret and apply the tests which have been developed regarding governmental justification of infringements of aboriginal rights.

This chapter focuses primarily on the first of these legal issues by asking whether the doctrine of aboriginal title[6] could be applied to the seabed.[7] How Canadian courts might reconcile claims of aboriginal title in the seabed with common law rights of fishing and navigation, as well as the international right of innocent passage, is uncertain; however, while it is difficult to offer definitive conclusions, it is possible that the doctrine of aboriginal title could be applied at least with regard to the seabed beneath Canada's territorial waters. The possible application of aboriginal title to an area of seabed, when taken in conjunction with the collaboration already required by the *Oceans Act*, makes it clear that the participation of First Nations will be an important element of the implementation of an oceans strategy.

Aboriginal title

The starting point for any discussion of aboriginal rights in Canada is s. 35(1) of the *Constitution Act*, which states that "[e]xisting treaty and aboriginal rights are hereby recognized and affirmed." Canadian courts have made it clear that the term "aboriginal rights" encompasses a range of different rights, including:

- title to the land itself;
- site-specific aboriginal rights where exercise of the right is tied to a particular piece of land, although the community does not hold aboriginal title to that land (for instance, the right to hunt or fish in a particular area); and,
- aboriginal rights to carry out certain activities that are not linked to any particular area.[8]

Source of aboriginal title

The first decision in Canadian law that commented on the source of aboriginal title (or "Indian title" as it was referred to at that time) was *St. Catharine's Milling and Lumber Co.* v. *The Queen*.[9] The issue in that case was whether, on the surrender of Indian title, the underlying fee simple lay with the federal or provincial Crown. In the course of deciding in favor of the provincial Crown, the Privy Council referred to the Royal Proclamation of 1763[10] as the basis of Indian title.

Modern Canadian jurisprudence on aboriginal title begins with the 1973 Supreme Court of Canada decision in *Calder* v. *British Columbia (Attorney-General)*.[11] Six of the seven justices of the Supreme Court who heard *Calder* addressed the substantive issues, and all of these six accepted that the area claimed had been inhabited by the claimants and their ancestors "since time immemorial."[12] The Court made it clear that aboriginal title did not have its primary source in any document or agreement and instead characterized aboriginal title as flowing from historic occupation and use of the land: "when settlers came, the Indians were there, organized in societies and occupying the land as their forefathers had done for centuries."[13]

Subsequent cases have accepted the *Calder* principle that aboriginal title is inherent. The most recent and comprehensive Supreme Court analysis on aboriginal title is the 1997 decision in *Delgamuukw* v. *British Columbia*.[14] In that case, two First Nations claimed aboriginal title to over 58,000 square miles in British Columbia. Because of deficiencies in the pleadings and errors in the trial judge's assessment of the evidence, the Court held that a new trial would be needed to determine the claim for aboriginal title; however, the Court went on to describe, among other things, the nature of aboriginal title, and the tests for establishing and justifying an infringement of aboriginal title, so as to give guidance to the lower courts. In *Delgamuukw*, Lamer CJC confirmed that aboriginal title flows from "the prior occupation of Canada by aboriginal peoples"[15] but also identified a second source for aboriginal rights; that is, "the relationship between common law and pre-existing systems of aboriginal law."[16]

Nature of aboriginal title

The Canadian law on the nature of aboriginal title has also evolved since *St. Catharine's Milling*. In that case, the Privy Council made its now-famous remark that "the tenure of the Indians was a personal and usufructuary right, dependent on the goodwill of the Sovereign,"[17] thus indicating that aboriginal title is not a right in the land itself. This is no longer the law today, however. In *Calder*, neither Judson nor Hall JJ. attempted to describe in any detail the exact nature of aboriginal title, but Judson J. suggested that ". . . it does not help one in the solution to this problem to call it a 'personal and usufructuary right'."[18] Justice Hall did set out three characteristics that are still part of the doctrine of aboriginal title: aboriginal title is not the same as a title in fee simple; aboriginal title exists in conjunction with the underlying "paramount" title of the Crown; and aboriginal title is inalienable except to the Crown.[19] In the 1983 Supreme Court of Canada decision of *R.* v. *Guerin*,[20] Dickson J. described aboriginal title as a "legal right to occupy and possess certain lands, the ultimate title to which is in the Crown,"[21] and as a *sui generis* title which cannot be transferred to anyone except the Crown.[22]

Delgamuukw contains an extensive discussion of the nature of aboriginal title. The claimants in *Delgamuukw* characterized aboriginal title as "tantamount to an inalienable fee simple"[23] while the provincial government characterized it as no more than a collection of aboriginal rights to engage in specific activities, or, at most "the right to exclusive use and occupation of land in order to engage in those activities which are aboriginal rights themselves. . . ."[24] According to Lamer CJC, who wrote the leading opinion:[25]

> The content of aboriginal title, in fact, lies somewhere between these positions. Aboriginal title is a right in the land, and as such, is more than the right to engage in specific activities which may be themselves aboriginal rights. Rather, it confers the right to use land for a variety of activities, not all of which need be aspects of practices, customs and traditions which are integral to distinctive cultures of aboriginal societies.[26]

Elsewhere in *Delgamuukw*, Lamer CJC reiterated that aboriginal title is a proprietary interest in land rather than a mere license to use and occupy, and that "the Privy Council's choice of terminology in *St. Catharine's Milling* is not particularly helpful to explain the various dimensions of aboriginal title."[27] In keeping with previous case law, *Delgamuukw* also held that aboriginal title exists in conjunction with an underlying fee simple in the Crown. Lamer CJC went on to describe aboriginal title as *sui generis* in a number of ways:

- The source of aboriginal title lies in the "prior occupation of Canada by aboriginal peoples"[28] and in "the relationship between common law and pre-existing systems of aboriginal law."[29]
- "[I]ts characteristics cannot be completely explained by reference either to the common law rules of real property or to the rules of property found in aboriginal legal systems. As with other aboriginal rights, it must be understood by references to both common law and aboriginal perspectives."[30]
- Aboriginal title is inalienable except to the Crown.[31]
- Aboriginal title is held communally.[32]
- The activities that may be carried out on aboriginal title land are not limited to traditional uses; however, such land cannot be used in ways that are "irreconcilable with the nature of the community's attachment to the land."[33] This was described by Lamer CJC as an "inherent" limit, which distinguishes aboriginal title from fee simple. On this point, Lamer CJC added that if a First Nation community wished to use its land in a way that is incompatible with this restriction, it must surrender the land to the Crown in exchange for valuable consideration.[34]

Proof of aboriginal title

The decision of Lamer CJC in *Delgamuukw* also discussed how the existence of aboriginal title is to be proved. The test was set out by Lamer CJC as follows:

> In order to establish a claim to aboriginal title, the aboriginal group asserting the claim must satisfy the following criteria:
>
> i the land must have been occupied prior to sovereignty,
> ii if present occupation is relied on as proof of occupation pre-sovereignty, there must be a continuity between present and pre-sovereignty occupation, and
> iii at sovereignty that occupation must have been exclusive.[35]

Extinguishment of aboriginal title

Once aboriginal title is established, the next issue becomes whether the claimants still hold aboriginal title to the area in question, or whether that title has been lawfully extinguished, either unilaterally by the Crown or bilaterally through a treaty or land claims agreement. The burden of proving extinguishment falls on the party disputing the claim to title.[36] The assumption that aboriginal title could be extinguished is clear in *St. Catharine's Milling*, with its description of aboriginal title as "dependent on the goodwill of the Sovereign."[37] In *Delgamuukw*, the Supreme Court held that since the advent of s. 35(1) of the *Constitution Act, 1982*, aboriginal title can no longer be unilaterally extinguished, although it can still be surrendered to the federal Crown in a land claims agreement. From the time of Confederation until 1982, only the federal Crown had the power to extinguish aboriginal title.[38] This flows from the fact that s. 91(24) of the *Constitution Act, 1867*[39] provides exclusive legislative jurisdiction with regard to "Indians and lands reserved for the Indians" to the federal government. In the pre-Confederation period, the authority to extinguish would have rested with the British Crown, whether that authority was exercised directly or delegated to colonial governments.

A clear and plain intent to extinguish must be shown[40] before a court will accept that a particular piece of pre-1982 federal legislation or particular pre-1982 actions of the federal government had the effect of unilaterally extinguishing aboriginal title. Specifically, Lamer CJC stated in *Delgamuukw* that "[w]hile the requirement of clear and plain intent does not, perhaps, require that the Crown 'use language which refers expressly to its extinguishment of aboriginal rights' . . . the standard is still quite high."[41]

Regulation and justification

The recognition of aboriginal rights in s. 35(1) of the *Constitution Act, 1982* has been interpreted by the Supreme Court to mean that government actions can only limit aboriginal rights when those actions can be justified. The analysis regarding which limitations will be allowed is a two-step process, involving tests for infringement and tests for justification.

If a First Nation wishes to argue that a particular governmental action infringes an aboriginal right, the First Nation must first show that the action interferes with that right. The court will consider such factors as whether the limitation is unreasonable, whether the limitation imposes "undue hardship" on the aboriginal people affected, and whether the limitation denies the holder of the right the "preferred means" of exercising the right.[42] Where an interference is established, a *prima facie* infringement exists and the onus shifts to the Crown to demonstrate that the infringement is justified. In an earlier case involving fishing rights, *R.* v. *Sparrow*, the Supreme Court of Canada set out a two-part test for justification: the legislative objective must be valid, and applying the legislation must be in keeping with the honor of the Crown. The Court also noted that the justification may place a "heavy burden" on the Crown.[43]

In *Delgamuukw*, Lamer CJC discussed the *Sparrow* test in the context of aboriginal title. Lamer CJC made it clear that provincial as well as federal legislation may be seen by the courts as justifiably infringing aboriginal title.[44] He also stated that in order for legislation which limits aboriginal title to be considered compelling and substantial, the legislative objective must be related to "the recognition of the prior occupation of North America by Aboriginal peoples or . . . the reconciliation of aboriginal prior occupation with the assertion of the sovereignty of the Crown."[45] It was suggested by Lamer CJC that:

> the range of legislative objectives that can justify the infringement of aboriginal title is fairly broad . . . In my opinion, the development of agriculture, forestry, mining, and hydroelectric power, the general economic development of the interior of British Columbia, protection of the environment or endangered species, the building of infrastructure and the settlement of foreign populations to support those aims, are the kinds of objectives that are consistent with this purpose, and in principle, can justify the infringement of aboriginal title. Whether a particular measure or government act can be explained by reference to one of those objectives, however, is ultimately a question of fact that will have to be examined on a case-by-case basis.[46]

The degree and kind of consultation required will be very dependent on the circumstances, and could range all the way from showing that the government had the "full consent of an aboriginal nation"[47] to merely showing that it had discussed decisions affecting aboriginal title land with the relevant First Nation. There are a range of possibilities between these two extremes, and Lamer CJC gave an example of an intermediate point on the spectrum: a requirement that First Nations be involved in making decisions regarding the land. It was also stated by Lamer CJC in *Delgamuukw* that because there is an "inescapable economic aspect" to aboriginal title, the Crown's "duty of honour and good faith" means "compensation will ordinarily be required where aboriginal title is infringed."[48]

Oceans management policy

Application of the doctrine of aboriginal title to the seabed

Having outlined the basic contours of the doctrine of aboriginal title, it is now possible to consider the applicability of that doctrine in the context of the seabed. Where the fee simple to subaquatic land lies with the Crown, it seems logical that this land, like terrestrial areas, could be subject to aboriginal title; in fact one of the *sui generis* aspects of aboriginal title is that it exists as a burden or limitation on the underlying Crown title. Therefore it must first be determined in Canada whether the Crown itself is seen as holding title to the land beneath its territorial waters.

Title below the low water mark

In the 1876 English case of *R.* v. *Keyn*, it was held the Court did not have jurisdiction over an offence committed on a foreign ship in waters within three miles of the English coast. According to the majority, the "dominion" of the common law

> extends no further than the limits of the realm. In the reign of Richard II the realm consisted of the land within the bodies of the counties. All beyond the low water mark was part of the high seas. At that period the three-mile radius had not been thought of. International law . . . cannot enlarge the area of our municipal law, nor could treaties with all the nations of the world have that effect. That can only be done by an Act of Parliament. As no such Act has been passed, it follows that what was out of the realm then is out of the realm now . . . Therefore, although, as between nation and nation, these waters are British territory, as being under the exclusive dominion of Great Britain, in judicial language they are out of the realm and any exercise of criminal jurisdiction over a foreign ship must in my judgment be authorized by an Act of Parliament.[49]

Although *Keyn* was not universally accepted at the time as having settled the issue of title to the seabed,[50] in *Reference Re: Ownership of Offshore Mineral Rights*, the Supreme Court of Canada accepted *R.* v. *Keyn* as representing the law in Canada. As set out in the passage quoted above, however, the Court in *Keyn* acknowledged that the Crown could, by way of legislation, extend its title beyond the low water mark.

Arguably, this is what Canada has done by way of the *Oceans Act*, which states:

> s.7 For greater certainty, the internal waters of Canada and the territorial sea of Canada form part of Canada.

> s.8(1) For greater certainty, in any area of the sea not within a province, the seabed and subsoil below the internal waters of Canada and the territorial sea of Canada are vested in Her Majesty in right of Canada.

Certainly this language of vesting is very much the language of real property law.

If sections 7 and 8 of the *Oceans Act* vest the radical fee simple in the federal Crown, this may permit the recognition of aboriginal title in the seabed. Aboriginal title has been described by the Supreme Court of Canada as flowing from continued use and occupation of the land since the time of British sovereignty, and from the relationship between aboriginal systems of law and the common law. On its face, nothing about this aspect of the doctrine of aboriginal title is inconsistent with a First Nation being able to claim aboriginal title in the seabed. The *Delgamuukw* decision did not, however, deal directly, or even indirectly, with the question of whether the test enunciated for establishing aboriginal title applies equally to submerged lands. In addition to evidentiary hurdles relating to physical possession, the courts will have to determine whether the notion of aboriginal title to the seabed can co-exist with the right of innocent passage at international law, or with established common law principles, such as the public rights of fishing and navigation.

Although the applicability of aboriginal title to the seabed has not been decided by a Canadian Court, the High Court of Australia has held that an aboriginal title claim could not be made to the seabed,[51] noting that recognition of exclusive aboriginal title in the seabed would conflict with the international right of innocent passage and domestic public rights of fishing and navigation.[52] More recently, however, the New Zealand Court of Appeal has acknowledged the potential for aboriginal title to exist in the foreshore and adjacent seabed.[53]

International law

One might argue that the *Oceans Act* cannot be read as vesting title in the Crown (therefore precluding an aboriginal title claim), because any notion of title in the seabed is irreconcilable with the fact that, in international law, foreign vessels have a right of innocent passage through the territorial seas of other nations. Thus in *Yarmirr*, the High Court of Australia held that "[a]s a matter of international law, the right of innocent passage is inconsistent with any international recognition of a right of ownership by the coastal state of territorial waters."[54] International law, however, has come to view the land beneath a coast's territorial waters as forming part of that state's territory in the same manner as does the land above water.[55]

If international law is willing to accept radical Crown title in the seabed as co-existing with the right of innocent passage, then presumably aboriginal title could also co-exist with the right of innocent passage, given that aboriginal title encompasses a more limited set of rights than does Crown title. True, in *Delgamuukw*, Lamer CJC describes aboriginal title as including the right to exclude others; however, this characterization of aboriginal title is not necessarily irreconcilable with the right of innocent passage. The fact that a form of title is seen as encompassing a right to exclude others cannot be automatically incompatible with the international right of innocent passage, since Crown title itself carries with it notions of exclusivity. A key component of both state and private ownership is the power to exclude others except where that power is otherwise limited by law.[56] Where this right to exclude is curtailed with regard to the underlying radical title of the Crown, it seems logical that any aboriginal title that exists as a burden on that radical title would be similarly curtailed.

When the international law perspective is taken in conjunction with the wording of sections 7 and 8 of Canada's *Oceans Act*, there seems to be an argument for saying that both international and Canadian law recognize the federal Crown as holding title to the seabed beneath Canada's territorial waters, although this title is subject to the international right of innocent passage. If that is so, then we are one step closer to saying that the doctrine of aboriginal title (subject to the right of innocent passage) could be applied to the seabed.

While the existence of underlying Crown title and the possibility of reconciling ownership with the international right of innocent passage are necessary prerequisites, these alone do not, however, allow for the conclusion that aboriginal title could exist in the seabed. There is a further argument, which also relates to Lamer CJC's description of aboriginal title as "exclusive," that must be explored; that is, can exclusivity be reconciled with common law rights of navigation and fishing?

Common law rights of public navigation and fishing

Since the time of the Magna Carta, English common law has recognized public rights of fishing and navigation in tidal waters. Further, in *Gladstone* the Supreme Court of Canada stated, with regard to aboriginal fishing rights, that:

> the aboriginal rights recognized and affirmed by s. 35(1) exist within a legal context in which, since the time of the Magna Carta, there has been a common law right to fish in tidal waters that can only be abrogated by the competent legislation . . . While the elevation of common law rights to constitutional status obviously has an impact on the public common law rights to fish in tidal waters, it was surely not intended that, by the enactment of s. 35(1), those common law rights would be extinguished in cases where an aboriginal right to harvest fish commercially existed . . . (I)t was not contemplated by *Sparrow* that the recognition and affirmation of aboriginal rights should result in the common law right of public access in the fishery ceasing to exist with respect to all those fisheries in respect of which exist an aboriginal right to sell fish commercially. As a common law, not constitutional, right, the right of public access to the fishery must clearly be second in priority to aboriginal rights; however, the recognition of aboriginal rights should not be interpreted as extinguishing the right of public access to the fishery.[57]

Does this mean that the existence of public rights of fishing and navigation preclude the possibility that aboriginal title could be established in the seabed? Arguably not, on both constitutional and common law grounds, which are discussed below. It may mean that the public rights would to some degree limit the rights which would otherwise be encompassed by aboriginal title, but that is substantially different than concluding that there can be no aboriginal title in areas where the common law would recognize rights of fishing and navigation.

The constitutional law argument is as follows: Lamer CJC's comments must be read narrowly, as simply meaning that the constitutionally recognized right of aboriginal fishing is not so extensive as to extinguish completely the common law right to fish. His comments cannot be interpreted as meaning that the common law right will overcome the constitutionally recognized aboriginal right such that the aboriginal right cannot exist, which would be the effect of saying that aboriginal title cannot exist in the seabed because of public rights of fishing or navigation. It was acknowledged by Lamer CJC in *Gladstone* that common law rights must be "second in priority" to constitutional rights. Common law rights can be curtailed or extinguished by the legislature, which is in turn curtailed by the constitution; thus, it is doubtful whether a court would find that common law rights of navigation and fishing could so completely trump constitutionally recognized rights that the doctrine of aboriginal title would be inapplicable to the seabed.

Even leaving aside the constitution and considering only the common law, it seems clear that the existence of title in submerged lands is not negated by public rights of fishing or navigation in the water over those lands. The common law did not resolve the tension between title and public rights by holding that there can be no ownership of the land beneath public waters, but by holding that where there is conflict between the rights associated with title and the public rights, the latter are paramount.[58] At English common law, land beneath tidal waters is held by the Crown, unless the Crown has granted that title to another.[59] Thus a privately held fee simple or even the radical title held by the Crown can co-exist with (but would be subject to) the common law rights of fishing and navigation. Presumably so too could aboriginal title co-exist with public rights.

Building on Lamer CJC's comments in *Gladstone* regarding aboriginal fishing rights, if aboriginal title were found to exist in an area where the common law would recognize public rights of fishing or navigation, the court might have to configure the rights associated with aboriginal title in such as way that the common law public rights were not completely extinguished, but that also recognized the priority of constitutional rights over those grounded only in the common law.

Conclusion

Thus, on the question of whether aboriginal title could be established in the seabed beneath Canada's territorial sea, the following tentative conclusions can be drawn:

- Given that aboriginal title is a "burden" on the underlying Crown title, aboriginal title cannot exist in the absence of that underlying title.

- By way of sections 7 and 8 of the *Oceans Act*, Canada has extended its territory to include the seabed beneath its territorial waters so that the federal Crown holds title to that portion of the seabed.
- Crown title in the seabed co-exists with and is subject to the international right of innocent passage.
- Aboriginal title cannot be more extensive than the Crown title on which it is overlaid; therefore, aboriginal title in the seabed could also co-exist with, but would also be subject to, the international right of innocent passage.
- At common law, the rights of public fishing and navigation are not seen as negating ownership of land beneath tidal waters.
- Therefore even at common law, these public rights should not be seen as negating the possibility of aboriginal title.
- However, at common law, the rights associated with ownership of the land beneath tidal waters are subservient to the public rights of fishing and navigation.
- Given the comments by Lamer CJC in *Gladstone*, it seems unlikely that aboriginal title, once established, would completely extinguish the public rights recognized by the common law; however, it is not clear that the balancing between a constitutionally recognized right and a common law right should necessarily bring the same outcome as the balancing of a common law right against a common law right. In fact, it seems unlikely that a constitutionally protected right should be wholly subservient to a common law right.

Obviously, there would be crucial evidentiary issues involved in any aboriginal title claim to the seabed. Leaving those aside however, and looking solely at the doctrine of aboriginal title, there seems to be a strong argument that the doctrine could apply to the seabed beneath Canada's territorial waters. Such title, if recognized, would be subject to the international right of innocent passage and might also be limited in some way by public rights of fishing or navigation.

Impact on oceans policy

The *Oceans Act* already mandates collaboration with certain aboriginal stakeholders and provides an opportunity for aboriginal participation on advisory and management bodies. The question which policy makers would have to consider is whether recognition of aboriginal title in the seabed would require more accommodation than is already provided in the *Oceans Act*. Canada's oceans policy involves three broad themes: declaring Canada's rights with regard to its territorial waters, the contiguous zone and the exclusive economic zone; providing for the development of a national oceans strategy based on the principles of sustainable development, integrated management and the precautionary principle; and consolidating various aspects of federal responsibility regarding oceans. Looking at each of these in turn, and considering whether there might be conflict between federal activities based on the themes of the *Oceans Act* and rights arising from aboriginal title:

- The declaration that the seabed beneath Canada's territorial waters is vested in the federal Crown strengthens rather than diminishes the possibility that the doctrine of aboriginal title could be applied to the seabed, since such title always exists in conjunction with underlying Crown title and Crown sovereignty.
- With regard to Canada's oceans strategy, conflict with aboriginal title is less likely to be identified at the level of generality with which the principles of that strategy is articulated in the Act. If aboriginal title were recognized in a portion of the seabed, the issue would be whether the implementation of a particular aspect of the oceans strategy infringed rights flowing from that aboriginal title. For instance, if a First Nation intended to use its aboriginal title area for hunting, fishing and gathering, would the federal government be able to establish a marine protected area within the area held by aboriginal title and restrict the First Nation in its hunting, fishing and gathering activities? According to *Sparrow*, this would constitute a *prima facie* violation of aboriginal title, but both *Sparrow* and *Delgamuukw* make it clear that such an infringement might, depending on the circumstances, be upheld as valid.
- Similarly, potential conflict between aboriginal title and other aspects of the exercise of federal jurisdiction in the *Oceans Act* would require a case-by-case analysis to determine whether federal legislation or activities infringed aboriginal rights and whether such infringement could be justified.

Even allowing for the fact that some limits on aboriginal title would likely be accepted as justified, recognition of aboriginal title in the seabed would significantly affect both how oceans policy is developed in the future and how it is applied. In the development and application of such policy, government will have to school itself to ask the same kinds of questions as it should now be asking with relation to terrestrial land, regarding the possible existence of aboriginal title in the relevant area, the possibility that legislation or policy initiatives might infringe such title, and whether or not such infringements would be seen as justified. Thus, what will be required is an awareness of the law on aboriginal title and a nuanced and contextual analysis of the potential interplay between aboriginal title and the implementation of federal oceans policy.

Acknowledgments

The author acknowledges the assistance of Cheryl Webb, Senior Counsel for Fisheries and Oceans Canada, for her helpful and perceptive comments on this chapter. The research assistance of the following students is also acknowledged: Russ Weinger (LLB, Dalhousie, 2003); David Steeves (LLB, Dalhousie, 2003) and Anne Tardiff (LLB, Dalhousie, 2005).

Notes

1 S.C. 1996, C-31.
2 *Oceans Act*, note 1, s. 29.
3 *Constitution Act 1982*, s. 35, being Schedule B to the *Canada Act 1982 (U.K.)*, 1982, ch. 11.
4 The Act refers to "affected aboriginal organizations" and "entities established under land claims agreements." See for example *Oceans Act*, ss. 29, 31, 32 and 33.
5 *Oceans Act*, note 1, s. 33(2).
6 Other aboriginal rights besides aboriginal title might also be relevant to policy on ocean management. For instance, a First Nation might argue that it holds an unextinguished aboriginal right of fishing, hunting or gathering in a sea area. Furthermore, First Nations who have signed treaties (whether historic or modern day land claims agreements) might argue that the terms of a particular treaty give them rights that would have to be respected by oceans management policy. The focus of this chapter is on aboriginal title claims; rights accruing from treaties or from aboriginal rights other than title are not considered here.
7 Throughout this chapter, when speaking of title, reference is made to the seabed, rather than to the marine resource as a whole.
8 *Delgamuukw* v. *British Columbia* [1997] 3 S.C.R. 1010 [*Delgamuukw*]. See also *R.* v. *Adams*, [1996] 3 S.C.R. 101 [*Adams*], and Brian Slattery "Making sense of aboriginal and treaty rights" (2000) 79(2) *Canadian Bar Review* 196–224. Slattery suggests that cultural rights (for instance, the right to speak one's language, or a community's right to be associated with certain traditional stories or songs) would be an example of aboriginal rights not associated with a particular land area.
9 (1888) 14 AC 46 [*St. Catharine's Milling*]. It should be noted that this case did not involve an aboriginal title claim, and no First Nation was party to the litigation.
10 R.S.C. 1985, Appendix II, No. 1.
11 [1973] S.C.R. 313 [*Calder*]. Although the existence of inherent aboriginal title was recognized in *Calder*, the Nishga'a lost the case. The six justices who dealt with the substantive issues split 3-3 on the question of whether the Nishga'a's title had been extinguished, and Mr. Justice Pigeon dismissed the appeal on other grounds unrelated to aboriginal title. Thus only three of the seven justices were willing to find existing aboriginal title in the area claimed. However, the recognition by the Supreme Court of Canada of inherent aboriginal title, which was seen to continue in existence until extinguished by the Crown, was sufficient to cause the federal government to establish the Office of Comprehensive Land Claims.
12 Ibid., at 317.
13 Ibid., at 328.
14 [1997] 3 S.C.R. 1010.
15 Ibid., at para. 114.
16 Ibid. See discussion below for comments regarding this second source for aboriginal rights.
17 *St. Catharine's Milling*, note 9 at 47.

18 *Calder*, note 11 at 328.
19 Ibid., at 352 where Hall J. stated:

> This is not a claim for title in fee but is in the nature of an equitable title or interest . . . a usufructuary right and a right to occupy the lands and to enjoy the fruits of the soil, of the forests and the rivers and the streams which does not in any way deny the Crown's paramount title as it is recognized by the law of nations . . . The Nishga'a do not claim to be able to sell or alienate their right except to the Crown.

20 [1984] 2 S.C.R. 335 [*Guerin*]. The issue before the Supreme Court of Canada in *Guerin* was whether the *Indian Act*, R.S.C. 1952, c.149, imposed a fiduciary duty on the federal government in dealing with surplus reserve land that had been surrendered to the Crown for the purpose of leasing it. Although *Guerin* did not involve a claim of aboriginal title, Dickson's comments are on point, given his statement: "[I]t does not matter, in my opinion, that the present case is concerned with the interest of an Indian Band in a reserve rather than unrecognized aboriginal title in traditional tribal lands. The Indian interest in the land is the same in both cases." [1984] 2 S.C.R. 379.
21 Ibid., at 382.
22 *Guerin*, note 20 at 336, where Dickson J. stated that this limit on alienability places a fiduciary duty on the Crown in dealing with surrendered lands.
23 *Delgamuukw*, note 8 at para. 110. Since one of the key characteristics of a title in fee simple is its alienability, by describing aboriginal title an inalienable fee simple, the claimants must have been using the analogy to argue that aboriginal title possesses other attributes of the fee simple, such as the fact that there is no limit on the use of fee simple land inherent in the title itself (although of course there are significant limitations on use found at common law and in statutes).
24 Ibid.
25 Decisions were also written by La Forest J. and L'Heureux-Dubé J., both of whom agreed with the outcome. L'Heureux-Dubé J.'s judgment was very brief; La Forest J.'s was somewhat more lengthy. The chief difference between Lamer CJC and La Forest J. seems to be as to the scope of aboriginal title, and in particular, the uses to which aboriginal title land could be put, with La Forest J. "taking a somewhat more restrictive approach to the content of aboriginal title." K. McNeil *Emerging Justice? Essays on Indigenous Rights in Canada and Australia* (Native Law Centre, Saskatoon: 2001) 65.
26 *Delgamuukw*, note 8 at para. 111.
27 Ibid., at para. 112.
28 Ibid., at para. 114. This is the first time that the Supreme Court of Canada explicitly identified this as a source of aboriginal title. Previous Supreme Court of Canada jurisprudence referred to the occupation of the land, prior to the British claim of sovereignty, as the principal source of aboriginal title, with some cases (such as *Calder*) speaking of the Royal Proclamation 1763 as either a secondary source or simply as confirmation of the inherent title flowing from occupation. The comments in *Delgamuukw* would seem to build on discussion in *R.* v. *Van der Pet* [1996] 2 S.C.R. 507 regarding the function of the *Constitution Act 1982*, s. 35(1).
29 Ibid.
30 Ibid., at para. 112.
31 Ibid., at para. 113.
32 Ibid., at para. 115.
33 Later, Lamer CJC likened this limitation to the doctrine of equitable waste, which prevents the holder of a life estate from carrying out "wanton or extravagant acts of destruction;" for commentary on this restriction, with different

commentators giving varying readings to the scope of Lamer CJC's limitation, see Nigel Bankes "*Delgamuukw*, division of powers, provincial land and resource laws: Some implications for provincial resource rights" (1998) 32 *University of British Columbia Law Review* 317–51; Richard H. Bartlett "The content of aboriginal title and equality before the law" (1998) 61 *Saskatchewan Law Review* 377–91; Brian J. Burke "Left out in the cold: The problem with aboriginal title under section 35(1) of the *Constitution Act, 1982* for historically nomadic Aboriginal peoples" (Spring 2000) 38 *Osgoode Hall Law Journal* 1–37; William F. Flanagan "Piercing the veil of real property law: *Delgamuukw* v. *British Columbia*" (1998) 24 *Queen's Law Journal* 279–326.

34 *Delgamuukw*, note 8 at para. 131.
35 Ibid., at para. 143; see also *R.* v. *Marshall* [2002] NSJ No. 98.
36 This is made explicit in *Calder*, note 11; *Delgamuukw*, note 8; and *R.* v. *Sparrow* [1990] 1 S.C.R. 1075 [*Sparrow*].
37 *St. Catharine's Milling*, note 9 at 47.
38 *Delgamuukw*, note 8 at para. 173.
39 *Constitution Act 1867* (U.K.), 30 & 31 Vict. C. 3, reprinted in R.S.C. 1985, App. II, No. 5.
40 *Delgamuukw*, note 8 at para. 180.
41 Ibid.
42 In *R.* v. *Nikal* [1996] 1 S.C.R. 1013, the Supreme Court of Canada elaborated on the concept of infringement, holding that the mere requirement to have a license to fish is not an infringement of an aboriginal right, although the conditions attached to a license may constitute an infringement. The Court emphasized the need to have one central authority regulating access to the fishery and stated that "a license is the essential first step in the preservation and management of this fragile resource" [1996] 1 S.C.R. 1013 at para. 102. The Court further stated that aboriginal rights do not exist in a vacuum and that there must be a balanced approach to limitations placed on the exercise of these rights.
43 *Sparrow*, note 36 at 186.
44 Ibid. This aspect of the decision has been criticized, see Kent McNeil "Aboriginal title and the division of powers: Rethinking federal and provincial jurisdiction" (1998) 61 *Saskatchewan Law Review* 431 at 435.
45 *Delgamuukw*, note 8 at para. 161, quoting *Gladstone* [1996] 2 S.C.R. 723 [*Gladstone*].
46 *Delgamuukw*, ibid., at para. 165. For commentary on this list of objectives, see Kent McNeil *Defining Aboriginal Title in the 90's: Has the Supreme Court Finally Got It Right?* (Robarts Centre for Canadian Studies, Toronto: 1998) 26; Lisa Dufraimont "From regulation to recolonization" (2000) 58 *University of Toronto Faculty of Law Review* 1.
47 *Delgamuukw*, ibid., at para. 168.
48 Ibid., at para. 169.
49 *Reg. v. Keyn* (1876), 2 Ex. D. 63 [*Keyn*], as quoted in *Reference Re: Ownership of Offshore Mineral Rights (British Columbia)* [1967] S.C.R. 792.
50 See *The Secretary of State for Indian Council v. Sri Raja Chellikani Rama Bao and Others* (1916) 32 *Times Law Reports* 652.
51 *Commonwealth of Australia* v. *Yarmirr & Ors; Yarmirr & Ors* v. *Northern Territory of Australia & Ors* [2001] HCA 56; 184 ALR 113 [*Yarmirr*].
52 The High Court of Australia in *Yarmirr*, ibid., endorsing the decision at first instance of Olney J in *Yarmirr* v. *Northern Territory (No. 2)* (1998) 156 ALR 370 did however recognize the claimants as having a right to fish, hunt and gather in the claimed area, to travel through it, and to "visit and protect places within the claimed area which are of cultural or spiritual importance [and] to safeguard

the cultural and spiritual knowledge of the common law holders" [2001] HCA 56 at para. 2.

53 See *Ngati Apa* v. *Ki Te Tau Ihu Trust* [2003] NZCA 117.

54 *Yarmirr*, note 51 at para. 57.

55 See R.R. Churchill and A.V. Lowe *The Law of the Sea* 3rd edition (Manchester University Press, Manchester: 1999) at 76.

56 See Bruce Ziff *Principles of Property Law* 3rd edition (Carswell, Scarborough: 2000) 7; F. Cohen "Dialogue on private property" (1954) 9 *Rutgers Law Review* 357 at 379.

57 Gladstone, note 45 at para. 67.

58 Gerard V. LaForest *Water Resources Study of the Atlantic Provinces* (Atlantic Development Board, Ottawa: 1968) at 14 states that "the public right of navigation is a paramount right; whenever it conflicts with a right of the owner of the bed or of a riparian owner, it will prevail." The same would appear to be true for the public fishery.

59 Ibid., at 27.

12 Canada's seas and her First Nations

A colonial paradigm revisited

Russ Jones

Introduction

Canada's claims to the seas surrounding her – the Pacific, Atlantic and Arctic oceans – have expanded to include land and resources from three to 200 nautical miles offshore over the short span of the past 40 years. But First Nations[1] occupied and utilized Canada's shorelines and ocean spaces long before Canada's most recent colonization. This chapter illustrates historic relationships of several First Nations in Canada to ocean spaces and discusses Canada's current approach to ocean management and sea title from a First Nation perspective. Canada's oceans policies are cautious about acknowledging specific aboriginal or treaty rights or prescribing a role for First Nations in ocean management and planning. This lack of direction is likely to lead to delays in implementing the *Oceans Act, 1996*[2] or uncertainty about the validity of resulting management plans if the government proceeds without substantive First Nations' input.

The aboriginal peoples of Canada, comprising Indian, Inuit and Métis, are made up of many individual First Nations with their own distinct history and culture. European advances and colonization in what is now known as Canada began on Atlantic shores in the early 1500s, the Pacific in the late 1700s, and the Arctic in the mid-1800s. First Nations often played a pivotal role in conflicts between Old and New World powers in the early settlement era. Britain eventually colonized what is now known as Canada, and relationships with many First Nations in eastern Canada and the Prairies were defined by means of treaties.[3] However, few treaties were struck on the Pacific coast due to disputes with British authorities in the late 1800s over who should provide treaty monies and policies of British Columbia leaders to deny aboriginal title.[4] Except for one disputed treaty in the MacKenzie River Valley, no treaties were signed in the Yukon and Northwest Territories up until the 1990s.[5] Canada's foundation myths include "fairness" to First Nation peoples and that land pre-empted for settlement was not being utilized by First Nations.[6] Colonizers in the end achieved their goals of preventing First Nations from impeding settlement or standing in the way of economic development. Dispossession of land and resources may have been less violent than other colonial episodes because First Nations may have believed that they could work with the newcomers and find ways to share the land and its resources.[7] Instead, First Nations have been effectively marginalized in one of the most prosperous countries in the world.

Processes are currently underway on all three coasts to define modern relationships between Canada and First Nations where these do not exist or are lacking,[8] with elevated attention to ocean issues in modern negotiations.[9] Yet, despite encouragement by courts to negotiate issues of aboriginal title and rights, the same pattern of dispossession is now being repeated with Canada's oceans.

Canada's approach to ocean management has been to pass legislation first and then develop policies and plans. The *Oceans Act* authorizes the Minister of Fisheries and Oceans to lead the development of a national oceans management strategy. Implementation of this legislation is in the early stages and a largely unresolved issue is how First Nations' ocean interests will be accommodated in planning and development. The legislation is likely sufficiently broad to encompass a variety of outcomes, but there is little clear direction set on how to proceed with difficult issues involving First Nations where issues are not already explicitly addressed in existing treaties.

Examples of First Nations' ocean interests

Canada's First Nations are diverse, with a wide range of ocean interests and practices adapted to specific environments. This section provides examples of First Nations' ocean interests drawn mainly from the Pacific Northwest coast. First Nations in this region were generally ocean-going peoples who utilized a wide variety of ocean resources and developed sophisticated technology, including canoes for hunting and fishing and general transportation. There were a variety of styles and sizes of canoe depending on purposes and traditions, ranging from less than six meters up to 20 meters in length.

As an example, attention will be given to First Nation offshore utilization of halibut (*Hippoglotus stenolepsis*) and blackcod (also known as sablefish or *Anoplopona fimbria*) on the Pacific coast. These fish are abundant on offshore fishing grounds at certain times of the year. First Nation peoples would have learned this through experience and passed this knowledge on from generation to generation.[10] First Nations had the means and technology to travel offshore to catch these fish and utilized these fishing grounds long before the arrival of Europeans.

The Nuu-chah-nulth and Ditidaht First Nations from the west coast of Vancouver Island in British Columbia fished for halibut at banks ten to 25 miles offshore.[11] To the north, the Haida from Haida Gwaii,[12] or the Queen Charlotte Islands, similarly used landmarks to locate fishing grounds that they called *gyu*.[13] Another fish that was caught offshore was *skil* or blackcod. Blackcod are commonly found at depths of 1,000 feet or more.[14] The Makah and Haida fished blackcod using fishing gear consisting of kelp lines with

20 to 100 bent-wood hooks attached, the latter made from steamed hemlock knots. The fishing line was said to be equal in strength to the best hemp cod-lines.[15] There is also evidence of proprietary interests in offshore fishing banks. In 1868, Sproat wrote, "The fishing tribes on both sides of the Straits of Fuca would drive away any other tribes which had not been accustomed to fish on the halibut banks."[16] Niblack, in 1890, observed that stretches of shorelines belonged to Haida lineages. Further, "Nor is this boundary confined to the strip of coast, but extends well out to sea, carrying with it the right to shoot seals and gather birds' eggs on outlying rocks, hunt sea otter, and to fish on well-known halibut or cod banks."[17] Similarly, in 1905, Swanton wrote of the Haida, "The halibut fishing grounds were all named and were owned by certain families."[18]

Halibut and blackcod had economic importance for tribes that fished offshore. Sproat notes that the Nuu-chah-nulth traded dried halibut, herring and cedar bark baskets to the Salish tribes of southern Vancouver Island.[19] Swan observed that the Makah traded dried halibut as well as whale oil and blubber to the Nuu-chah-nulth.[20] The Haida also traded halibut with the Tsimshian.[21] Blackcod oil was a valuable trade item for the Haida, not only with the mainland First Nations, but with Haida from other areas of Haida Gwaii who did not have access to blackcod.[22] Despite historic trade there is presently only minimal First Nation participation in commercial halibut and blackcod fisheries due in large part to government policies such as limited entry licensing.[23]

First Nations connections with the landscape and seascape were spiritual as well as physical. Haida beliefs about their origin and relationship to the natural world give an intrinsic spiritual value to the natural world and all its elements, including fish, sea mammals, birds, land animals, creeks and places.[24] First Nations in Canada have been marginalized in most economic pursuits, including the fisheries, and are struggling to regain access. At the same time, First Nation interests in ocean spaces have broadened from traditional activities such as fishing and hunting to a variety of modern economic and environmental ocean interests. Many new developments in the ocean sector such as oil and gas, aquaculture and marine tourism are seen as threatening traditional pursuits and encroaching on ocean spaces that are in the traditional territories of First Nations.

Canada's emerging ocean management framework and First Nations

Sea title – a new legal frontier

First Nations have a clear sense of their ownership of ocean spaces but the recognition of rights and title to ocean spaces remains a source of debate in Canadian law.[25] As yet there are no defining cases in Canada about sea title and, as a result, government policy has been exceedingly cautious. For First Nations, the existence of sea title is as obvious as aboriginal title to dry land[26] as there are no dividing lines when you step out your front door onto the beach or when you board a canoe to travel to a nearby fishing ground. The Government of Canada on the other hand sees the ocean as a common property resource and a frontier that must be developed and managed for the prosperity of all Canadians. These opposing views create a situation that encourages confrontation and litigation unless First Nations are willing to accept the minimal role offered in oceans stewardship through proposed ocean development processes.

Haida, Tsimshian and Nuu-chah-nulth First Nations recently initiated court proceedings concerning their sea title and fishing rights, but these claims are still at an early stage. The Haida case is a broad land and sea claim to the lands and inland and offshore waters surrounding Haida Gwaii. The statement of claim, filed in November 2002, claims infringement of the aboriginal rights and title of the Haida Nation and seeks a declaration of Haida aboriginal title and rights to Haida Gwaii, damages, and a quashing of licenses, leases, permits and tenures over Haida Gwaii. This would include foreshore leases, oil and gas tenures, and fishing licenses. The Haida suit was followed in September 2003 by an offer of 200,000 hectares of land by the British Columbia government mainly comprising of areas declared by the Haida as protected areas from logging.[27] There was no offer from the federal government regarding the offshore component of the claim. The Haida rejected the offer.[28] The Lax kw'alaams Indian Band and nine allied Tsimshian tribes filed a statement of claim in January 2003 that was specific to fishing rights and title to fisheries resource harvesting sites. Harvesting sites would include the mouth of the Skeena River, a prime commercial fishing area for salmon. The plaintiffs are seeking a declaration of Tsimshian aboriginal rights to harvest fish for commercial purposes and aboriginal title to resource harvesting sites, that Canada and British Columbia have infringed those rights, and damages. In June 2003, ten Nuu-chah-nulth First Nations filed a writ of summons claiming aboriginal rights and title to a large portion of the land and offshore waters on the west coast of Vancouver Island from Kuyoquot to Port Renfrew. The writ claims aboriginal rights to harvest all species of fish, the right to sell these fish on a commercial scale, aboriginal title to the fishing territory and fishing sites, and infringement of these rights by Canada and British Columbia. It is anticipated that these claims may take some time to settle.

Legislative provisions relating to First Nations

The main provision relating to First Nations in the *Oceans Act* is a non-derogation clause found in s. 2(1). It provides:

> For greater certainty, nothing in this Act shall be construed so as to abrogate or derogate from any existing aboriginal or treaty rights of the First Nations of Canada under Section 35 of the *Constitution Act, 1982*.

Similar clauses are repeated in other legislation.

There are several provisions for involving First Nations, among others, in ocean management activities. The *Oceans Act* refers to collaborating, cooperating and consulting with affected First Nations' organizations and bodies established under land claims agreements. In addition, the *National Marine Conservation Areas Act, 2002* adds aboriginal governments to the list and has several provisions not included in the *Oceans Act*. One provision allows the Minister of Canadian Heritage to make management agreements with other entities including First Nations. Another provision provides for the designation of marine conservation area reserves where:

> (An) area or a portion of an area proposed for a marine conservation area is subject to a claim in respect of aboriginal rights that has been accepted for negotiation by the Government of Canada.[29]

This designation allows for possible changes to be made to reserve boundaries once land claims agreements (or treaties) are negotiated.[30]

The *Migratory Bird Conventions Act, 1994* ratifies a 1995 Protocol[31] to the 1916 Convention between the United Kingdom and the United States for the Protection of Migratory Birds in Canada and the USA.[32] The protocol adds "use of aboriginal and indigenous knowledge, institutions and practices" as a conservation principle and recognizes regulatory and conservation regimes for harvesting migratory birds and eggs in Canada as defined in relevant treaties, land claims agreements and co-management agreements.

Assessment

The foregoing legislation enables Canada to proceed with development of ocean resources in a seemingly orderly manner that may involve First Nations, at least where there may be provisions in existing treaties or agreements. There was little consultation with First Nations about the foregoing legislation.[33] While provisions relating to First Nations do no harm to aboriginal and treaty rights, they are more likely enacted in an effort to prevent legislation from being struck down, rather than being proactive about addressing First Nations' ocean interests.

Policy provisions relating to First Nations

Policy documents and recent land claims agreements provide further insight into Canada's approach to addressing First Nation marine issues. As mandated in the *Oceans Act*, the federal government released *Canada's Oceans Strategy: Our Oceans, Our Future* (*Oceans Strategy*) in August 2002 that "sets out the policy direction for ocean management in Canada."[34] The *Oceans Strategy* repeats commitments not to override aboriginal and treaty rights and to provide opportunities for First Nation involvement in integrated management planning and ocean management decision making. The *Oceans Strategy* identifies "collaboration" as the governance model for integrated management. But it indicates that this may extend to "co-management" in cases such as settled land claims agreements where structures established through land claims agreements might be applied.[35] Two of the three examples of integrated management initiatives cited in the *Oceans Strategy* involve First Nations. These are the Beaufort Sea Integrated Management Planning Initiative (BSIMPI) in the Arctic Ocean that involves committees established under the 1984 Inuvaliat Final Agreement and a marine planning process on the British Columbia central coast.[36]

The 1994 Parks Canada *National Marine Conservation Areas Policy* predates the *National Marine Conservation Areas Act* but provides little policy direction different from the legislation.[37] While Parks Canada's means may not be explicit, their ends are described in a system plan that sets a goal of establishing National Marine Conservation Areas in 29 distinct marine regions of Canada.[38]

Assessment

A major deficiency of existing oceans legislation and policy is that there is little or no policy guidance on the role of First Nations or mechanisms for involving them in oceans planning and management except where treaties and specific processes are already in place. Parks Canada's approach to creation of protected areas signals some flexibility in dealing with First Nations through temporary designations of marine conservation areas in advance of treaties with the potential to negotiate management agreements. However, it is too early to say how satisfied First Nations will be with that approach. Another difficulty with implementation is that until June 2005, when the Oceans Action Plan Phase 1 was announced, there was little in the way of new government resources identified for implementing the *Oceans Act*. This has affected the ability of government to consult with and effectively involve First Nations in ocean planning initiatives, although this may improve in future.

Marine protected areas

Marine protected areas (MPAs) can be created under the *Oceans Act* and related legislation.[39] A key lesson learned from MPAs worldwide is that "Local people need to be deeply involved from the earliest possible stage in any MPA that is to succeed."[40] Indigenous peoples are recognized as having a special role to play in protected area management because of their unique relationship to their environment. In Canada, First Nations may exercise aboriginal and treaty rights in protected areas unless prohibited for valid reasons such as conservation.[41] Thus a major motive for government seeking agreements with First Nations about protected areas is to define and limit aboriginal and treaty rights. From Canada's perspective, agreements provide certainty about First

Nation activities in protected areas and benefit First Nations by giving them a say in protected area development and management.

The *Oceans Act* gives the Minister of Fisheries and Oceans power to establish MPAs without any special limitations or requirements to consult or collaborate with others. In contrast, provisions of the *National Marine Conservation Areas Act* and the Parks Canada *National Marine Conservation Areas Policy* allow for agreements with First Nations and interim designations of protected areas where treaties are not yet in place. Parks Canada has entered into a variety of co-management agreements with First Nations regarding terrestrial parks. One example is the Gwaii Haanas National Park Reserve – Haida Heritage Site that is co-managed by an Archipelago Management Board comprised of two Haida and two Parks Canada representatives.[42] Despite the lack of a clear policy regarding First Nations, the Department of Fisheries and Oceans (DFO) seems to be following a similar approach with MPAs on the Pacific coast. These MPAs provide opportunities for First Nations and Canada to work together on common issues, but if rushed could lead to confrontations or court challenges resulting in the creation of MPAs that are ineffective.

Self-government

Self-government has been a prominent First Nation issue in recent years that has yet to be realized despite political commitments. In a 1995 policy statement, the Government of Canada recognized the inherent right of aboriginal self-government as an existing right within the *Constitution Act, 1982*, s. 35. Canadian examples of modern self-government include recognition of the Nisga'a government in British Columbia and establishment of Nunavut Territory. Self-government was directly implemented for the first time in the 1998 Nisga'a Final Agreement that came into effect in 2000.[43] Nunavut, which is larger than any province in Canada, consists of Inuit and non-Inuit lands and has its own elected government. Although the Nunavut Lands Claims Agreement and federal legislation established a new self-governing territory, Nunavut has a public rather than an Inuit-exclusive government structure, which does not benefit from constitutional protection.[44]

Traditional knowledge

Both the *Oceans Strategy* and the *National Marine Conservation Areas Act* acknowledge the value of First Nations' knowledge although traditional knowledge is not necessarily applied exclusively to First Nations in the latter. Despite recent recognition of the benefit of First Nations traditional knowledge, there are few examples of successful melding of traditional knowledge and science in natural resource management. Establishment of co-management committees through treaties provides a venue for exchange of information and makes it more likely that traditional knowledge will be used in decision making. In the Arctic, at least, traditional knowledge has come into play in establishing management plans for wildlife. In much of Canada, century-old policies of assimilation of First Nations have been far too successful. Fewer and fewer First Nation languages are being spoken in the home. And as a result of a variety of social and economically-driven changes, young First Nations peoples are no longer able to spend as much time on the land and waters as past generations. Other factors are the political nature of information and that there are limited resources available to assist with recovering or documenting First Nations traditional knowledge.

Non-commercial fishing and hunting

First Nation fishing and hunting rights have been a source of controversy in Canadian society. But it is now understood that existing hunting, fishing and trapping rights are constitutionally protected in s. 35(1) of the *Constitution Act, 1982* and, as such, prevail over federal, provincial and territorial legislation unless that legislation is justifiable.[45] Current policies also recognize the priority of First Nations fishing for food, social and ceremonial purposes over commercial and recreational users.

These and other requirements are incorporated into modern land claims agreements through explicit allocations and the establishment of fish and wildlife co-management boards. For instance, the Nisga'a Final Agreement[46] establishes a Joint Fisheries Management Committee and a Wildlife Committee to provide advice to Canada and the Nisga'a (through the Nisga'a Lisims Government) on fisheries and wildlife management. The committees generally aim to make consensus recommendations which helps to create a climate of cooperation in decision making. Similarly, in the Arctic, the Nunavut Land Claims Agreement establishes a Nunavut Wildlife Management Board responsible for managing and regulating access to wildlife in the Nunavut Settlement Area.[47] Even though the legal groundwork has been established, there are still frequent court cases that have helped to define Canada's ability to regulate food, social and ceremonial fisheries when there are conservation issues.

Commercial fishing and hunting

Fishing and hunting are traditional activities commonly addressed in early treaties and have continued to be important in modern treaties. However, government polices have tended to limit and restrict these First Nations activities as they gained economic value in Canadian society.[48] Modern treaties contain explicit references to fish and wildlife, but commercial access remains controversial. For example, the Nisga'a have a commercial allocation of salmon returning to the Nass River. However, the allocation is an agreement separate from the treaty that is not constitutionally protected due to provincial objections to including commercial allocations in the treaty. In addition, Nisga's organizations hold commercial fishing licenses that are utilized in the regular commercial fishery. In the northeast Arctic, the Nunavut Land Claims Agreement establishes co-management boards and gives priority to Inuit organizations in developing sports and commercial ventures with the purpose of developing a long-term, healthy renewable resource economy.

Recent court decisions are causing major changes in government policy. In *Marshall* v. *The Queen*, the Supreme Court ruled that certain Mi'kmaq Indians had a treaty right to catch and sell eels without a license and during a closed season.[49] It affirmed the right of the Mi'kmaq to sell the products of their hunting, fishing and other gathering activities to achieve a moderate livelihood. The decision led to confrontations between Mi'kmaq exercising their rights in local lobster fisheries and the federal fisheries department. The government response has been short-term measures such as increasing participation of affected First Nations in the commercial fishery and steps to establish negotiating processes to address long-term issues involving First Nations, federal and provincial players. The Supreme Court of Canada also found in *R.* v. *Gladstone* that the Heiltsuk people of British Columbia possessed an aboriginal right, protected under s. 35(1) of the Constitution, to trade in herring spawn on kelp because this activity was a central and significant feature of their society.[50] However, the court said that these were not blanket rights but would have to be proven on a case-by-case basis. The federal fisheries department has allocated new commercial spawn on kelp licenses to the Heiltsuk on an interim basis, although the amount has not been to the Heiltsuk's satisfaction.

Commercial fish allocations are expected to be a major part of British Columbia treaty settlements with the federal government following a policy of purchasing commercial licenses from willing buyers and transferring them to First Nations. The allocation transfer program is one component of the *Aboriginal Fisheries Strategy* that was initiated in 1992 in response to a court decision on aboriginal rights to fish for food, social and ceremonial purposes.[51] Three pilot sales initiatives were also initiated in 1992 in which First Nations received fixed allocations of salmon that could then be sold. The sales initiatives have been controversial but were recently upheld by British Columbia Appeal courts after civil disobedience and court challenges by commercial fishermen.[52]

Assessment

Unless there is adequate policy intervention, disputes over rights to fish commercially in Canada are likely to continue and may resemble the "fish wars" in the state of Washington. The American dispute continued even after the courts allocated up to 50 percent of the catch of salmon to treaty tribes. A policy of "circling the wagons" in the face of fundamental change did not work in Washington and is unlikely to work north of the border in Canada.

Offshore oil and gas and minerals

In Canada, development of offshore resources and exploration was halted in the Arctic and the Pacific in the 1980s due to environmental risks, land claims and a poor economic climate. But development in the past decade of the Hibernia oilfields offshore of Newfoundland signals that environmental risks of offshore development are acceptable to Canadians if there is political support, including royalties for the provincial government. In the Arctic, where treaties have recently been concluded, there has been a shift in the position of First Nations on development of oil and gas and mineral resources. The Aboriginal Pipeline Group, representing the ownership interest of aboriginal peoples of the Northwest Territories of the proposed Mackenzie Valley Pipeline, recently reached a funding agreement with oil companies and the project is moving ahead with regulatory reviews.[53] While the reserves in this case are located onshore, the announcement signals a fundamental change in approach by First Nations.

In British Columbia, virtually all First Nations oppose oil and gas development, but the provincial government has requested that the federal government lift a moratorium on exploration and development. In response, the federal government announced the creation of three task groups to investigate issues concerned with lifting the moratorium. But oil companies have not been interested in investing in areas where there is uncertainty about title. While there is little interest in offshore mineral deposits at present, this is an emerging issue that will get more attention as less costly sources become depleted.

First Nation interests in development of offshore non-renewable resources have varied. Settlement of treaties in northern Canada recently resulted in a change in the position of some First Nations to onshore development. An important factor has likely been economic incentives provided through recent treaties. However, environmental risks of offshore development are still a major concern, particularly on the Pacific coast. In the end, development of oil and gas is likely to be a political decision rather than one that can be resolved by integrated ocean planning and one that is closely tied to the resolution of aboriginal rights in offshore areas.

Aquaculture

Aquaculture is another developing industry that can significantly affect First Nation utilization of foreshore and nearshore areas.[54] However, new aquaculture tenures alienate foreshore and marine areas and may adversely affect traditional First Nation practices. In addition, there may be environmental effects such as introduction of exotic species and waste generation. Aquaculture is an area of shared federal–provincial responsibility that has only recently started to receive serious federal attention. The provinces have a significant role in aquaculture, including siting, economic development and environmental protection. DFO's role according to its 2003 *Aquaculture Policy Framework* is to create policy, program and regulatory conditions in its areas of responsibility that will enable aquaculture development. There are ten policy principles in the Framework including:

> DFO will respect constitutionally protected Aboriginal and treaty rights and will work with interested and affected Aboriginal communities to facilitate their participation in aquaculture development.

In British Columbia, salmon farming has been contentious. A moratorium on new operations was lifted in April 2002 after being in place for seven years. Many First Nations opposed the lifting of the moratorium. Current issues for resolution include sea lice infestation of wild salmon and Atlantic salmon escapes.[55]

Assessment

Federal government policies recognize that aquaculture is a significant issue for First Nations. Implicit in this is that First Nations may have power to stop or at least slow development in their territory. The government approach appears to encourage First Nation participation in development by trying to put First Nations inside the tent with government and business. These policies appear to be making inroads with several salmon farms proceeding in remote First Nation communities with the consent and involvement of the affected First Nations.

Protecting the marine environment

Like many other countries, Canada is experiencing the negative effects of coastal development. The *Oceans Strategy* supports improved collaboration between federal and provincial governments and measures such as the creation of a national network of marine protected areas and establishment of marine environmental quality guidelines. In addition, Canada has recently passed legislation, including the *Species at Risk Act, 2002* that imposes mandatory restrictions on taking threatened or endangered species. First Nations can play a significant role in species recovery plans as they are collectively a major owner of undeveloped lands in Canada. Increased environmental protection is clearly in the interest of First Nations due to their reliance on wild fish and game and dependence on a healthy ecosystem for cultural survival.

Conclusion

This chapter had provided a glimpse of First Nations' ocean interests and practices and a survey of Canada's current approach to ocean management and development. The Canadian government's approach resembles a fortress mentality where issues such as aboriginal and treaty rights are denied until explicitly defined in treaty negotiations or proven in court. As shown through a few examples, First Nations exercised proprietary rights to ocean spaces and resources and developed sophisticated skills and technology for exploiting many offshore resources such as groundfish long before the arrival of Europeans. Historic treaties focused to a large extent on depriving First Nations of land and limiting interference with development activities such as agriculture, mining or other resource development. Modern treaties and emerging oceans policies have a similar focus by denying aboriginal title to ocean spaces and encouraging non-interference and even offering limited participation by First Nations in ocean development.

Canada's ocean policies are still emerging, but the general direction has been set by the *Oceans Act* and *Oceans Strategy*, and legislation and policies concerning marine conservation areas and migratory bird refuges. In general, aboriginal and treaty rights are recognized as a burden on the Crown, and the role of First Nations in oceans management or planning is not defined except where management arrangements (relating to fish and wildlife) are established under existing treaties or agreements. Canada's position of denying aboriginal title to ocean spaces is unlikely to change as shown by recent conflicts over aboriginal and treaty rights to fish commercially. A different approach is being followed with co-management of fish and wildlife where a role has been spelled out in treaties and marine conservation areas and allows for agreements with First Nations (and others) in advance of treaties.

Based on the scope of recent treaties, it is unlikely that the Canadian approach to ocean areas will change through negotiation except in a limited way. Some First Nations describe the *Constitution Act*, s. 35 as an empty box – a box that is being filled little by little through "negotiated" aboriginal and treaty rights in modern treaties and rights defined through court decisions. The broader Canadian society has been reluctant to accept differentiated citizenship for First Nations or to recognize broader ocean interests than hunting and fishing. So far it has been possible for Canada to be successful at limiting aboriginal rights to ocean spaces through interest-based negotiations where access to land-based resources are traded for extinguishment of marine interests. But the broader issue of sea title is about to be tested in Canadian courts, and the result is likely to require fundamental changes to Canada's ocean policies.

Acknowledgments

The author acknowledges the assistance of several colleagues including Robert Galois, Don Hall and Nancy Morgan who reviewed a draft of this chapter.

Notes

1 First Nations is used here as synonymous with aboriginal peoples or indigenous peoples.
2 *Oceans Act* S.C. 1996, C-31.
3 Treaties in Canada's Maritime (Atlantic) provinces generally dealt with peace and friendship while those in the prairies dealt with surrender of land.
4 Reasons behind these actions are discussed in Paul Tennant "Reaching Just Settlements: Land Claims in British Columbia" in Frank Cassidy (ed.) *Reaching Just Settlements: Land Claims in British Columbia, Proceedings of a conference held February 21–22, 1990* (Oolichan Books and The Institute for Research on Public Policy, Lantzville: 1991) 153. A few treaties (the Douglas treaties) were concluded in southern Vancouver Island in 1850–54, but the only modern treaty in British Columbia is the Nisga'a Final Agreement that was ratified in 2000.
5 See Bradford W. Morse (ed.) *Aboriginal People and the Law: Indian, Métis and Inuit Rights in Canada* (Carleton University Press, Ottawa: 1991) at 720 discusses the 1911 treaty. Several modern land claims agreements have been reached in the Arctic including the Inuvialiat Final Agreement and Nunavut Final Agreement.
6 See for instance Ken Coates "The 'Gentle' Occupation: The Settlement of Canada and the Dispossession of the First Nations" in Paul Havemann (ed.) *Indigenous People's Rights in Australia, Canada and New Zealand* (Oxford University Press, Auckland: 1999) 141–61; and the introduction by Paul Havemann "Settling the Anglo-Commonwealth" in ibid., at 123–8.
7 Coates, ibid., at 156–7.
8 Current negotiations with marine components under Canada's Comprehensive Land Claims Policy include the Inuit of Quebec, Labrador and Newfoundland and the Cree of northern Quebec; British Columbia has a separate treaty process that involves individual negotiations with 53 First Nations. See, Indian and Northern Affairs Canada *Comprehensive Claims Policy and Status of Claims* (February 2003). Online at: www.ainc-inac.gc.ca/ps/clm/brief_e.pdf (accessed 2 February 2004).
9 For example, 22 out of 51 statements of intent filed by First Nations under the BC Treaty Process include ocean spaces within the described "traditional territory," see C. Rebecca Brown *Starboard or Port Tack? Navigating a Course to Recognition and Reconciliation of Aboriginal Title to Ocean Spaces* (Masters of Laws Thesis, University of British Columbia, Vancouver: 2000) at 8.
10 A common saying in Skidegate is "When the salmonberries are ripe the halibut are in the kelp." See R. Russ Jones and William Lefeaux-Valentine *Gwaii Haanas – South Moresby National Park Reserve Review of Vertebrate Fishery Resources* (Canadian Parks Service, Ottawa: 1991) at 234.
11 See Alan D. McMillan *Since the Time of the Transformers: The Ancient Heritage of Nuu-chah-nulth, Ditidaht, and Makah* (University of British Columbia Press, Vancouver: 1999) at 141.
12 Haida means "people" and Haida Gwaii means "islands of the people."
13 See accounts of Haida fishing for halibut and blackcod by Haida Solomon Wilson as recorded by David W. Ellis in Jones and Lefeaux-Valentine, note 10.

14 Blackcod are abundant at depths of 400–1,830 meters, G.A. McFarlane and R.J. Beamish "Biology of adult sablefish (*Anoplopoma fimbria*) in waters off western Canada" in *Proceedings of the International Sablefish Symposium, Second Lowell Wakefield Fisheries Symposium* (Alaska Sea Grant Report 83-8, Anchorage: 1983) 59 at 63.
15 James G. Swan *The Indians of Cape Flattery, at the Entrance to the Strait of Juan de Fuca, Washington Territory* (Smithsonian Contributions to Knowledge Volume 16, Smithsonian Institution, Washington, DC: 1870); according to Barbara Lane "Makah Marine Navigation and Traditional Makah Offshore Fisheries" report prepared for the Makah Indian Tribe (20 March 1977) (on file Oregon Historical Society, Portland) at 19.
16 Gilbert Malcolm Sproat *Scenes and Studies of Savage Life* (Smith, Elder and Co, London: 1868) 225 as cited in Lane, ibid., at 20, reprinted as Charles Lillard (ed.) *The Nootka: Scenes and Studies of Savage Life* (Sono Nis Press, Victora: 1987) at 151.
17 See Albert P. Niblack "The Coast Indians of Southern Alaska and Northern British Columbia" in *Annual Report of the U.S. National Museum for 1888* (U.S. National Museum, Washington: 1890) (reprinted Johnson Reprint Corporation, New York: 1970) at 335.
18 See John R. Swanton "Haida texts and myths, Skidegate dialect" (1905) 29 *Bureau of American Ethnology Bulletin* 31.
19 Sproat, note 16 at 79 as cited in McMillan, note 11 at 154, reprinted Lillard, note 16, at 58.
20 Swan, note 15.
21 Swanton, note 18 at 384. This is a transcription of Haida oral history: "Fights between the Tsimshian and Haida and Among the Northern Haida."
22 Jones and Lefeaux-Valentine, note 10 at 188 (from Haida Solomon Wilson as recorded by David W. Ellis).
23 See, for example, a brief discussion of the history of First Nations in the British Columbia halibut fishery in Diane Newell *Tangled Webs of History: Indians and the Law in Canada's Pacific Coast Fisheries* (University of Toronto Press, Toronto: 1993) 180–8. The blackcod fishery has a similar history.
24 See, for example, Russ Jones and Terri-Lynn Williams-Davidson "Applying Haida Ethics in Today's Fishery" in H. Coward, R. Ommer and T. Pitcher (eds) *Just Fish: Ethics and Canadian Marine Fisheries* (Social and Economic Papers 23, Institute of Social and Economic Research, Memorial University of Newfoundland, St Johns: 2000) 100–15.
25 See discussion by Ginn in Chapter 11 in this volume.
26 Title to dry land was denied by successive governments until the landmark *R.* v. *Calder* [1973] S.C.R. 313 affirmed the existence of aboriginal title as being guaranteed by the Canadian common law.
27 See Treaty Negotiations Office (British Columbia) "British Columbia Offers Treaty Land to Haida Nation" Treaty Negotiations Office, British Columbia, Information Bulletin 2003TNO0025-000784 (3 September 2003). Online at: www2.news.gov.bc.ca/nrm_news_releases/2003TNO0025-000784.htm (accessed 2 February 2004).
28 See "B.C. offers Haida control of 20 percent of islands, but Natives reject move, reaffirming their claim to all of the Queen Charlottes" *Globe and Mail* 4 September 2003.
29 *Canada National Marine Conservation Areas Act* 2002, c. 18, s. 4(2).
30 This designation also provides notice to the public that boundaries may change in the future.
31 *Migratory Birds Convention Act* 1994, c. 22; Protocol amending the 1916 Con-

vention for the Protection of Migratory Birds (Washington, DC, 14 December 1995). Online at: www.le.fws.gov/pdffiles/Canada_Mig_Bird_Treaty.pdf (accessed 2 February 2004).

32 Convention between his Majesty and the United States of America for the Protection of Migratory Birds in Canada and the United States (Washington, DC, 7 December 1916). Online at: www.lexum.umontreal.ca/ca_us/d_219_en.html (accessed 2 February 2004).

33 One change made to the *National Marine Conservation Areas Act* in the initial parliamentary review was to ensure consistency with existing land claims agreements through references to aboriginal governments. See Mollie Dunsmuir *Bill C-10: An Act Respecting the National Marine Conservation Areas of Canada* (Legislative Summary LS-396E, Parliamentary Research Branch, Library of Parliament, Ottawa: 2001).

34 *Canada's Oceans Strategy: Our Oceans, Our Future* (Oceans Directorate, Fisheries and Oceans Canada, Ottawa: 2002).

35 *Policy and Operational Framework for Integrated Management of Estuarine, Coastal and Marine Environments in Canada* (Fisheries and Oceans Canada, Oceans Directorate, Ottawa: 2002) at 15.

36 A major focus of BSIMPI is oil and gas development in the Mackenzie Delta and Beaufort Sea. Efforts in the Central Coast have been a joint federal, provincial and First Nation initiative that have been limited at this stage to gathering marine information.

37 Parks Canada *Guiding Principles and Operation Policies* (Minister of Supply and Service, Ottawa: 1994).

38 This includes five in the Pacific, nine in the Arctic, ten in the Atlantic and five in the Great Lakes. For further information, see Parks Canada *Sea to Sea to Sea: Canada's National Marine Conservation Areas System Plan* (Parks Canada, Hull: 1995).

39 See discussion in Chapter 2 by Chircop and Hildebrand in this volume.

40 Graeme Kelleher (ed.) *Guidelines for Marine Protected Areas* (The World Conservation Union/IUCN, Gland: 1999) at xii.

41 See R. Russ Jones and Sylvie Guénette "First Nation Issues and MPA Planning" in S. Bondrup-Nielson, N.W.P. Munro, G. Nelson, J.H.M. Willison, T.B. Herman and P. Eagles (eds) *Managing Protected Areas in a Changing World* (Proceedings of the Fourth International Conference on Science and Management of Protected Areas, 14–19 May 2000, University of Waterloo) (Science and Management of Protected Areas Association, Wolfville, NS: 2002) 1427–37.

42 The Gwaii Haanas co-management process is described in R. Russ Jones "Accounting for First Nation Interests in Marine Protected Areas" in Neil W.P. Munro and J.H. Martin Willison (eds) *Linking Protected Areas with Working Landscapes Conserving Biodiversity* (Proceedings of the Third International Conference on Science and Management of Protected Areas, 12–16 May 1997, Wolfville, Canada, Science and Management of Protected Areas Association, Wolfville, NS: 1998) 317–24; the Gwaii Haanas Agreement and the Gwaii Haanas Management Plan are available from Gwaii Haanas National Park Reserve – Haida Heritage Site, Queen Charlotte City, BC, Canada. Online at: www.pc.gc.ca/pn-np/bc/gwaiihaanas/plan/plan2a_E.asp (accessed 2 February 2004).

43 See Mary C. Hurley and Jill Wherrett *Aboriginal Self-Government* (PRB 99-19E Parliamentary Research Branch, Library of Parliament, Ottawa: 1999).

44 Ibid.

45 See Thomas Isaac *Aboriginal Law: Cases, Materials and Commentary* 2nd edition (Purich Publishing, Saskatoon: 1999) at 308.

46 Nisga'a Final Agreement. Online at: www.gov.bc.ca/tno/popt/final_agreements.htm (accessed 2 February 2004).
47 The committee comprises four Inuit representatives, four government appointments and one representing the public interest; *Inuit Land Claims Agreement* (DIAND, Ottawa: 1993).
48 See for instance Newell, note 23 at 3–27.
49 *Marshall* v. *The Queen* [1999] 4 CNLR 161 (S.C.C.).
50 *R.* v. *Gladstone* [1996] 4 CNLR 65 (S.C.C.). See also Isaac, note 45 at 400.
51 *R.* v. *Sparrow* [1990] 1 S.C.R. 1075.
52 *R.* v. *Kapp et al.* [2004] BCSC 958.
53 Shell Canada News Release "Funding Agreements Reached & Preliminary Information Package to be Submitted, Mackenzie Delta Producer Group Confirms" (18 June 2003). Online at: www.shell.ca/code/library/news/2003/03nr_jun18_mackenzie.html (accessed 2 February 2004).
54 Canada's aquaculture industry currently focuses on Atlantic salmon, Pacific salmon, oysters, clams, mussels and scallops. Production is expected to increase and is likely to expand to other species such as halibut, cod or sablefish. See DFO *Aquaculture Policy Framework.* Online at: www.dfo-mpo.gc.ca/aquaculture/library_e.htm (accessed 2 February 2004). See also "A new way to feed the world" *The Economist* 9 August 2003 at 9.
55 See the following publications by the Pacific Fisheries Resource Conservation Council (PFRCC), an independent agency that provides advice to the governments of Canada and British Columbia and the general public: PFRCC *2002 Advisory: The Protection of Broughton Archipelago Pink Salmon Stocks* (PFRCC, Vancouver: November 2002); PFRCC *Making Sense of the Salmon Aquaculture Debate: Analysis of Issues Related to Netcage Salmon Farming and Wild Salmon in British Columbia* (PFRCC, Vancouver: January 2003). Online at: www.fish.bc.ca (accessed 2 February 2004).

13 Indigenous rights in the sea

The law and practice of native title in Australia

Geoff Clark

Introduction

The seminal decision of the High Court of Australia in *Mabo (No. 2)* determined that the common law of Australia was capable of recognizing that indigenous rights and interests in land and sea could form part of the common law. This chapter examines the development or extension of native title below the high water mark through decisions of courts in interpreting the legislative schemata created by the *Native Title Act 1993* (Cth). The chapter goes on to examine the structural mechanisms for mediation between parties under the *Native Title Act*. It postulates that the legislative schemata offers significant opportunities for parties to the process (including indigenous parties) to reach significant agreements on native title and non-native title matters. The chapter concludes that, at least in Australia, the embryonic developed state of the case law, the structural provisions of the legislation, and the model of mediation available point to the need for matters in relation to the integration of indigenous customary marine tenure and Western maritime management systems be resolved through agreements rather than litigation.

The development of the common law and statute law of native title

The development of the common law

The starting point for contemporary legal recognition of indigenous rights in land and waters in Australia is the 1971 decision in *Milirrpum* v. *Nabalco Pty Ltd.*[1] In that case, Blackburn J declared:

> The evidence shows a subtle and elaborate system, highly adapted to the country in which the people led their lives, which provided a stable order of society and was remarkably free from the vagaries of personal whim or influence. If ever a system could be called "a government of laws, and not of men", it is shown in the evidence before me.

Nevertheless, Blackburn J found that, despite the existence of the subtle and elaborate system of law, the Aboriginal plaintiffs did not "own" land in property law terms.[2] This decision gave rise to a political response to Aboriginal needs for land. At a federal level, the *Aboriginal Land Rights (Northern Territory) Act 1976* (Cth) was enacted. State parliaments followed the Commonwealth example of statutory land rights in a rather haphazard fashion. Statutory land rights schemes were eventually established in South Australia, New South Wales and Queensland.

Mabo *and the development of the common law*

The High Court decision in *Mabo* v. *Queensland (No. 2)* was handed down on 3 June 1992[3] as a result of proceedings that lasted some ten years.[4] A majority of six to one of the judges agreed:

1 There was a concept of native title at common law;
2 The source of native title was a traditional connection to or occupation of the land;
3 The nature and content of native title was determined by the character of the connection or occupation under traditional laws or customs; and
4 Native title could be extinguished by the valid exercise of governmental powers provided a clear and plain intention to do so was manifest.[5]

Until the *Mabo (No. 2)* judgments, the general law of Australia did not recognize that indigenous people possessed any inherent or pre-existing legal rights to land. Post-*Mabo (No. 2)*, such an inherent right was recognized in areas where it was not extinguished. Native title is therefore inherently different from statutory land rights titles. Under land rights schemes, groups of indigenous people in Australia are granted a fee simple title or a lease by the Crown. On the other hand, native title is the recognition of something that groups of indigenous people already have. Native title laws exist to identify, recognize and protect what already exists. The Crown grants nothing, as native title is not the Crown's to grant.[6]

The legislative response to Mabo

The *Native Title Act 1993* (Cth) was the Commonwealth of Australia's response to *Mabo*. The protracted parliamentary debate over the Act, and the 1998 amendments to it, attracted considerable public interest and comment. The Act is long and detailed. For the purposes of the present chapter it is necessary only to outline the scheme of the Act.

The preamble of the Act "sets out considerations taken into account" by the federal Parliament in enacting it.[7] The policy considerations underlying the Act have been summarized as follows:

- the protection of the rights of indigenous peoples;
- the need to provide a special procedure for the just and proper ascertainment of native title rights and interests;
- the importance of ensuring that native titleholders are able to enjoy fully their rights and interests and the need to significantly supplement those rights;
- the requirements for certainty and the enforceability of acts which were potentially made invalid because of the existence of native title; and
- the importance of providing certainty to the broader Australian community that future acts that affect native title may be done validly.[8]

Key concepts

Two key concepts lie at the core of the *Native Title Act*: what is Native Title and what does the Federal Court of Australia have to consider in making a determination that native title exists? These two concepts permeate and subliminally direct Federal Court proceedings and the mediation process that usually occurs as part of those proceedings.[9] The definition of native title common law rights and interests is set at s. 223 of the Act. Essentially native title rights and interests are:

- the communal or group rights possessed under traditional laws and customs including rights to hunt, fish or gather;
- where the group seeking to assert the rights have had and maintained a connection to the land or waters; and
- the rights are capable of being recognized by the common law.

Section 225 of the Act provides that a determination of native title by the Federal Court must specify:

- how the group of native titleholders is composed;
- the rights and interests held by the group;
- any other rights and interests held by non-native titleholders in the claimed area; and
- how the native title rights and rights and interests held by others in the claimed area will co-exist.

The difficulty, in an operational sense, posed by s. 225 lies in ss 225(d). This requires a precise enumeration of how the native title rights are to operate beside or be integrated with the rights granted to parties by the Crown.

The schemata of the Native Title Act

The Act seeks to establish legislative schemata for resolution of native title claims. This "codification" of native title rights and interests and how they are to be determined distinguishes the Australian process from that which operates in Canada, New Zealand and the United States. The principal features of the statutory scheme are:

- claimant applications are lodged with the Federal Court of Australia;[10]
- the Federal Court sends the application to the Tribunal Registrar[11] who undertakes administrative procedures, including applying the registration test to the application;[12]
- the Tribunal Registrar is responsible for notifying the relevant persons and bodies and the public about each application;[13]
- persons who feel their rights may be affected by a determination of native title apply to the Federal Court to become parties to the application;[14]
- once the party list is settled, the Federal Court will usually refer the application to the Tribunal for mediation;[15]
- mediation must be presided over by a member of the Tribunal;[16]
- the Tribunal carries out mediation of each matter referred to it and reports to the Federal Court on the progress of mediation.[17]

Where native title exists, the Federal Court makes an appropriate determination of native title, either in or consistent with the terms agreed by the parties or as decided by the Court after a trial.[18]

Native title in the sea

Introduction[19]

Thus far the chapter has focused on the common law and statute law recognition of indigenous rights and interests in land and water. This part focuses on the development of common law and statute law native title rights and interests in the sea. The term "sea" is used to describe waters below the high water mark.

Ironically, the application for a determination of common law rights that resulted in the *Mabo (No. 2)* decision was brought by representatives of an island culture located on the eastern edge of the Torres Strait at the northern tip of Australia. The application was originally one for a determination that indigenous rights and interests could be found to exist in both the land and the surrounding seas.[20] In the course of proceedings, the application was amended to remove the sea component of the claim.[21] The judges in *Mabo (No. 2)* made fleeting reference to the question of whether native title could be found to exist in the sea. Similarly, the subsequent landmark decision of the High Court in *Wik*[22] made scant reference to the recognition of customary marine tenure by the common law. Prior to the 1998 and 1999 decisions of the Federal Court of Australia in *Yarmirr*[23] and the subsequent 2001 decision of the High Court,[24] the only judicial consideration of native title in the seas had arisen from defences raised to prosecutions for taking marine resources contrary to legislation. In those cases,[25] the court took the view that native title rights could be likely to include a right to fish offshore in relation to particular areas.

The legislative approach to native title rights in the sea

In framing their response to *Mabo (No. 2)*, the legislators sought to extend the application of the *Native Title Act* to the area below the high water mark. They took as their fundamental tenet the areas over which the Commonwealth sought to exercise sovereignty – even in a limited or qualified form. This approach is demonstrated in the *Native Title Act 1993* (Cth), s. 6, which extends operation of the Act to Australia's territorial sea, contiguous zone, exclusive economic zone and continental shelf.[26] Nevertheless, it would be wrong to think that, in practice, native title rights and interests would extend to the edge of the continental shelf. As the native title rights and interests must be sourced in traditional laws and customs and must be exercisable in a discrete area, it would be extremely difficult to prove that such rights had traditionally been exercised some 100 miles from a land base. However, it would not be difficult to prove the exercise of such rights closer to shore such as in the territorial sea.

Judicial application of the Native Title Act *– the Yarmirr cases*

The application of the *Native Title Act* to Australian offshore waters fell to be determined in the *Yarmirr* line of cases. These cases involved a claim for native title rights to the seas, including the seabed and its resources, in the vicinity of Croker Island, Northern Territory. At first instance, Justice Olney of the Federal Court[27] reached several conclusions including the following:[28]

- The *Native Title Act* contemplates recognition and protection of native title over the sea, at least to the 12 nautical mile limit of the territorial sea.
- Native title can only be understood as a combination of rights and interests, and the term "ownership" in the context of native title does not necessarily equate with the type of dominion normally associated with ownership in its absolute form.
- The *evidence* did not support a finding that the applicants enjoy exclusive possession, occupation, use and enjoyment of the waters in the claimed area, although a system of permission operates as between the applicants themselves and perhaps other Aboriginal people.
- A right of exclusive possession would be inconsistent with the right of innocent passage[29] recognized in international law, and with the public right of navigation and the public right to fish recognized by the common law, and therefore a right to exclusive possession could not be recognized by the common law.

- In relation to fisheries legislation, the scheme of such legislation was regulation, and there was no plain and clear intent to extinguish any non-exclusive, non-commercial native title fishing, nor to create any third party rights inconsistent with such native title. Rather, the native title rights are capable of co-existence with the regulatory regime.
- There was no *evidence* to support any control over any of the resources of the sub-soil, and any right to control the resources of the sea is necessarily co-extensive with the right to control access, which was not *established*.
- There was no *evidence* to suggest any traditional law or custom in relation to minerals, and no native title to minerals could be recognized. In any event, any native title right could not survive legislative enactments which vested beneficial ownership of minerals in the Crown.
- The rights to protect places of importance and to safeguard cultural knowledge were established.
- The *evidence* did not support any native title right to trade in the resources of the claimed area.

As noted by Olney J:

> The net result of all of the foregoing is that native title rights have been, and are now, regulated, but not extinguished, by prior legislative enactments or administrative action. However, to the extent that the scheme of regulation would otherwise require the applicants to obtain a licence, permit or other instrument under a law of the Northern Territory or of the Commonwealth to lawfully exercise their native title rights of hunting, fishing, gathering or to engage in any cultural or spiritual activity for the purpose of satisfying their personal, domestic or communal non-commercial needs, they are not required to have any such licence, permit or other instrument.[30]

Whilst the decision may not have pleased the applicants, it certainly established that native title rights can be found to exist in the sea, although it is unlikely that those rights could be found to be exclusive rights.

The appeal in the Federal Court[31]

The decision of Olney J was subject to appeal to the Full Court of the Federal Court. In its judgment,[32] a majority (Beaumont and von Doussa JJ, Merkel J dissenting) dismissed both appeals. The key points to note from this decision are that all of the judges agreed that the *Native Title Act* recognized and protected native title rights and interests in respect of offshore places. The majority also adopted Olney J's reasoning that as the right to trade in resources of the claimed area was inexorably linked with the right to exclusive possession, once the former could not be made out, the latter was unsustainable.[33]

At first instance, Olney J had also found that the native title rights were regulated but not extinguished by municipal and Commonwealth fishing legislation and administrative action. The majority essentially adopted this view and found that because of the legislative regime and administrative actions that existed over the claim area, any right of the public to fish for commercial purposes and any asserted traditional right to fish for commercial purposes were, at a minimum, regulated or partially extinguished by statute or executive act.

The appeal in the High Court

In October 2001, the High Court gave its decision on the appeal from the Full Federal Court[34] in which the essential tenets of the decision at both first instance and on appeal were upheld.[35] The main features of the Court's decision were:

- The Court rejected the Commonwealth argument that the common law did not "extend" beyond the low water mark.
- The feudal concept of "radical title" that had been used in native title cases to denote the Crown's ultimate ownership of land can extend to the offshore area. That concept does not have a "controlling role" but is no more than a "tool of analysis which is important in identifying that the Crown's rights and interests in relation to land can co-exist with native title rights and interests." The concept is not determinative but simply an aid that "reveals the nature of the rights and interests which the Crown obtained on its assertion of sovereignty over land"[36] (and waters).
- Native title rights in the sea cannot be absolutely exclusive. "There is a fundamental inconsistency between the asserted native title rights and interests and the common law public rights of navigation and fishing, as well as the right of innocent passage."[37]
- Accordingly, management regimes on resources imposed by the Commonwealth or the littoral state did not, of themselves, operate to extinguish native title.

Obviously, the High Court decision has disappointed some of the interests represented in the case. Nevertheless, the legal landscape is now clear. At the time of the decision there were approximately 120 native title claimant applications taking in sea areas around Australia and another 61 applications included areas in the inter-tidal zone between high and low water marks.

The *Yarmirr* decision is an important precedent that lays the foundation for successful claims to extensive areas of Australia's coastal waters. A major issue still to be determined is in what manner the existence of native title rights and interests in those areas will affect the management and use of those places. One major issue still to be worked through is the operation of the future act provisions of the *Native Title Act* in the sea. The Act establishes schemata whereby registered claimants are entitled to be consulted about proposed activities in the claimed area that may impact on their asserted native title rights and interests. This is known as the "right to negotiate" process. The registered claimants do not possess a veto right in relation to the proposed activity, but proponents are required to undertake a process of consultation. This is designed to promote agreements between the applicants and the proponent about the doing of the future act. In instances where the future act operates to extinguish native title, the native titleholders are entitled to compensation for that extinguishment. In addition, where native title rights and interests have been determined to exist, they co-exist with other rights and interests.

The current matrix of offshore native title

Coastal waters

Registered claimants enjoy a right to negotiate for future acts in onshore places (defined in s. 253 as effectively within three miles of the shore).[38] This is considerable in some parts of Australia where there are extensive islands close to the mainland.[39] Native titleholders will be entitled to compensation for future acts that impact adversely on their determined native title rights and interests.

Offshore places

With respect to waters beyond the three mile limit, registered claimants have no right to negotiate in respect of offshore places (defined in *Native Title Act*, s. 253) but native titleholders will be entitled to compensation for future acts that impact adversely on their determined native title rights and interests.[40]

Such a complex matrix points to the need for agreement about co-existence in onshore and offshore areas rather than for parties to seek to litigate each future act and to then conduct compensation litigation in respect of each future act or class of future act. The complex jurisdictional matrix, when coupled with the issues facing resource managers about the nature and extent of the level and kind of native title interests that can exist below the high water mark, demonstrate the difficulty of seeking to resolve both the existence of native title rights and interests and how those native title rights and interests will co-exist with statutory schemes for resource management through litigation. These are issues that are so complex as to be resolved through negotiation, rather than litigation. The complexity is enhanced by reference to s. 211 of the *Native Title Act* which effectively provides that municipal laws that regulate or control activities under government regulatory or management regimes do not and cannot operate to limit or prevent the right of native titleholders to take resources for communal, non-commercial purposes. The framers of the Act contemplated a situation where regulators and indigenous peoples would negotiate the interaction of the resource management regimes and the native title rights. The entire legislation is focused on mediation and agreement making.

It is important that those responsible for the management of offshore waters understand and appreciate the opportunities that are provided by the framework of the Act and the manner in which the National Native Title Tribunal carries out its mediation function. It is only when seized of that understanding that managers will be in a position to optimize the development of an integrated approach to resource management that provides durable agreements that afford of consensual management regimes that can operate without disruption and challenge.

Mediating sea claims: the powers and functions of the National Native Title Tribunal

Introduction

Faced with a narrowing "black letter law" approach to customary marine tenure by the courts, it is appropriate to look at how native title is mediated in Australia within the litigation framework. This involves an examination of the legislative focus on agreement making rather than on litigation, the role of the Tribunal in mediating claims,[41] and how the Tribunal operates in its agreement-making practice.

The National Native Title Tribunal is an administrative body established in accordance with the main objects of the *Native Title Act 1993*, with power to make determinations about whether certain future acts can be done and whether certain agreements concerning native title are to be covered by the Act, and to provide assistance or undertake mediation in other matters relating to native title.[42] The powers and functions, membership and administration of the Tribunal are governed by the Act and regulations made under the Act.

The Tribunal has certain functions in relation to Federal Court proceedings arising from native title determination applications (or claimant applications). When the Court refers proceedings to the Tribunal for mediation, the Tribunal holds conferences of the parties. The Act sets out how mediation conferences are to be conducted, who may attend or participate in conferences, the circumstances under which questions of fact or law may be referred to the Court, and the way in which reports about the mediation are to be made to the Court.

The Tribunal must pursue the objective of carrying out its functions in a fair, just, economical and prompt way.[43] In carrying out its functions, the Tribunal may take account of the customary and cultural concerns of indigenous peoples, but not in a way that unduly prejudices other parties to any proceedings.[44] The Tribunal is not bound by technicalities, legal forms or rules of evidence.[45] The Tribunal may sit in a wide range of places, including external territories.[46] The Tribunal may carry out research for the purposes of performing its functions, including research into anthropology, linguistics or the history of interests in relation to land or waters in Australia.[47]

How mediation occurs under the Native Title Act[48]

It has been recognized that mediation is at the center of the work of the Tribunal.[49] Generally speaking, the Federal Court refers every claimant application to the Tribunal for mediation[50] as soon as practicable after the end of the notification period for the application.[51] Further, the Court may, at any time in a proceeding, refer the whole or part of the proceeding to the Tribunal for mediation if the Court considers that the parties will be able to reach agreement on (or on the facts relevant to) any matter set out in the *Native Title Act*, s. 86A(1).[52]

In making Federal Court orders in relation to the case management of claimant applications, the Court has referred to the "central role" of the Tribunal in the mediation process. That role is apparent from the mandated terms of referral and is reinforced by the Federal Court's facility to request the Tribunal to provide reports to the Court on the progress of mediation. The various statutory provisions relating to mediation conferences convened by the Tribunal are "ancillary to the referral of applications" to the Tribunal for mediation and they "do not define the limits" of the Tribunal's role.[53]

A myriad of factors impact on and determine how mediation occurs in each particular case. To its credit, the Parliament, in developing the Act, made no attempt to dictate how mediation should be conducted. There is no prescribed way of proceeding. The Tribunal may "hold such conferences of the parties or their representatives as the Tribunal considers will help in resolving the matter."[54] The formal part of the mediation process is conducted at such "conferences." Each mediation conference must be presided over by a member.[55] The President may appoint a consultant to mediate.[56]

The presiding member at a mediation conference is provided with several statutory powers, including:

- to allow parties to participate in a conference in person, by telephone, or any other means of communication;[57]
- to hold conferences with only some of the parties;[58]
- to exclude persons from conferences at the member's discretion;[59]
- to permit (with the consent of parties present) non-parties to attend a conference as observers[60] or as participants;[61]
- to refer (at the member's own initiative) to the Federal Court a question of fact or law that arises during a mediation[62] and to continue the medi-

ation whilst the referred question of fact or law is being determined by the Federal Court;[63]

- to prohibit, limit or qualify the disclosure to persons (including to other parties or parties' constituencies) of any information given or statement made at or documents produced at a conference.[64]

The Tribunal member appointed by the President to conduct mediation in a particular application develops a process or framework for progressing the mediation from stage to stage towards the identification and resolution of the issues raised by the application. That program will usually include a series of time frames for each stage of the process. This is known as the design phase of the mediation.

The emphasis on agreement and consensus building

Having set out the legislative schemata for resolution of native title, it is appropriate to examine the statutory and practical emphasis on agreement making. The starting point of such an examination is the preamble to the Act that provides:

> A special procedure needs to be available for just and proper ascertainment of native title rights and interests which will ensure that, if possible, this is done by conciliation and, if not, in a manner that has due regard to their unique character.

Similarly, the main objects of the Act include providing for the recognition of native title and establishing the mechanism for determining claims of native title. Whilst the preamble to the Act uses the term "conciliation," it is interesting to note that (since the 1998 amendments) the Act is replete with the term "mediation."[65] Whatever words like "mediate," "negotiate" or "conciliation" mean to the different parties, the aim of the process is to resolve a range of issues by agreements that will endure based on relationships that will develop.[66]

It is difficult to judge how successful (in empirical terms) the agreement making emphasis in the Act has been. Optimists would point to the 31 determinations of native title that have been made by the Federal Court as at June 2003, of which over 25 have been by consent of all of the parties.[67] Optimists would also point to the 77 Indigenous Land Use Agreements already registered under the Act and the 24 such agreements currently in the process of being registered.[68]

The Tribunal practice in mediating native title claims

The interest-based negotiation model

The Tribunal has adopted the interest-based model of mediation as a basis for the manner in which it will conduct mediation of native title issues. The interest-based mediation model is based on the interest-based negotiation model. The underlying process for the interest-based negotiation model is that disputing parties attempt (without the help of another person) to determine each other's interests and to generate options that satisfy the interests of all of the parties. The discussions do not focus on each party's rights but on their interests. Unlike some other schools of negotiating, this model does not prescribe what the negotiator should say at every point in the negotiation. This model provides a framework for proponents to work through their issues and to explore their options.[69]

Interest-based mediation model

This term is used to describe principled or interest-based negotiation between parties with a negotiation expert, a mediator, who can assist the parties to overcome any obstacales during their negotiations. Although there are several models for interest-based mediation, all of the models attempt to determine the parties' interest in the hope that the disputants and the mediator will develop or create solutions to satisfy the interests. The classical model of interest-based mediation is a seven-stage model. The stages are:

- setting the table;
- storytelling;
- determining interests;
- determining the issues;
- brain-storming options;
- selecting an option;
- closure.

It will be seen that the stages in interest-based mediation reflect quite closely several elements in interest-based negotiation.

Moving through the stages

It would be wrong to imagine mediation as involving a linear transition from one stage of the mediation to the next. Inevitably, mediations involve jumping from one stage to the other and normally not in sequence. A typical pattern is for mediation to jump from storytelling to options, to interests, and back to storytelling as interests are jointly explored. Each mediation develops its own ebb and flow, and the wise mediator is always conscious of this and does not seek to be too prescriptive. The greatest facilitator of progress in mediation is storytelling. If the parties need to go back and tell the story from another angle or from a different perspective then the wise mediator lets them do that.

Sea claims – where to from here?

The Croker Island case was "the first significant opportunity for the courts to clarify complex legal issues regarding native title and the seas, including whether the common law is capable of recognizing indigenous sea rights, the relationship between native title and statutory fishing interests, and the relationship between native title and the public right to fish."[70] The law, at least on the basis of the evidence adduced, is now clear. Native title can exist in offshore waters, although the rights may not be exclusive.

Customary marine issues

It is argued that, "Following the *Mabo* judgment coastal communities are . . . seeking to exercise their sea rights through native title claims, co-management arrangements of conservation and joint venture economic developments."[71] This is consistent not only with the inherent interest of many Aboriginal peoples in Australia in marine areas,[72] but also as a result of the clarification of marine native title following the High Court's decision in *Yarmirr*.

In many parts of Australia, there are competing private, recreational and commercial uses of the traditional clan estate areas of native titleholders with the potential for conflict. Such conflict may be avoided or minimized by agreed resource sharing or by spatial separation of competing or inconsistent interests. Native titleholders may seek various mechanisms for involvement in the management of their traditional clan estates including some or all of the following arrangements:

- indigenous control of areas of particular resources through sea closures, traditional hunting and fishing zones;
- joint management of areas in conjunction with relevant government agencies and/or relevant industry bodies; and
- co-management arrangements where decision making is by a relevant government agency in a close working relationship with the relevant indigenous group.[73]

None of the above, or any variations thereof, can be directly determined by litigation. They are best resolved and structured through negotiation. Sparkes suggests[74] consideration should be given to using such tools as Indigenous Land Use Agreements[75] to provide interim and lasting solutions to meet the needs of all of the parties. Such agreements may be a precursor to consent determinations in relation to native title and amendment of relevant legislation to meet the needs and aspirations of indigenous peoples in relation to offshore areas.

Conclusion

The decisions of the Courts have established sufficient principles for applicants to commence proceedings for a determination of native title rights and interests in offshore waters. Given the state of the law, it is reasonable to assume that applicants will seek to negotiate with governments and others who have interests in the claimed waters. Those negotiations will usually occur as part of the mediation process established under the *Native Title Act*. In other words, applicants (and other parties) will decide that they can achieve more by negotiation than by litigation.

The mediation of negotiations about those rights and interests will require the Tribunal to manage cross-cultural, multi-party negotiations about rights and interests in the sea. The Tribunal will seek to carry out interest-based negotiation within the framework of a rights-based court process. Successful outcomes from those negotiations will be determined by a number of factors including:

- the ability of the Tribunal to create the appropriate environment in which parties will feel comfortable enough to set aside their bargaining positions and put their issues on the table;
- the ability of the parties to recognize that the negotiations present an opportunity to develop cooperative processes going forward that create a basis for sound resource and management arrangements for the claimed areas; and
- a common understanding of the parties that issues of customary marine tenure, at least in the coastal management zone will not simply "go away" but needs, wherever possible, to be integrated into a holistic management regime.

However, it is suggested that there has been sufficient judicial development of the legal principles involved. The parties now have a comprehensive statutory framework and sufficient certainty in the mediation model for them to prepare for and to proceed through the process with some degree of certainty.

Acknowledgments

I am grateful to the work of Dr Guy Wright and Dr Stephen Sparkes of the National Native Title Tribunal for their assistance in the provision of the international context of indigenous sea rights overview.

Notes

1 (1971) 17 FLR 141 at 267.
2 Ibid., at 273, see also 268–72.
3 *Mabo* v. *Queensland (No. 2)* 1992 (175 CLR 1).
4 For a wonderful insight into the history of the proceedings see B.A. Keon-Cohen "The Mabo litigation: A personal and procedure account" (2000) 24 *Melbourne University Law Review* 893–951.
5 See the comment by R.D. Lumb "The Mabo Case – Public Law Aspects" in M.A. Stephenson and Suri Ratnapala (eds) *Mabo: A Judicial Revolution* (University of Queensland Press, St Lucia, Qld.: 1993) 1 at 12 who notes that "the reasoning of the judges in Mabo appear to have been influenced by a thesis written by Dr G.S. Lester, entitled *The Territorial Rights of the Inuit of the Canadian North West Territory*; see G.S. Lester *The Territorial Rights of the Inuit of the Canadian North West Territories: A legal argument* (Ph.D.Jur. dissertation, York University, Toronto: 1981).
6 See G. Neate "Indigenous land rights and native title in Queensland: A decade in review" (2002) 11 *Griffith Law Review* 90 at 106.
7 See commentary on the Commonwealth Attorney General's Legal Practice *Native Title Act 1993* (Australian Government Publishing Service, Canberra: 1993) at C8 for a detailed history.
8 Policy considerations as well as the history of the development of legislation are set out in the second reading speech, Commonwealth *Hansard*, House of Representatives, 16 November 1993, 2877, by the then Prime Minister.
9 See *Native Title Act 1993* (Cth), s. 86B(1).
10 Ibid., at s. 62 sets out what must be contained in the application.
11 Ibid., at s. 63.
12 Ibid., Part 7 requires the Registrar to make a procedural and merits-based assessment of an application. For a discussion on the registration test see Greg McIntyre, David Ritter and Paul Sheiner "Administrative avalanche: The application of the registration test under the *Native Title Act 1993* (Cth)" (1999) 4, No. 20 *Indigenous Law Bulletin* 8. An application that passes the registration test vests certain procedural rights in the claimants (between the date of registration and the date of a determination of native title) in relation to future acts that may be proposed to be carried out on any part of the claimed area; *Native Title Act 1993* (Cth), Div. 3, SubDiv P.
13 *Native Title Act 1993* (Cth) s. 66. See also Federal Court Orders-Order 78, r 43.
14 Ibid., at s. 84.
15 Ibid., at s. 86B.
16 Ibid., at s. 136A(2).
17 Reports by members to the Federal Court on the progress of mediation can either be made in response to an Order of the Court, ibid., at s. 86E, or at the initiative of the member, ibid., at s. 136G(3).
18 Ibid., at ss. 86F, 87.
19 In preparing this part of the chapter I have been greatly assisted by the work of Dr Stephen Sparkes, Head, Legal Section, National Native Title Tribunal.
20 The original claim included seas that extended ten miles from Murray Island to the Great Barrier Reef. These seas were located within both Commonwealth and Queensland waters.
21 This approach was intended to divert the Commonwealth from adopting an antagonistic position to the claim and to ensure that the case focused on the principle that native title may be recognized by the common law. See G. McIntyre, "Mabo and Sea Rights: Public Property Rights or Pragmatism?" paper presented at Turning the Tide: Conference on Indigenous Peoples and Sea Rights,

Darwin, 14–16 July 1993 (Faculty of Law, Northern Territory University, Darwin: 1993) at 108. See also D. Haigh "Torres Strait and Customary Marine Tenure – A Legal Baseline" presented at Turning the Tide: Conference on Indigenous Peoples and Sea Rights, Darwin, 14–16 July 1993 (Faculty of Law, Northern Territory University, Darwin: 1993) at 135.

22 *Wik Peoples* v. *Queensland* (1996) 187 CLR 1.

23 *Yarmirr* v. *Northern Territory* (1998) 156 ALR 370 (Decision of Olney J) and *Commonwealth* v. *Yarmirr; Yarmirr* v. *Northern Territory* (1999) 168 ALR 426 (Decision of Beaumont, von Doussa and Merkel JJ) [*Yarmirr* 1999].

24 *Commonwealth* v. *Yarmirr* (2001) 184 ALR 113 [*Yarmirr* 2001].

25 *Mason* v. *Tritton* (1994) 34 NSWLR 572 (especially Kirby J at 582) and *Derschaw anors* v. *Sutton*, Full Court, Supreme Court of Western Australia, unreported 16 August 1996.

26 Consistent with the provisions of the *Seas and Submerged Lands Act 1973* (Cth).

27 *Yarmirr* v. *Northern Territory* (1988) 156 ALR 370 [*Yarmirr* 1988].

28 The summary of issues is based on Neville Henwood "The Croker Island case – A landmark decision in native title" (1998) 3, No. 10 *Native Title News* 146. See also Graham Hiley "Croker Island – Final orders and appeals" (1998) 3, No. 11 *Native Title News* 177. Mr Henwood was an instructing solicitor in the case and Mr. Hiley was a Senior Counsel. Note that the highlight of words such as "*evidence*" and "*established*" is mine. It is designed to demonstrate that the Court did not find that those things were not capable of existing as native title rights. Rather, it emphasizes that the evidence was not of a sufficient standard to convince the Court in this case that the particular right (on the evidence) could be found to exist.

29 See United Nations Convention on the Law of the Sea (1982) 1833 UNTS 396, arts. 19 and 31.

30 *Yarmirr* 1988, note 27, at para. 156. See also *Native Title Act 1993* (Cth), s. 211 which addresses certain native title rights and decisions on this provision: *Dillon* v. *Davies* [1998] TASSC (20 May 1998); *Underwood* v. *Gayfer* [1999] WASCA 56 (15 June 1999); *Wilkes* v. *Johnsen* [1999] WASC 74 (23 June 1999); *Yanner* v. *Eaton* [1999] HCA 53 (7 October 1999).

31 In preparing the commentary on the appeal, I have been greatly assisted by Gordon Kennedy "Croker Island appeals – No "watershed" for application of the Native Title Act" (1999) 4, No. 6 *Native Title News* 109.

32 *Yarmirr* 1999, note 23.

33 Although Olney J had found some evidence that the applicant's ancestors had engaged in an exchange of goods, the evidence did not permit him to find that the objects traded could be characterized as subsistence resources derived from either the waters or land.

34 *Yarmirr* 2001, note 24.

35 See comment by Pat Brazil "The Croker Island appeals to the High Court" (2001) *Australian Mining and Petroleum Law Journal* 206–311, esp. table at 311.

36 *Yarmirr* 2001, note 24.

37 Ibid., interestingly, the High Court upheld the trial judge's determination that, although a right to possess (which implicitly meant exclusion rights) the sea country claimed as against non-members of the claimant group had been extinguished in that case, the right to grant permission to use the country as against non-members of the claimant group had survived. See ibid., at paras 86–93.

38 See the operation of *Coastal Waters (State Title) Act 1980* (Cth) s. 4(1) which vested in each state "the same right and title to the property in the seabed beneath the coastal waters of a State" as each state would have in the seabed beneath the waters of the sea within the limits of the state. Note *Coastal Waters*

(State Title) Act 1980 (Cth), s. 8 (a) which makes it clear that the Act does not operate to extend the limits of the state, merely its jurisdictional reach.

39 For an indication of the extent of onshore places in areas of sea where there are extensive island systems, see Stuart B. Kaye "Jurisdictional patchwork: Law of the sea and native title issues in the Torres Strait" (2001) 2 *Melbourne Journal of International Law* 381.

40 *Native Title Act 1993* (Cth), s. 24NA (4) and (6).

41 The major difference in the role of the Tribunal in mediating claims – as distinct from its arbitral role in dealing with "future acts" – is that it is not a determinative body. Unlike the Waitangi Tribunal in New Zealand, it cannot make decisions about whether native title exists, but its role is more than that of "keeper of the process" as in the British Columbia model. See British Columbia Treaty Commission *Why treaties?* (British Columbia Treaty Commission, Vancouver: 2004). Online at: www.bctreaty.net (accessed 5 February 2004).

42 *Native Title Act 1993* (Cth) ss 4 (2), 4 (7).

43 Ibid., at s. 109(1).

44 Ibid., at s. 109(2). It is worth noting that a similar provision is made for the Federal Court's way of operating: ibid., at s. 82(2).

45 Ibid., at s. 109(3). However, the Federal Court is bound by the rules of evidence, except to the extent that the Court otherwise orders. Ibid., at s. 82(1).

46 Ibid., at s. 127 of the Act and read with s. 6.

47 Ibid., at ss. 108(2), 108(3).

48 See Graeme Neate "Meeting the Challenges of Native Title Mediation" presented at LEADR 2000: ADR International Conference, Sydney (2000) at 17–18.

49 See National Native Title Tribunal *Annual Report 1998–99* (Australian Government Publishing Service, Canberra: 1999) 11; see also *North Ganalanja Aboriginal Corporation* v. *Queensland* (1996) 185 CLR 595 at 657 per Kirby J.

50 *Native Title Act 1993* (Cth) s. 86B (1). See *Bropho* v. *Western Australia* (2000) 96 FCR 453 at 445, 456, 461–2 per French J.

51 On occasions, the Court has referred matters to the Tribunal for mediation prior to notification occurring. This is usually where there are overlapping claims.

52 See *Smith* v. *Western Australia* (2000) 104 FCR 494 at 498 per Madgwick J.

53 *Frazer* v. *Western Australia* [2003] FCA 351 at para. 26.

54 *Native Title Act 1993* (Cth), s. 136A (1).

55 Ibid, at s. 136A (2). The current practice is that the President appoints a member to preside over and conduct the mediation for each application. *Native Title Act 1993* (Cth) s. 123 (1) (b).

56 Ibid., at s. 131A.

57 Ibid., at s. 136A (6).

58 Ibid., at s. 136B (1).

59 Ibid., at s. 136B (2).

60 Ibid., at s. 136C (a).

61 Ibid., at s. 136C (b).

62 Ibid., at s. 136D.

63 Ibid., at s.136D (4).

64 Ibid., at s. 136F.

65 See references to mediation in Ibid., at ss. 4, 43A, 44F, 44G, 79A, 86A, 86B, 86C, 86D, 86E, 108, 123, 131A, 131B, 136A, 136D, 136G, 136H, 183.

66 Neate, note 48 at 18.

67 Most of those cases that do not involve consent determinations have been "test" cases designed to answer difficult questions of law or have occurred where negotiations have broken down.

68 Indigenous Land Use Agreements (ILUAs) are agreements between claimants and other parties to a claim that typically deal with the carrying out of activity on or in relation to some or all of the claimed area.

69 William L. Ury, Jeanne N. Brett and Stephen B. Goldberg *Getting Disputes Resolved* (Jossey-Bass, San Francisco: 1988) describe three approaches to resolving disputes: a power-based approach, a rights-based approach and an interest-based approach. It is obvious that a power-based approach would be inappropriate in the native title context as the applicants have little or no power. The power rests with government and industry. Similarly, a rights-based approach would be an inappropriate model as the law in relation to native title rights is still in its infancy and is being constantly tested, re-stated and modified by courts and legislature (the 1998 amendments to the Act). On this reasoning, the Tribunal's decision to adopt the interest-based model may well have been because the alternatives were either unattractive or inappropriate – or both.

70 Ron Levy "Native title – A catalyst for sea change – Part 1" (2001) 3, No. 9 *Native Title News* 120.

71 Sue Jackson "Sea country" (June–July 1995) *Arena* 24.

72 See discussion by Dillon in Chapter 14 in this volume.

73 Particularly arrangements where the native titleholders are not simply treated merely as another "stakeholder" and attributed the same status as general members of the public.

74 Stephen Sparkes "Native Title – All at Sea?" paper presented at the Native Title Representative Bodies Legal Conference, 28–30 August 2001 (updated 12 October 2001) at 30.

75 *Native Title Act 1993* (Cth), Pt. 2, Div. 3, subdiv. B, C and D.

14 Aboriginal peoples and oceans policy in Australia

An indigenous perspective

Rodney Dillon

Introduction

In a recent publication on aboriginal customary marine tenure, it was noted that even in the field of anthropology, Aboriginal peoples' relationships to the sea have been misunderstood and neglected in a manner that "has resulted in the indigenous relationship to the sea being seen only in terms of resource usage and in the many and complex indigenous systems of near-shore marine tenure worldwide becoming invisible."[1] One of the reasons proffered for this "blind spot"[2] was the manner in which Western relationships to the sea, including views that the seas were open to all, blinkered the way in which indigenous cultures were understood.

This "invisibility" has been more than just a "blind spot" in the field of anthropology. It has been a matter of great convenience to governments and industry groups who, by ignoring Aboriginal peoples' interests in marine environments, have been able to exploit the resources that Aboriginal peoples have always managed. It has also served to deny us a right to make a livelihood from those resources. While the myth of invisibility is being increasingly dispelled, the desire of governments and industry to continue to reap the benefits of that historical injustice remains largely unchanged. Despite being the subject of numerous reports and policy statements espousing principles of increased participation in both resource management and industry participation, tangible benefits for Aboriginal people have yet to be realized.[3]

Australia's Oceans Policy was released on 23 December 1998[4] and creates a new policy and political framework for the management of Australia's oceans and seas. This chapter will review the impact of the *Oceans Policy* upon the interests of Aboriginal people in Australia from an indigenous perspective. Reference will be made to how Aboriginal people's rights and interests are addressed in the *Oceans Policy*, and the challenges to be faced in implementing *Australia's Oceans Policy* consistently with Aboriginal people's rights and interests.

Aboriginal peoples' interests in the marine environment

The special relationship between Aboriginal peoples and marine environments is at least becoming increasingly understood by non-indigenous people. For Aboriginal peoples living in coastal regions, the marine environment has always constituted a fundamental economic and cultural resource. It is central to the spiritual well-being of Aboriginal peoples. Indeed, in some parts of the country, it is not unusual for Aboriginal people to have stories and Dreaming tracks that travel through the sea, imbuing those areas with the cultural identity of the communities to which they belong. The travels of Dreaming creatures and the songs associated with them are sign-posted by numerous sites of significance that can be located several kilometers offshore.[5]

The coastal environment is also important to Aboriginal peoples in other ways. Activities such as hunting and fishing have been passed down over the centuries and are carried out with the knowledge and pride that they represent the continuing of a tradition. It is a means of being on country, of enjoying our home, and making a living from a country that has always been ours. In these ways, hunting and fishing by Aboriginal people are not mere physical activities, they are cultural activities that are central to our identity. Understanding Aboriginal hunting and fishing in these ways makes it easy to understand why Aboriginal people are angered[6] when governments and policy-makers treat them as just another category of recreational user of marine resources.

Aboriginal people have an understanding of our coastal environment that has assisted us to manage coastal resources for generations. We understand the life cycles and seasonal movements of many species. We have knowledge of the times of the year when it is appropriate to fish certain species. There are areas where we do not fish at particular times of the year because they are known breeding areas. We know how to look for, and understand, indicators from the environment that tell when a particular resource is under threat and should be left alone. When we see those signs we do not fish in those areas. That is how we have traditionally ensured that the species is properly managed.

For Aboriginal people, the marine environment is not only an important source of food, it remains an important economic resource. It constitutes a resource that has always been exploited to the extent necessary to satisfy the needs that were required to be satisfied in order to sustain our existence.

The requirement that Aboriginal peoples' interests in coastal environments be protected is no longer solely in the realm of the goodwill of governments. The entitlement of Aboriginal peoples to have our interests in coastal waters protected arises from a range of international instruments and the decisions of Australian courts. The need for the involvement of Aboriginal people in the management of resources and our active participation in resource industries also arises as a matter of social justice. This was recognized by the *Commonwealth Coastal Policy*, which states:

> As a matter of social justice, Aboriginal and Torres Strait Islander peoples should be recognised as participants in the coastal management process, and should be able to derive social, cultural and economic benefit from the use of coastal environments in which they have an interest.[7]

Native title

The recognition of native title in the High Court of Australia's decision in *Mabo*[8] was a significant step for Aboriginal people and is a means to obtain limited recognition and protection of Aboriginal people's relationships to land and sea, including some hunting and fishing activities. However, native title is a white man's law. Yet it also sanctioned the dispossession of Aboriginal people by recognizing the authority of Australian governments to extinguish Aboriginal people's interests in land and sea.[9] Such concepts of "extinguishments" are foreign to our cultures. Furthermore, the recognition of native title itself only occurs on limited terms that have nothing to do with the full and meaningful recognition of Aboriginal people's interests in land and sea. Despite the fact that many Aboriginal people's cultures consider that they own their sea country as much as their land, the Australian courts will not recognize the existence of those interests.[10] The courts have also introduced arbitrary requirements in relation to continuing connection, which have meant that although Aboriginal people themselves maintain their laws and customs, the courts will no longer recognize them.[11] This is done through criteria that pay no regard to the views of indigenous people in relation to such matters.

The manner in which Aboriginal people's cultures are dealt with under the Australian legal system was highlighted in *Yarmirr* v. *Northern Territory*.[12] First, the court held that the recognition of aboriginal inter-tidal rights would be inconsistent with the common law.[13] Second, exclusive fishery rights were held to not exist.[14] Third, in relation to the requirements of permission to fish in the sea, the court held that although the system of permission applied to Aboriginal people, it did not apply to white people.[15] This has in effect meant that white people are not bound by that system of law and custom. To have a system of law and custom that can be freely ignored by a section of the community is little different from having that section of the community saying it does not exist at all.

The *Native Title Act 1993* (Cth) (NTA) affords little protection to native title rights and interests even in the limited form in which they have been recognized. Any future act in relation to the management of waters and living resources in those waters is valid.[16] Although native title holders are afforded a right to comment in relation to those acts,[17] that right has been read to provide for only minimal involvement in the management processes,[18] and even if the procedural right is not afforded to native title-holders, the future act will be valid anyway. Furthermore, there is no right to negotiate in relation to the inter-tidal zone or areas below low water mark.[19] Any act in relation to an offshore place affecting native title is also valid regardless of whether the appropriate procedural rights[20] are afforded to native titleholders.[21] While Aboriginal people are entitled to be compensated for such acts,[22] no such compensation has been paid to date. While the NTA protects aboriginal hunting and fishing rights,[23] that protection does not extend to the protection of those rights carried out for commercial purposes, even where they are recognized.

Finally, one unintended and negative impact of the recognition of native title has been the increasing tendency of Australian governments to consider Aboriginal peoples' interests in marine resources only in terms of native title rights and interests. It is, however, a flawed approach to see the need to recognize the rights of Aboriginal people only in terms of the limited and artificial confines of native title. Aboriginal people's involvement in resource management, the continuation of customary fishing rights, and the need to be able to obtain a livelihood from our country is a human right as well as an issue of social justice.

Aboriginal people and *Australia's Oceans Policy*

Australia's Oceans Policy identifies the concerns of the coastal Aboriginal peoples as including the equitable and secure access to resources, direct involvement in resource planning, management and allocation processes and decisions, formal recognition of traditional patterns of resource use and access, traditional management practices and customary law and conservation of the oceans and its resources, intellectual property, and ownership. In its response to those concerns, the Commonwealth has committed itself to a number of goals in relation to several different areas of aboriginal concern.

The Oceans Policy *and the recognition of Aboriginal sea titles*

While the legal recognition of pre-existing Aboriginal peoples' interests in *Mabo* increased focus on Aboriginal peoples' interests in the marine environments, it should not have taken that recognition to trigger such an interest. The 1993 Coastal Zone Inquiry noted the perceived inadequacy of land tenure arrangements and marine estates commensurate with the status of Aboriginal peoples as the original owners of the coastal zone.[24] The *Oceans Policy* does not satisfactorily address this issue. The absence of an adequate legislative response to the non-recognition of aboriginal sea titles and the limitations of native title means that there will be an ongoing grievance on the part of Aboriginal people in relation to this lack of recognition.

The Oceans Policy *customary fishing rights*

The recognition of aboriginal hunting and fishing rights has been the subject of numerous reports including the Law Reform Commission's inquiry into the Recognition of Aboriginal Customary Law[25] and the Coastal Zone Inquiry.[26] What is clear from the numerous reports and the recommendations contained therein is that, in recent years at least, it is not through a lack of any identification of Aboriginal peoples' interests that they have not been recognized. It is through a lack of willpower, and in some instances outright antagonism, on the part of governments. Indeed, it has often been more convenient for governments merely to have another report or inquiry than meaningfully to implement the recommendations of the previous report.

For most Aboriginal people, the protection of aboriginal hunting and fishing rights has generally only taken the form of exempting Aboriginal people from licenses and otherwise allowing Aboriginal people to take flora and fauna that the general public are prohibited from taking.[27] Very little regard has been given to protecting Aboriginal people from the effects of the intrusion caused by the exploitation of our traditional sea country by commercial fishers, tourist operators and recreational users.

Australia's Oceans Policy provides that the government will "continue" to "remove barriers to indigenous groups practicing subsistence fishing on a sustainable yield basis consistent with conservation of species."[28] Several points can be made in relation to this aspect of the strategy. First, any measures that will increase the ability of Aboriginal peoples to exercise traditional hunting and fishing are welcomed. The second is that the measure is limited to "subsistence" hunting and fishing. It is regrettable that a similar goal was not made in relation to aboriginal hunting and fishing activities generally. Third, the goal is limited to a "sustainable yield basis." While this is in principle a matter of general support, in practice it is a matter of considerable controversy. Unless the government is willing to address all matters relevant to a particular resource, it is wrong simply to point the finger at Aboriginal people for the impact on the species and judge sustainability by that criteria without first reducing the impact on species by other users and coastal development generally.

Finally, it is doubtful just how committed the Commonwealth is in relation to this matter anyway. At the time of the release of the *Oceans Policy*, which undertook to "remove barriers to indigenous groups practicing subsistence fishing," the Commonwealth was actively arguing in Australian courts that the Australian legal system should not recognize the traditional native title rights of Aboriginal people below low water mark.[29] Far from trying to remove barriers, it was actively seeking to create them. With such a two-faced approach to such a fundamental issue, is it any wonder that Aboriginal people are skeptical that the other measures in the *Oceans Policy* will be approached in good faith?

Oceans Policy *and Aboriginal cultural heritage*

Nowhere is the tension between being seen to protect Indigenous peoples' interests and the need to protect the convenience of the "invisibility" of

Aboriginal peoples' interests more apparent than in the area of aboriginal cultural heritage. White Australians are happy to hang the works of Aboriginal peoples on their wall or use them in opening ceremonies for sporting events. However, when it comes to a choice between protecting sites of significance and other cultural interests which conflict with the interests of industry, aboriginal cultural heritage is seldom the winner.

Commonwealth and state heritage protection legislation is as inadequate to protect Aboriginal peoples' interests in coastal areas as it is in relation to other areas.[30] Some state heritage legislation tends to emphasize the protection of aboriginal cultural heritage of archaeological value, such as "artifacts" and "relics," but is inadequate to protect sites of spiritual significance.[31] Furthermore, in some legislation, part of the protection mechanism is to deem those "artifacts" and "relics" the property of the Crown.[32] This is offensive to Aboriginal people. Far from providing protection, it often is no more than a means of regulating the destruction of that material, with government agencies being provided with authority to give consent for its destruction.[33] While the *Aboriginal and Torres Strait Islander Heritage Protection Act 1984* (Cth) is supposed to provide protection where state regimes are inadequate, that protection is afforded at the discretion of the government.[34] Not only has it been sparingly used,[35] but when convenient the government has also passed specific legislation removing the protection afforded by the Act.[36]

The Coastal Zone Inquiry noted the concerns of Aboriginal people that their cultural heritage, including sites of cultural significance in coastal areas, was not under aboriginal control, was inadequately protected or managed by government agencies and that such heritage should be under the control of Indigenous people.[37] There has been little change to heritage legislation since that report.

Australia's Oceans Policy now provides that the government will address the threats of impacts posed by activities on fishery resources and marine sites valued by Aboriginal communities.[38] While this is welcomed, it is in many respects an example of the government being seen to be doing something without really doing anything. If the government was serious about addressing our concerns it could have enacted legislation improving the protection of aboriginal cultural heritage and implemented the recommendations of the "Review into the Aboriginal and Torres Strait Islander Heritage Protection Act" which took place in 1996.[39] In any event, given the reluctance of the Commonwealth to use its own legislation to protect sites of significance, Aboriginal people can have little comfort that this will be meaningfully implemented. Indeed, at the same time that the *Oceans Policy* was being developed, the Commonwealth was removing the protection of that legislation in relation to one coastal region to allow the construction of the Hindmarsh Island bridge in South Australia.[40]

Oceans Policy *and the recognition of Aboriginal marine knowledge*

There is presently an inadequate recognition of aboriginal cultural knowledge in the marine environment. It receives inadequate protection under copyright laws and as a native title right.[41] The courts have recently also stated that native title does not extend to the protection of knowledge of sites and country because such matters are not regarded as interests in land and waters as defined in the NTA.[42] It is also inadequately protected under aboriginal cultural heritage legislation. In relation to aboriginal cultural knowledge, the *Oceans Policy* states that the government "will continue to implement the National Aboriginal and Torres Strait Islander Cultural Industry Strategy as it is applicable to the natural and cultural heritage values of Australia's marine areas."[43] That strategy is aimed at providing economic empowerment, access and coordination and various support networks to enable indigenous communities to use their cultural resources to obtain a degree of self-sufficiency.[44]

Oceans Policy *and the involvement of Aboriginal people in management of marine resources*

Along with the rejection of the stereotype of hunter-gatherers as passive "food-collectors" in opposition to "active, food producing agriculturalists"[45] has come a greater recognition of the benefits of Aboriginal people's participation in resource management,[46] including coastal resource management. Despite this, in most areas, governments have refrained from permitting Aboriginal people to enjoy meaningful partnerships or management of coastal environments. The *Oceans Policy* provides that the Commonwealth government will:

- provide for Aboriginal and Torres Strait Islander representation on the National Oceans Advisory Group and on Regional Marine Plan Steering Committees;
- provide for Aboriginal and Torres Strait Islander participation at the National Oceans Forum;
- consult with Indigenous groups on the requirements for establishing a national consultative mechanism, such as an annual forum; and
- continue to develop and implement principles and guidelines for co-management or relevant marine areas and resources.[47]

Making a statement in relation to Aboriginal people's involvement in the management of coastal resources, and even making provision for such consultation to occur, does not guarantee effective involvement. Aboriginal people will be unable to benefit from measures designed to increase their involvement in those processes unless they are adequately resourced to participate effectively in those processes.[48] Furthermore, creating requirements for Aboriginal people to be consulted or to have representation on various committees does not mean that their views will be accommodated. Meaningful recognition of a role in marine resources management means being able to make decisions in relation to the management of resources. There are now, across the country, numerous policies that require consultation with Aboriginal people in relation to resource management. However, merely being able to put a point of view across, which may or may not be taken into account, does not constitute management of resources.

Furthermore, Aboriginal people's involvement in the management of resources, while important, should not be seen as a means of avoiding the need to address other matters of importance, such as the need to increase the involvement of Aboriginal people in the fishing industry.[49] Again, the government's commitment to meaningful involvement in the management of resources is questionable. Around the same time that the Commonwealth was formulating the *Oceans Policy* it was enacting the amendments to the *Native Title Act 1993* (Cth). These amendments ensured that native titleholders would only have "a right to comment" in relation to any future act affecting native title that is done under legislation in relation to the management of waters and marine resources.[50]

Oceans Policy *and Aboriginal peoples' involvement in the commercial fishing industry*

Aboriginal people in coastal areas have always made their livelihood from the sea. The disruptions caused by white settlement have meant that there are new pressures and demands on Aboriginal communities. Accordingly, our use of sea resources has been required to adapt in order to deal with those new pressures and demands. In many areas, Aboriginal people have had a history of involvement in the fishing industry. That involvement has been both with Aboriginal people themselves participating in those industries as well as comprising a labor force for the industry.[51] With more complicated licensing regimes being introduced and the introduction of share quota management systems, it is Aboriginal people who are increasingly being denied a livelihood from the exploitation of resources that they have always exploited. Similarly, with greater mechanization of the fishing industry, the industry has become less labor intensive and the employment opportunities for Aboriginal people have decreased.

Aboriginal people continue to be disadvantaged by being increasingly isolated from the commercial fishing industry. The emphasis on regional management plans tends to focus on existing uses rather than a reconsideration of the equitable distribution of resources. The nature of the use that Aboriginal people should be entitled to is not properly considered in the preparation of those plans. This can itself form a barrier to the future involvement of Aboriginal people because remedial measures for Aboriginal people, which may be considered at a later date, may be inconsistent with the plan and this can be used as an excuse not to take any further action.

A similar problem arises in the context of the increasing trend towards share quota management fisheries. Under share-managed fisheries, the allocation of quotas is often worked out by existing commercial operations and the current catches under various licenses. In circumstances where we have already been excluded, such a criteria effectively excludes us from the industry permanently. In effect, such criteria rely on the historical invisibility of Aboriginal people and entrench that injustice into legislative regimes.

The need for a greater level of Aboriginal people's involvement in the fishing industry has been the subject of numerous reports in the last 20 years. The Coastal Zone Inquiry recommended that an indigenous fishing strategy should be developed that includes measures to improve economic development and employment opportunities for indigenous communities in fisheries and mariculture ventures. Options included the reservation of a proportion of fishing or other licenses for indigenous communities, the purchase of such licenses on behalf of indigenous communities by the Aboriginal and Torres Strait Islander Commission, and the establishment of fishing zones adjacent to land owned or controlled by indigenous people in which communities could operate their own commercial enterprises, participate in joint ventures, or license access by other marine resource users.[52] Despite such recommendations, there have not been any significant moves to involve Aboriginal people in the commercial fishing industry.

Australia's Oceans Policy does not take any significant step forward on this issue either. The *Oceans Policy* provides that Australia will enable increased opportunities for Aboriginal and Torres Strait Islander people to be involved in commercial fishing and will implement the National Aboriginal and Torres Strait Islander Rural Industry Strategy as it is relevant to ocean-based industries.[53] That Strategy relevantly states that the Commonwealth is only to:

- encourage the extension of preferential licensing to indigenous people for collection of abalone, trochus, beche-de-mer and mud crabs in appropriate locations;
- support the reservation and buy back of fishing licenses where Aboriginal people have been excluded from the local commercial fishing industry; and
- assess market opportunities for increased production and value adding by indigenous communities in relation to abalone, trochus, beche-de-mer, shark fins, rock lobster and mud crabs.[54]

The goals of the *Oceans Policy* are not sufficiently clear to ensure their effective and meaningful implementation. The Policy focuses on increasing opportunities within the existing regime. Where detailed measures could have been implemented, the Policy instead speaks in terms of providing "increased opportunities." The nature and extent of the increase remains undefined.

If the requisite goodwill was forthcoming, the *Oceans Policy* could have been much more precise in its standard setting in this regard. Not only could it have identified tangible steps to ensure a more equitable distribution of the commercial fishery, it could have also set out principles for inclusion of Aboriginal people where new share quotas or developmental fisheries are established. The absence of such measures is unfortunate because both these last mentioned matters represent clear opportunities to provide equity to Aboriginal people in the context of either a broader restructuring of the industry or otherwise with minimal interference with other existing commercial operations. Yet even in these circumstances it seems that the disadvantage to Aboriginal people is difficult to address.

The Oceans Policy *and Aquaculture*

Aquaculture is one area where there has been increasing interest in developing partnerships with Aboriginal people, and it represents a valuable opportunity for a different outcome for Aboriginal people from our isolation from traditional fish resources. The "National Aquaculture Development Strategy for Indigenous Communities"[55] at least identifies several steps that may be undertaken to provide Aboriginal people with the means to enter into that industry. The undertaking in the *Oceans Policy* to implement the "National Aboriginal Torres Strait Island Rural Industry Strategy"[56] includes the references in that Strategy to measures in relation to aquaculture. Those measures include that the Commonwealth is to:

Action 2.10: Recognise the interests of Indigenous communities within the National Aquaculture Strategy.

Action 2.11: Provide technical support to Indigenous communities wishing to plan for and establish aquaculture enterprise for community food supplies or for external sales.

Action 2.12: Assist the planning and establishment of aquaculture enterprises where they are likely to achieve significant economic benefits for Indigenous communities, either in their own right or as a component of diversified production.

As this is a developing area of the fishing industry it is perhaps too early to pass judgment on how meaningfully these measures will be introduced and how effectively the Strategy will lead to the ongoing involvement of Aboriginal people in the industry. It is hoped that the Strategy will be pursued with more conviction by Australian governments than other areas of the *Oceans Policy* relating to Aboriginal people.

Conclusion

Smyth has succinctly summarized the frustration of Aboriginal people over the inaction by governments in relation to Aboriginal participation in the management of marine resources and the commercial fishing industry.[57] The *Oceans Policy* does not contain structural change in the management of coastal resources or Aboriginal involvement in the commercial fishing industry. Even with those aspects of the *Oceans Policy* that may deliver real change to Aboriginal people, there has been no rush to commit to genuine reform. Instead, we have been left wallowing in a morass of reports and "feasibility studies" containing recommendations calling for "the examination of options," the "development of protocols," the further investigation of "options," and the identification of "further strategies." While the various policies nearly always represent an improvement on what was previously available, the fundamental issues are not addressed. In this way the disadvantage for Aboriginal people, which arises from the historical invisibility of our interests, is entrenched for future generations. The extent to which the *Oceans Policy* can deliver outcomes where previous reports and policies have failed remains to be seen, but current indications are that it will be unlikely to deliver real change to how indigenous peoples' interests are recognized and protected.

Acknowledgment

The assistance of the Jumbunna Indigenous House of Learning, University of Technology, Sydney, Australia, in the preparation of this paper is acknowledged.

Notes

1 N. Peterson and B. Rigsby (eds) *Customary Marine Tenure in Australia* (Oceania Monograph 48, University of Sydney, Sydney: 1998) at 1.
2 Ibid., at 2.
3 For a description of some of those varying reports, see M. Tsamenyi and K. Mfodwo *Towards Greater Indigenous Participation in Australian Commercial Fisheries: Some Policy Issues* (Aboriginal and Torres Strait Islander Commission, Canberra: 2000); and D. Smyth "Fishing for recognition: The search for an indigenous fisheries policy in Australia" (2000) 4 No. 29 *Indigenous Law Bulletin* 8–10.
4 Commonwealth of Australia *Australia's Ocean Policy* Volumes I and II (Environment Australia, Canberra: 1998) (*Australia's Ocean Policy*).
5 See, for example, Dick Roughsey's account of the flood-making ceremony on Sydney Island in the Gulf of Carpentaria in D. Roughsey *Moon and Rainbow: The Autobiography of an Aboriginal* (A.H. & A.W. Reed, Sydney: 1971) 63–9; P. Memmott and D. Trigger "Marine Tenure in the Wellesley Islands Region, Gulf of Carpentaria" in Peterson and Rigsby, note 1 at 109–24, especially 119–21.
6 See, for example, M. Leon "NSW indigenous fisheries strategy: Friend or foe" (2001) 5, No. 9 *Indigenous Law Bulletin* 13.

7 Department of Environment, Sport and Territories (DEST) *Living on the Coast: The Commonwealth Coastal Policy* (DEST, Canberra: 1995) as quoted in J. Sutherland *Fisheries, Aquaculture and Aboriginal and Torres Strait Islander Peoples: Studies, Policies and Legislation* (Environment Australia, Canberra: 1996) at 57.
8 *Mabo* v. *State of Queensland (No. 2)* (1992) 175 CLR 1.
9 Ibid., per Brennan J at 63–9. See also *Fejo* v. *Northern Territory* (1998) 195 CLR 96; *Western Australia* v. *Ward* (2002) 191 ALR 1; *Wilson* v. *Anderson* (2002) 190 ALR 31; *Fourmile* v. *Selpam Pty Ltd* (1997) 80 FCR 151.
10 *Commonwealth* v. *Yarmirr* (2001) 208 CLR 1 at paras. 97–100.
11 See, for example, *Members of the Yorta Yorta Aboriginal Community* v. *State of Victoria* [2002] HCA 58 and *De Rose* v. *South Australia* [2002] FCA 1342.
12 *Yarmirr* v. *Northern Territory* (1998) 82 FCR 533 (at trial) and on appeal to the High Court, *Commonwealth* v. *Yarmirr* (2001) 208 CLR 1; see also discussion by Clark in Chapter 13 in this volume.
13 *Commonwealth* v. *Yarmirr* (2001) 208 CLR 1 at paras. 97–100 per Gleeson CJ, Gaudron Gummow and Hayne JJ.
14 See at trial *Yarmirr* v. *Northern Territory* (1998) 82 FCR 533 at 593. It was also implicit in the majority judgment in the High Court, ibid. See also *State of Western Australia* v. *Ward* (2002) 191 ALR 1 at para. 388 per Gleeson CJ, Gaudron, Gummow and Hayne JJ.
15 *Yarmirr* v. *Northern Territory* (1998) 82 FCR 533 at 585. That finding was not disturbed on appeal to the High Court of Australia. Part of the reasoning was based on an answer given by an Aboriginal witness who had poor English skills. For an account of some of the evidence in this matter and how it was misconstrued in that case, see N. Evans "Country and the Word: Linguistic Evidence in the Croker Sea Claim" in J. Henderson and D. Nash (eds) *Language in Native Title* (Australian Institute of Aboriginal and Torres Strait Islander Studies, Canberra: 2002) 53–99 esp. 86–93.
16 See *Native Title Act 1993* (Cth), s. 24HA (3).
17 Ibid., at s. 24HA (7).
18 See generally F. Anggadi "The Ambit and Nature of Claimant Rights Under S.24HA of the Native Title Act: *Harris* v. *Great Barrier Reef Marine Park Authority* [2000] FCA 603" (2000) 5, No. 2 *Indigenous Law Bulletin* 18.
19 *Native Title Act 1993* (Cth), s. 26(3).
20 In relation to offshore places, native titleholders have the same procedural rights as if they instead held corresponding non-native title rights in those waters: *Native Title Act 1993* (Cth), s. 24NA (8). In relation to coastal waters which would be regarded as an onshore place, native titleholders have the same rights as other titleholders, although in relation to a compulsory acquisition a right to negotiate may apply: *Native Title Act 1993* (Cth), ss. 24MB, 24MD. Neither of these rights apply if the act relates to the management of water and airspace because such acts would then be covered by *Native Title Act 1993* (Cth), s. 24HA.
21 The act will be valid regardless of the procedural rights being afforded: *Lardil Peoples & Ors* v. *State of Queensland & Ors* [2001] 108 FCR 453.
22 In relation to acts constituting management of water and airspace, see *Native Title Act 1993* (Cth), s. 24HA (5) and in relation to offshore places see *Native Title Act 1993* (Cth), ss. 24NA (5), 24NA (6).
23 *Native Title Act 1993* (Cth), s. 211.
24 Resource Assessment Commission *Coastal Zone Inquiry: Final Report* (Resource Assessment Commission, Canberra: 1993) at 177.
25 Australian Law Reform Commission *The Recognition of Aboriginal Customary Laws* (Report No. 31, Vol. 2, Australian Law Reform Commission, Sydney: 1986) at 200.

26 Resource Assessment Commission, note 24, ch. 10.
27 See, for example, *Living Marine Resources Management Act 1995* (Tas) s. 60(2); *Fish Resources Management Act 1994* (WA) s. 6; *Fisheries Act 1995* (NT) s. 53; *Fisheries Act 1994* (Qld) s. 14. For an overview, see Sutherland, note 7 at 25–36.
28 *Australia's Oceans Policy: Specific Sectoral Measures*, note 4 (Part 2.11) at 24.
29 Indeed the Commonwealth lodged an appeal in the Crocker Island case on this point, see *Lardil, Kaiadilt, Yangkaal and Gangalidda Peoples* v. *State of Queensland & Ors,* FCA Proceedings QG207, Written Submissions of the Second Respondent: The Commonwealth of Australia (1 November 2002), paras 85–6.
30 For a comparative overview of state and territory aboriginal heritage legislation, see E. Evatt "Overview of state and territory aboriginal heritage legislation" (1998) 4, No. 16 *Indigenous Law Bulletin* 4–8.
31 E. Evatt *Review of the Aboriginal and Torres Strait Islander Heritage Protection Act 1984* (Minister for Aboriginal and Torres Strait Islanders Affairs, Canberra: 1996) 77–81.
32 See, for example, *National Parks and Wildlife Act 1974* (NSW) s. 83; *Aboriginal Relics Act 1975* (Tas) s. 11; *Cultural Record (Landscape Queensland and Queensland Estate) Act 1987* (Qld), s. 33.
33 See, for example, *Aboriginal Relics Act 1974* (Tas), s. 14(1); *National Parks and Wildlife Act 1974* (NSW), s. 87.
34 *Aboriginal and Torres Strait Islander Heritage Protection Act 1984* (Cth), ss. 9, 10.
35 For a complete analysis of the *Aboriginal and Torres Strait Islander Protection Act 1984* (Cth), see Evatt, note 31.
36 See, for example, the *Hindmarsh Island Bridge Act 1997* (Cth) and *Kartinyeri* v. *Commonwealth* (1998) 195 CLR 337.
37 Resource Assessment Commission, note 24 at 178.
38 *Australia's Ocean Policy: Specific Sectoral Measures*, note 4 (Part 2.11) at 24.
39 Evatt, note 31.
40 See *Hindmarsh Island Bridge Act 1997* (Cth) and *Kartinyeri* v. *Commonwealth* (1998) 195 CLR 337.
41 *Bulun Bulun* v. *R&T Textiles Pty Ltd* (1998) 86 FCR 244 at 256 per Van Doussa J.
42 *State of Western Australia* v. *Ward (2002)* 191 ALR 1 at paras 58–60 per Gleeson CJ, Gaudron, Gummow and Hayne JJ.
43 *Australia's Oceans Policy: Specific Sectoral Measures*, note 4 (Part 2.11) at 24.
44 Aboriginal and Torres Strait Islander Commission (ATSIC) *National Aboriginal and Torres Strait Islander Cultural Industry Strategy – A Summary* (ATSIC, Canberra: July 1997).
45 Nancy M. Williams and Eugene S. Hunn (eds) *Resource Managers: North American and Australian Hunter-Gatherers* (Westview Press, Boulder: 1982) 1. See also Aboriginal and Torres Strait Islander Social Justice Commissioner *Native Title Report: January–June 1994* (Aboriginal and Torres Strait Islander Social Justice Commissioner, Canberra: 1994) at 145.
46 Aboriginal and Torres Strait Islander Social Justice Commissioner, ibid., at 145.
47 *Australia's Ocean Policy: Specific Sectoral Measures*, note 4 (Part 2.11) at 24.
48 Resource Assessment Commission, note 24 at 181 at 10.44. See also Tsamenyi and Mfodwo, note 3 at 9–12.
49 A similar point has been made in Tsamenyi and Mfodwo, note 3 at 6.
50 See *Native Title Act 1993* (Cth), s. 24 HA (7).
51 For one account of a history of Aboriginal peoples' involvement in the fishing industry in one community, see S. Cane "Aboriginal Fishing Rights on the New South Wales South Coast: A Court Case" in Peterson and Rigsby, note 1 at 66–88.

52 Resource Assessment Commission, note 24 at 187.
53 *Australia's Ocean Policy: Specific Sectoral Measures*, note 4 (Part 2.11) at 24.
54 See Actions 2.4–2.6, the ATSIC *Aboriginal and Torres Strait Islander Rural Industry Strategy* (ATSIC, Canberra: 1997).
55 B. Lee and S. Nel *A National Aquaculture Development Strategy for Indigenous Communities in Australia: Final Report* (Department of Agriculture, Fisheries and Forestry – Australia, Canberra: March 2001).
56 *Australia's Ocean Policy: Specific Sectoral Measures*, note 4 (Part 2.11) at 24.
57 Smyth, note 3 at 10.

Part VII

Future directions

Theoretical and practical challenges

15 The challenge of international ocean governance

Institutional, ethical and conceptual dilemmas

Douglas M. Johnston

Introduction

The concept of "governance," distinguished from "government," envisages a broad array of legitimate inputs into public policy making under the changing conditions of contemporary society.[1] At the international level, the governance concept is the latest intellectual response to the "anarchic" state of world society,[2] and reflects a general awareness of the inadequacy of government systems and intergovernmental organizations to deal with complex problems. The broadened governance view of legitimacy, embracing economic and societal as well as political values, coincides with a new determination to achieve a higher degree of effectiveness in the highly complex field of ocean management. The need for a higher degree of sophistication in "ocean governance" opens up challenging opportunities for cross-disciplinary and cross-sectoral interaction, whether or not one espouses the vision of "comprehensive ocean management,"[3] or subscribes to the educational goal of "consilience."[4]

We are living in an age of ambitious hopes for the future of international society. With a view to evaluating the case for optimism in the field of ocean management, it may be useful to identify some institutional, ethical and conceptual dilemmas that seem to obstruct progress. This chapter concludes with reflections on the future role of maritime regime building as the best approach to effective ocean governance.

Institutional dilemmas

The Westphalian framework and UNCLOS III

The global governance concept has become popular in many circles because of the general perception of the limitations inherent in state institutions. Yet among specialists in the sub-discipline of international relations theory, the Westphalian model still occupies a central position. It postulates a "system of territorially organized states operating in an anarchic environment,"[5] each state being "capable of defining its own goals and cultural mission."[6] Although the goal of ultimate control within a specific territory remains elusive to many actors, it is still widely assumed by them that their potential should be defined in terms of "an autonomous state, nationally unified and in control of its own economic policy."[7]

Traditionalists in the sub-discipline of public international law have reinforced the Westphalian view of the world by insisting on the centrality of the principles of state sovereignty and formal equality among states. By assuming that sovereignty must reside in "final authority,"[8] they have imposed a strait-jacket of dichotomous (either–or), sovereignty-related concepts such as autonomy, jurisdiction, control and territoriality.[9] The traditional principle of territorial sovereignty, so long and so widely regarded as the *grundnorm* of the Westphalian system of nation-states,[10] has been subject to "reduction" within the United Nations (UN) system since 1945, originally on the basis of overriding human development values invoked in the Charter[11] and more recently within the expanding framework of human rights.[12] On the other hand, very few would argue, as a matter of law, that the international community's interest in environmental protection gives it an entitlement to intervene in the territory of a state against the will of the sovereignty-holder, although some multilateral interventions, as in the case of Kuwait, have been responses to crises with several dimensions, including the threat of environmental disaster. Yet it can certainly be argued that neighboring states sharing a vulnerable transboundary ecosystem or migratory range have special responsibilities to each other, responsibilities that would limit the sovereignty-holder's autonomy within its own territory.[13] The special case for transboundary responsibility might be said to rest on "soft law" obligations to consult, notify, negotiate and cooperate that necessarily reduce the degree of autonomy inherent in the traditional concept of territorial sovereignty. In certain circumstances, such an obligation might take the form of a duty to participate in the building of a cooperative environmental regime designed to serve protective or problem-solving purposes.[14]

In the modern age of environmentalism, which places unprecedented emphasis on transboundary cooperation, wholly rational ocean management is virtually inconceivable within the Westphalian framework of sovereign nation-states. At the Third UN Conference on the Law of the Sea (UNCLOS III) ingenious efforts were made to cultivate more equitable statist concepts, such as "sovereign rights," that would be free of the rigidities associated with absolute sovereignty and yet provide the assurance demanded by developing, and (mostly) newly independent, coastal states that their nation-building would be assisted by a huge seaward extension of their land economy without the threat of obstruction by technologically superior foreign states and enterprises. Indeed the establishment of the exclusive economic zone (EEZ) and the re-definition of the continental shelf accomplished at UNCLOS III have helped a number of these states to develop their own national fishing and offshore oil industries, and others to acquire new sources of revenue from foreign states or enterprises permitted to operate in their waters. But the seaward extension of the Westphalian order at UNCLOS III has failed to achieve an effective system of ocean management, as distinct from an equitable system of state entitlement.[15]

Despite the progress achieved at UNCLOS III in the development of "functional jurisdiction," which is intended to de-territorialize the concept of extended coastal state jurisdiction, it remains difficult to promote the goal of cooperative ocean management on the foundation of national entitlement that was solidified at UNCLOS III. Adjustments would have to be made within the Westphalian framework of state autonomy principles and analogues, through the application of relativist notions such as modified or conditional sovereignty, shared territory, or degrees of authority to the holders of EEZ and continental shelf entitlements. The future of such relativist concepts is taken seriously by some political scientists, and there are some historical analogies.[16] Yet in retrospect, it seems that the neo-Westphalian ethic of nation-building that pervaded the UNCLOS III restructuring of the law of the sea may have raised, rather than lowered, the barriers to international cooperation in the oceans, overweighting national entitlement at the price of transnational responsibility. In most countries around the world, the goal of truly effective national ocean management remains remote, and will become less so only if the entitled state is prepared to sacrifice some degree of national autonomy, *de facto* if not *de jure*. The "structural dilemma" is how to persuade sovereignty-holders to comply with globally negotiated provisions that can be implemented effectively only through resort to genuinely cooperative ocean management arrangements.

The world community arenas

Since the 1920s, and especially since the creation of the United Nations in 1945, the traditional "transactional" model of international law, based on (mostly) bilateral inter-state relationships, has been complemented by an "organizational" model based on a proliferation of intergovernmental agencies. Today we have inherited a myriad of arenas for official interactions and cooperative initiatives in a wide variety of complex problem contexts. At the global level, both inside and outside the UN system, these arenas serve several sectors of ocean management: fishery management and conservation (the Food and Agricultural Organization – FAO), whaling regulation (International Whaling Commission – IWC), vessel-based marine pollution control (International Maritime Organization – IMO), marine scientific research (International Oceanographic Commission – IOC), environmental protection (United Nations Environment Programme – UNEP), and the prospective regulation of deep ocean mining (International Seabed Authority – ISA). Noticeably missing is one economically important ocean use sector, the management of offshore hydrocarbon development, which from its infancy in the 1930s has remained exclusively within the jurisdiction and control of coastal states and been permitted a high degree of autonomy.

It was understood at UNCLOS III that all of these international organizations, and several others, would have a major role in the implementation of the numerous provisions in the 1982 United Nations Law of the Sea Convention (LOS Convention) that invoke the need for "international cooperation."[17] Indeed it was foreseen that the effectiveness of much of the new normative system would depend on the potentiality for institutional development at the regional as well as the global level. But these tasks of implementation are enormous.[18]

The fatal flaw in the Westphalian model of world affairs is the "problem of difference": states are very far from equal except in legal fiction. Participation in international organizations is assumed to be the key to reducing disparities that tend to limit the functionality of cooperative management regimes. In theory such participation is voluntary; in practice it is virtually obligatory. Yet, in reality, all of these organizations are sadly deficient in funds and other resources. Despite the numerous programs and projects generated at the global level, most developing coastal states are making only glacial progress toward the Westphalian goal of national autonomy in contexts as difficult as ocean management.

Specialists in the sub-discipline of international organization generally emphasize the limitations inherent in large-scale structures at the global or macro-regional level – arguably all those structural, hierarchical, political and financial limitations inherent in all large-scale national government bureaucracies, overlaid by inter-cultural, strategic and even ideological constraints that operate at a high "international community" level.[19] Even in relatively favorable circumstances, where an international organization functions with relatively minor friction and infrequent eruptions over fundamental issues, the politics of conference diplomacy at the global or macro-regional level is rarely simple. In most of those organizations devoted to ocean development and management, the national strategies of the United States, Western Europe, Japan, and a few other well-endowed member states, such as Canada and Australia, tend to carry the day more often than not, but only if they are consistent with one another, as demonstrated by climate change diplomacy.[20] It is by no means certain that the American electorate will support a US government policy favoring more generous funding of international organizations than in the past, except perhaps in situations where the use of these funds is seen to be in line with US foreign policy. Since the terrorist attacks on the American homeland in September 2001, the specter of US unilateralism, if not neo-isolationism, has become a more disturbing threat to the attainment of ambitious global goals, such as international ocean and environmental management, especially if the undertaking is seen to depend on intergovernmental organizations.

The dilemma here, the "organizational dilemma," is whether it is realistic to adhere to the orthodox UNCLOS III view that international organizations are central to national efforts to achieve ambitious ocean management goals.

In the theory of international relations, the "globalization debate" has re-introduced the question of why states cooperate at all. In the past, utilitarians such as Hobbes, Locke, Madison and Mill regarded institutions as necessary to provide solutions to problems, because without them chaos would reign.[21] Modern political realists have scant regard for international institutions, which they consider to have no capacity to enforce compliance, and therefore no power. For modern institutionalists, on the other hand, the power of an institution arises not from its capacity to use physical force, but from the benefits its members derive from participation.[22] By this neo-utilitarian reasoning, international organizations in the field of ocean management can play a central role only if they are able to generate sufficient benefits for their members in the form of "mutual gains."[23] The modest scale of such benefits in that field is a further reminder to governance advocates of the inadequacy of organizations consisting of state institutions.

Development assistance strategies

The prospect of accelerating progress toward the goal of global ocean governance might be enhanced if the development assistance agencies, charitable foundations and other institutions that fund (and often conceptualize) programs and projects were to assign a higher priority to ocean affairs. But these institutions are divided over the best way to allocate development assistance and administer funds. Some development assistance agencies, for example, are beginning to favor increased allocations to global and regional funding outlets (e.g. World Bank, the Global Environment Facility – GEF, favorite UN agencies like the UN High Commissioner for Refugees and UNICEF, and the regional development banks); others, in a minority, suggest that trust should be reposed in the supervisory capacities of at least the more effective and reliable regional political organizations (e.g. the Association of Southeast Asian Nations – ASEAN, South Pacific Forum, the Caribbean Community – CARICOM) as vehicles for the delivery of ocean-related programs and projects within their region.[24]

It seems to be generally agreed that there is, and ought to be, a gradual (or not-so-gradual) movement away from the traditional emphasis on direct, bilateral (donor-to-donee) assistance. It is often argued that this emphasis made sense only in the era, now passing, when primacy had to be given to basic infrastructural deficiencies in public works such as housing, hospitals, roads, bridges, railways and communications. Although these deficiencies are still severe in many countries, especially in Africa, a growing number of countries have reached a stage of development where greater weight should be given to their need for technical and professional expertise in dealing with problems of great complexity.

Not so long ago much of the funding allocated to "ocean affairs" was still seen to contribute to the highly developed, relatively homogeneous, though increasingly cross-disciplinary, field of the "law of the sea." By the early

1980s much of that funding, applied to developing regions such as Asia-Pacific, was justified chiefly by the practical need to provide a forum for regional law of the sea specialists to meet and discuss their countries' common problems associated with the implementation of the LOS Convention.[25] But very quickly after the completion of UNCLOS III, law of the sea issues lost much of their appeal to funding sources,[26] and it became increasingly necessary for project proponents to gear their applications for funds to the training and human resource development requirements of developing coastal states.[27] By the late 1980s the focus was on the more ambitious tasks of institutional capacity building,[28] with special attention to the nurturing of indigenous institutions.[29] Since then, and especially since the 1992 Earth Summit (the UN Conference on Environment and Development), funding in "ocean affairs" has often focused on the implementation of such documents as Agenda 21 framed around ambitious goals such as integrated coastal zone management.[30] In the eye of hard-headed project officers, this surely represents a weakening of focus, creating administrative problems of project evaluation that are aggravated by demands for stricter accountability.[31] Moreover, those responsible for executing ambitiously conceived, broadly designed, projects find it increasingly difficult to show impressive results within the normal five-year time frame.

Given the trend to increasingly ambitious objectives, a number of crucial dilemmas have to be addressed. Is the goal of effective ocean management of such transcending priority as to justify the increasing use of "leverage" by such institutions as agents for the "international community"?[32] How many kinds of equal or non-equal partnerships between donors and recipients can be usefully invented? Would the participation of regional political organizations as "intermediary" between donor and beneficiary be likely to accelerate or delay progress toward the twin international goals of effective and equitable ocean management? Would conditionality imposed by a regional organization be more acceptable than that imposed by a global organization? Is the growing concern with the fact or prospect of climate change the most reliable means of raising worldwide consciousness of the need for a rational and equitable use of the oceans? How much will non-governmental organization (NGO) assistance to developing countries contribute to that objective without undermining the statist objectives and methods of international organization?[33]

Diplomatic options

The foreign policy approach to ocean management does perhaps bring in some of the brightest and best of those in government service; many with a flair for flexibility and an internationalist view of the world. The danger has always been that diplomats, dealing much of their professional lives with counterparts from other foreign ministries, may bring home an ingenious and forward-looking text that cannot, however, be "sold" to the managers of foreign policy, much less to the managers of the various cognate sectors of domestic policy. To reduce this risk, international negotiations in complex situations are increasingly preceded by careful, and often protracted, "internal diplomacy" at national and sub-national levels. Intra-national consultations often result in the appointment of national delegations composed of representatives of divergent, but potentially convergent, viewpoints, in the hope that intra-national compromises can be reached through exposure to the greater complexity of divergences, and the greater need for compromise, within the international arena.

Perhaps the chief danger in global diplomacy today is the tendency to generate "over-expectations" through ambitious and ostensibly successful negotiations. In the specific context of ocean management, idealized goals such as "integrated coastal or ocean management" set an exciting agenda for both kinds of diplomacy, intra-national as well as inter-national.[34] Selling such an ambitious, and potentially sophisticated, goal is a stirring challenge to the art of salesmanship. But the international marketing of this vision may be resisted by ocean managers of less developed countries who interpret the new thrust as unduly ambitious, creating educational challenges that cannot be met by their institutions.

Resistance of this kind might be more easily overcome if the "foreign policy challenge" is brought down to the regional level. Arguably, the global goal of integrated ocean or coastal management is a degree less intimidating if it becomes the focus of regional diplomacy designed to adapt global commitments to local conditions and realistic expectations at that level. Moreover, a regional organization (e.g. ASEAN, the South Pacific Forum, or CARICOM) may be more likely than individual member states to be able to engage in successful fundraising outside the region with a view to initiating research networks, programs and projects designed around globally approved goals.[35]

By this argument, the foreign policy approach to ocean management, begun at the global level, requires follow-up diplomacy at the regional level in order to make the goals attainable. But it is certainly true that few, if any, of the existing political organizations at the regional level are capable of more than the *promotion* of ocean management goals and priorities in the region, and perhaps the *supervision or coordination* of research, training and other modes of capacity-building among the members. So the "operational dilemma" is whether it is feasible for regional political organizations to assume responsibilities for facilitating operational cooperation among its members in the field of ocean governance.

The "political regions" associated with such organizations rarely, if ever, correspond with what would be regarded as appropriate units of ocean management.[36] The "ecological imperative" that has had so much impact on post-UNCLOS thinking about the future of ocean management has furnished a new kind of idealism around the concept of "large marine ecosystems" (LMEs).[37] Marine ecologists in particular advocate the need to pursue lines of research that would demonstrate both the feasibility and desirability of cooperative ocean management regimes structured and designed for each LME.[38] However, even the most enthusiastic recognize the range of political, legal and perhaps even cultural barriers that would have to be circumvented before the vision of a global system of LME regimes could be realized.

Given these institutional difficulties, it might be more sensible to focus instead on developing cooperative ocean management arrangements at the *sub-regional* level, whereby smaller groupings of neighboring coastal states would be required to participate in the day-to-day tasks of building and operating an appropriate regime. The most obvious example of sub-regional cooperation of this kind would be in *semi-enclosed waters*, where the neighboring littoral states have a shared interest in the negotiation, formation and maintenance of such a regime.[39]

The role of social institutions

Even with the political energies of intergovernmental organizations, the resources of international development institutions and the skills of sophisticated diplomats, state institutions in most developing countries find the challenges of ocean management to be beyond their capacity. Increasingly they find themselves dependent on the growing array of "civil society" institutions that offer funds and talents to supplement the resources available in the public sector. Through the combined contributions of state and social institutions, in various kinds of coalitions and partnerships or separately, it becomes possible to bring much more significant capacities to focus on the complex tasks of ocean management under the banner of "ocean governance."

The involvement of NGOs in international ocean affairs began modestly at the time of the 1972 UN Conference on the Human Environment, the first of many UN mega-conferences organized for designated fields of particular complexity.[40] The 15-year law-making process of UNCLOS III presented more technical issues to be grasped and more stubborn diplomatic hurdles to be circumvented. Yet by the end of the process in 1982, over 100 NGOs had been accredited as observers, and the history of external penetration of the diplomatic arena had begun. Since then it has become increasingly common for government delegations with ocean management mandates to be composed of at least a limited number of non-governmental participants, and the global-level initiatives in ocean-related conference diplomacy are now frequently accompanied by parallel initiatives on the part of NGOs.[41] With the accessibility of huge amounts of information through instant electronic communication and the explosion of "civil society" institutions, the traditional "knowledge gap" between state and social institutions has very nearly disappeared.

The impact of NGOs on the shaping of ocean management is, of course, particularly evident in the treatment of environmental and resource management issues. Since the end of UNCLOS III, environmental NGOs have become extraordinarily successful in influencing, or even determining, the outcome of inter-state ocean-related conferences and meetings in such varied contexts as whaling, fur seal harvesting, overfishing, by-catch, ocean dumping, ocean transportation of radioactive and other hazardous materials, offshore hydrocarbon production, river discharge, marine protected areas, coastal habitat protection and other aspects of coastal management.[42] Despite philosophical, strategic and tactical differences among them, they all tend to be critical of governmental efforts to deal with such issues. Many of them are also highly critical of the community of professional marine scientists for its perceived obsession with excessive rigor in quantification, measurement and prediction. Since these scientists are mostly government-employed, they are perceived as fatal flaws in the Westphalian system, tending to delay or obstruct preventive action.[43] Frequently, the preference of environmental NGOs is to substitute prohibitions for regulatory mechanisms, which they perceive to be government-driven, amenable to industrial pressures and intellectually sustained by a tradition of "scientism."

Above all, the outlook of activist organizations is ethical, in contrast to the more "neutral" influences of bureaucracy and the modern industrial economy. The legitimacy of NGO opinion seems to have been conceded in Article 71 of the UN Charter.[44] However, the sheer size and power of the "transnational ethical community" that now exists has caused alarm and resentment in many countries, and there is probably no bureaucracy or extractive or manufacturing industry that feels entirely secure in the face of this army of dissenters. Resistance to these social pressures takes different forms, but chiefly it requires diplomats, bureaucrats and industrial representatives to "discover among themselves a consensus on how to distinguish reasonable demands from unreasonable ones, realistic from utopian, and popular from radical, in the hope of separating the moderates from the extremists."[45]

These conflicts reveal a deeply rooted "societal dilemma": whether state and social institutions can find an accommodation in the common interest based on a generally shared perception of what constitutes reasonableness in public policy. As argued below,[46] a struggle for power-sharing of this kind might be welcomed if it were to result in a wider sharing of responsibilities for developing a more effective system of guardianship of the world's oceans.

The proliferation of "soft law" norms and instruments

Over the last two or three decades it has become evident to political scientists and international lawyers that formal, legally binding agreements of the classical kind that create legal obligations are now outnumbered by a rapidly growing proliferation of instruments characterized as "informal" or "non-binding."[47] Moreover, many international instruments, both bilateral and multilateral, that seem to pass the normal test of what constitutes a legally binding agreement contain provisions that are too loosely worded to possess clear obligatory force. Thus, their binding effect is not much greater, from a functional point of view, than the content of clearly non-binding or non-legal instruments.[48] Increasingly, the normative language negotiated at global and even regional levels is, at least partly, in the form of "soft law."[49]

In the past, international lawyers have been inclined to view the corpus of the law of the sea, including especially the LOS Convention, as "hard law" in its essence. In contrast, much of the newer domain of international environmental law is thought of as evolving juridically from a normative foundation that consists primarily of "soft law" principles and commitments that generate political expectations or ethical commitments rather than strictly legal obligations. But this distinction seems simplistic. Even the LOS Convention, which is certainly a formally binding treaty in the fullest sense of that term, contains a lot of generally or loosely worded provisions that seem to await more specific or more clearly mandatory language in order to qualify as obligatory and enforceable "hard law."[50] Conversely, many of the principles of international environmental law are now quite deeply rooted in juridical soil, nourished by successive textual incorporations. Moreover – to pursue the organic metaphor – we are all witnesses to numerous recent efforts by the diplomatic community to "graft" "hard law" and "soft law" together in areas of overlap between the law of the sea and international environmental law. So the new, post-UNCLOS legal framework for international ocean governance becomes an increasingly miscellaneous mixture of "hard law" and "soft law."

A pointer in this direction was provided by the Brundtland Commission, whose report in 1985 paved the way to endorsement of a number of "principles" such as those that address the need for sustainable development, pollution prevention, and precaution and recognize that liability for environmental damage should be fixed on the polluter.[51] But jurists differ on the juridical status of these norms, and indeed the concept of a "principle" has a rather broad "penumbra of uncertainty."[52] Despite controversy over its legal status, the need for precautionary restraint – whether principle, policy or goal – has been incorporated into several important instruments, both binding and non-binding, straddling the undemarcated boundary between the law of the sea and international environmental law.[53] Scholars differ also on whether the European "polluter pays principle" has now passed into customary international law, and has thus acquired universally binding effect on all states.[54] Similarly, it is debatable whether the unquestionable need for environmental sustainability and prevention (or at least reduction) of pollution generates automatically some kind of responsibility in customary international law.

So this trend poses the "normative dilemma" of whether such norms at the center of the "soft," but gradually "hardening," environmental law of the sea should be regarded as imposing operationally binding responsibilities on state institutions. This question that might seem scholastic has practical implications not only for national governments, which drive the Westphalian system, but also for social institutions now challenging the system's legitimacy.

Order versus anarchy in ocean management

Underlying this set of institutional dilemmas is a serious concern that our institutions do not have, and may not acquire, the necessary will or capacity to deal effectively with the complex problems and expensive tasks of international ocean management. At the national level, state institutions with the coercive capacity to bring "order" into the field of national ocean management rarely display a determination to do so. Normally, national governments are reflectors of sovereignty sentiment, striving for "autonomy" and usually reluctant to yield to the need for operationally effective regimes of cooperative ocean management with neighboring and other states. At the international level, intergovernmental organizations have the will to assist member governments with ocean management problems, to the extent it falls within their mandate to do so, but are chronically underfunded and understaffed. Most of the more affluent donor states display an unwillingness to increase their international development contributions in a field such as ocean management, perhaps because its technical complexity limits its political attractiveness. Modern conference diplomacy is quite impressively creative, perhaps more so than ever before, but chiefly at the global level, far removed from the regional and sub-regional levels at which it may be realistic to develop effective cooperative ocean management regimes or more modest arrangements.

The best hope may be that the rising awareness of the "sustainability crisis" in ocean affairs is convincing more and more governing elites of the urgent necessity for collaboration with responsible social institutions, and with ocean industries and coastal communities. Yet, even armed by the latest technologies of worldwide communication, the "transnational ethical community" is likely to remain highly heterogeneous in a field as difficult as ocean management. Even though they are becoming more influential on governing elites in many countries, INGOs (international non-governmental organizations) will continue to lack coercive power so as to ensure appropriate official action, even if a consensus could emerge on such a requirement.

The problem of reducing the "anarchy" of international society is compounded by the mixture of "hard law" and "soft law" norms, instruments, institutions and expectations in the evolving environmental law of the sea. On the one hand, a "softer" or more relativist normativity in global ocean governance seems to be contributing to the formation of "apparent consensus" reflected in a wider range of (mostly unthreatening) instruments and commitments. On the other hand, the soft–hard mixture of negotiated texts creates confusion over the "musts" and "oughts" of ocean governance and over the limits of attainability.

International law has always been on the cusp between idealism, in one form or another, and a practical understanding of reality of one kind or another. It has survived down the ages because all elites have agreed on the need for maintaining such a balance, and nothing else can take its place. Many of the new, bold, visionary principles of the environmental law of the sea are derived from the tradition of legal idealism. "Ah," we might exclaim with Browning, "But a man's reach should exceed his grasp, Or what's a heaven for?" Yet the realist, looking out, sees the challenge as a thousand tiny steps.

Ethical dilemmas

Elitism, participatory democracy and the quest for consensus

In the scientific literature, it is generally assumed, idealistically, that effective ocean management must be "rational," and to that extent must be controlled – or at least strongly influenced – by the scientific community. In short, it is assumed that hope for effective ocean management rests with the appropriate scientific elite, or with an intellectual elite whose nucleus resides in the marine science community. Ethical dilemmas in international ocean governance might, therefore, be said to begin with the alleged inevitability of elitism in modern society.

In the history of liberal democratic theory, the utilitarians, led by Jeremy Bentham and James Mill, argued for a "protective" brand of democracy in response to the perceived reality of a threatening mass of politically ignorant or apathetic citizens.[55] This model of democracy was soon to be challenged by John Stuart Mill's more optimistic "moral vision" of human improvement, which according to one well-known analysis gave rise to the model of "developmental democracy."[56] The apparent failure of this second model during the totalitarian challenges of the 1930s brought a more somber theory of "equilibrium democracy" that was dependent on a kind of enlightened elitism designed to balance out claims to general interest within a pluralistic framework of political interactions.[57] By the 1960s, however, an alternative model of "participatory democracy" had begun to evolve, responding to some successes in employees' shareholding and workers' control arrangements in certain sectors of industry, but also coinciding with the idea that there should be substantial citizen participation in governmental decision making.[58] Most today are likely to agree "the hope of a more participatory society and system of government has come to stay."[59]

In a context as complex as ocean management, it is likely that most specialists believe that professional expertise is essential to understanding the problems and issues within their field. Yet these experts will certainly concede that policy making in virtually all ocean-related sectors must be "democratic" in some sense. Probably they would begin with the premise that "elite consensus" is essential for all sophisticated societies, and for progress in world community development, especially in an environment dominated by bureaucracy such as that of ocean management. Nervousness about the dangers of manipulation by an established "power elite" may result in the reservation that the latter must be responsive to "the humanizing force of an intellectual elite."[60] At this point the argument for democracy in public policy formation turns on the question of how far to support a broadening of the concept of the elite whose consensus is necessary to humanize bureaucratic decision making without risking the civic chaos or anarchy that might arise from "mass participation."

Since the "transformational decade" of the 1960s, it has become common among political theorists and ethicists to support experiments in *participatory democracy* that center not on the participation of the people or the ordinary citizen, but on that of minority elites.[61] In general, participatory democracy, in contrast to representative (or parliamentary) democracy, is associated with "those decision-making structures that adhere to basic democratic procedural norms, such as equality and majority rule, yet tend to extend equality by some sort of 'grass roots' decision-making of an authoritative nature."[62] Radical theorists advocate the "self-determination" model of participatory democracy, whereby "amateurs" pursue lines of political action independent of those lines followed by "non-amateurs" (i.e. formally trained experts and regularly elected officials). More moderate theorists, on the other hand, tend to prescribe the "co-determination" model, which involves cooperation between amateurs and non-amateurs.[63] Advocates of this second model of participatory democracy invoke the need for civic goodwill or responsibility and for the reconciliation of rationality and democracy.[64]

In the ocean management context, the first ethical dilemma under contemporary conditions of democracy is how widely to seek for *consensus* on issues that have both technical and social components. Political and social ethicists differ among themselves in the identification of the "enemy" they are challenging: the bureaucracy, the state, the political establishment, the military complex, capitalism, industry, the science community, experts in general or central government seen to reflect the values of the urban middle class. Yet within the "transnational ethical community," despite the wide dispersion of its energies and emotions, there is widespread agreement on the legitimacy of their challenge to established authority and the institutions seen to support the distrusted status quo, however defined.

Sustainability and the obligation to future generations

In a world seemingly threatened with overpopulation, nothing appears more likely to win general acceptance than the notion that the finite natural resources of the planet need to be wisely conserved. Significantly, the conservation movement first surfaced in the final quarter of the nineteenth century, coinciding with sharply rising levels of human population. Indeed, to the extent conservationism originated in the fear of overpopulation, it can be said to have found its first apostle in Thomas Malthus (1766–1834), who as early as 1798 published an anonymous pamphlet arguing that population will always tend to outrun the growth of production, expanding to the limit of subsistence but checked there by the ravages of famine, disease and warfare.[65]

Over the two centuries since that original publication, economists have been sharply divided over the Malthusian legacy of pessimism. Critics have found it easy to demonstrate the shallowness of Malthus' analysis, and his lack of understanding of the many factors that are now seen to contribute to economic growth, such as technological innovation, entrepreneurship and product substitution. Yet in recent times neo-Malthusian environmentalists such as Lester Brown and Paul Ehrlich have staked their reputations on the essential correctness of economic "catastrophism." The fear of impending planetary catastrophe is shared so widely among so many contemporary environmentalists (and indeed voters) that *resource sustainability* is now accepted by almost all national governments as a goal of the highest priority. This global priority was set through worldwide adoption of Agenda 21 at the Earth Summit in 1992 and the subsequent establishment of the UN Commission on Sustainable Development to monitor and guide its implementation.

On the other hand, there is a great deal of reason to question the claims of catastrophists and to counter their influence on policy makers. Indeed the "eco-skeptical" challenge extends to all four of the principal environmental fears that underlie contemporary environmental ethics: (i) natural resources are running down; (ii) population is growing out of control; (iii) species are becoming extinct; and (iv) air and water are increasingly polluted. Since the publication of *The Limits of Growth* by the Club of Rome in 1972 a lot of counter-evidence has been presented. First, energy and other resources have become more abundant, not less so, over the last 30 years. Second, starvation is decreasing steadily, as global *per capita* food production today is higher than ever before. Third, although many species are indeed threatened with extinction, only about 0.7 percent of them are expected to disappear in the next half-century. Fourth, most forms of environmental pollution seem less severe or more transient than claimed by catastrophists, except global warming and greenhouse gases, which seem likely to prove to be a serious long-term phenomenon.[66]

Moreover, *The Limits of Growth* has been strongly criticized for its faulty methodology, its factual distortions and false assumptions. Given the ideological popularity of catastrophism on the political left, as a weapon to be used on behalf of the poor against market capitalism, technocracy and the forces of globalization, it is to be noted that *The Limits of Growth* has been the target of leftist as well as rightist critics. On the right, the criticism stems from belief that economic growth is necessary to end poverty; on the left from the belief that catastrophism is an expression of pessimism generated by the contradictions inherent in imperialism.[67]

At a more technical level, economists are divided over the usefulness of the concept of sustainability as a sophisticated tool of economic policy. Much of the technical debate focuses inevitably on the question of operationally useful "indicators": that is, how to measure progress toward such an elusive goal. To be operationally useful, it is argued by many economists and biologists, the indicators must be quantifiable.[68] In response to this, many environmentalists question whether it is possible to reconcile the goal of sustainability with that of economic growth to the extent they represent two distinct perceptions of human values.[69]

The ethical dilemma over sustainability turns, then, on value priorities as much as on factual interpretations. "Weak sustainability," it might be said, rests on the perceived value of maintaining economic output and the lifestyle associated with it, whereas "strong sustainability" gives the ultimate priority to sustaining the "ecosystem services of the natural world."[70] Whereas optimists may still put faith in the possibility of "consilience," the integration of disciplines and the paradigmatic values inherent in them, pessimists are more likely to conclude that adherents of weak and strong sustainability goals "each have their own specific valuation approach and thus different ethical perspectives."[71]

The ethics of sustainability is buttressed by further considerations. Not the least of these is the question of obligation to future generations.[72] The issue of *intergenerational equity* has surfaced not only in the domain of moral philosophers (or professional ethicists) but also in that of economists and lawyers. Philosophers with concerns about "environmental justice" are divided over whether, how and why to distinguish present and future rights. Some are prepared to defend the rights of the unborn, but others resist this notion to the extent that the "right to life" seems to limit a pregnant woman's freedom of action and to downgrade, if not eliminate, her right to choose an abortion. It is not obvious why unconceived and unborn generations, as a future collective, have the right to a healthy environment, as so often argued,[73] while a conceived but unborn fetus, as a potential individual, has no right to life itself. It seems anomalous, moreover, to assign precedence to collective rights over individual rights, especially in cultures that traditionally place a premium on the rights of the individual. Discomfort with this anomaly may displace the focus of debate from the human species of the present generation to ecosystems that are expected to survive indefinitely, but not many will be easily persuaded that an ecosystem's right to survive, or to remain undamaged, transcends the rights of the present generation of humans to a better quality of life, if that is what is promised through economic growth. To argue that all species, human and non-human, have the same rights to life, both in the present and future, is surely logically unsustainable, and impossible to convert into behavioral norms.

Most economists willing to concede the rights of the unborn do so from the perspective of "weak" sustainability, whereby a development is sustainable if it is "non-diminishing from generation to generation."[74] In this way economic growth theory translates weak sustainability into intergenerational equity, whereby a reduction of human welfare is allowed "as long as the level of consumption exceeds some subsistence level."[75] Lawyers are familiar with the difficult concept of individual "future interests" in the fields of property and inheritance, and are becoming accustomed to the recent idea of cross-generational collective rights in such contexts as that of cultural property. In international environmental law an elegant case has been made for an obligation to respect the rights of the unborn to a healthy environment, no less than the same entitlement of the living human population.[76] But it is arbitrary to stipulate "how much" of an ecosystem should be protected for the benefit of future generations. So it has been suggested by environmental economists that we move away from the future-regarding concept of sustainability to the future-regarding concept of "social bequests," which puts emphasis on "what" to leave to our descendants rather than on "how much."[77]

In the context of ocean management, it is the collapse or disappearance of major commercial fishery stocks that seems to make the strongest case for the principle of conservation or sustainability. The reservations of "eco-skeptics" regarding allegedly dwindling world food supplies cannot be allowed to discount the economic and social distress suffered in many regions as a result of such problems. The argument for "sustainable fisheries" has been accepted widely for a long time by conservationists, long before the coinage of the Rio terminology. Yet it is true, at the least, that the concept of "sustainable fisheries," with its ethical connotation, gives fishery management priorities a sharper political edge than was possible under the rubric of the more limited concept of fishery conservation.

Both the FAO's 1995 Code of Conduct for Responsible Fisheries[78] and the 1995 Fish Stocks Agreement[79] reflect acceptance by the UN system of the need for a comprehensive, ecosystem-based approach to sustainable development. These soft-law and hard-law instruments, respectively, have stimulated FAO to undertake new and potentially innovative initiatives such as the development of a "sustainable development reference system" (SDRS) for fisheries based on specific indicators of sustainability.[80] If plans of action such as the 2001 International Plan of Action to Prevent, Deter and Eliminate Illegal, Unreported and Unregulated Fishing (IPOA-IUU)[81] can achieve operational significance at the national level, credit would have to be given to the mobilizational influences of the goal of sustainable fisheries. IUU fishing is believed to account for about one-third of the actual world fishing catch.[82] Moreover, the IPOA-IUU represents, among other things, an effort to give port states a role in monitoring IUU fishing, so as to reduce the problem of overdependence on flag states to enforce their fishery treaty obligations.[83]

It might be supposed, then, that the emphasis on sustainable fisheries is bound to be progressive. But it remains to be seen how successful the Code of Conduct and the Fish Stocks Agreement will be in operation.[84] The ecosystem-based approach to fishery management associated with both of these instruments has its critics. Diplomatic efforts at the global level to advocate a LME approach to fishery management have run into difficulties because of its perceived threat both to coastal state authority within the limits of the EEZ and to international high seas fishery commissions with management responsibilities immediately adjacent to such limits.[85] In short, ecosystem-based fishery management may prove to be unworkable because of recently extended institutional investment in "politically defined" ocean spaces.[86]

Obligation to nature

Of the intellectual and political developments of the 1970s, one of the most significant was the rise of environmental ethics, arguably derived from the earlier work of Aldo Leopold and his conception of "land ethics."[87] Prior to the 1970s, nature conservation and the extent of human responsibility for the preservation of ecological balance were largely the concerns of humanists, biologists and certain nature-nurturing religions, sects and tribes.[88] The preparations for the 1972 UN Conference on the Human Environment might be regarded as the first globally organized effort in the diplomatic arena to articulate the need for a world community ethic in support of the environment, conceived as embracing the entire spectrum of species and habitat areas: the "biosphere" or "ecosphere."[89] Today nothing receives more attention as a potential "world ethic" than environmentalism, with its imperatives purporting to transcend all cultures, societies, economic systems and political regimes.

At the heart of environmentalism is the deeply felt sentiment or conviction – held as a matter of faith as much as reason – that, above all, humans have a fundamental obligation to "nature." When feelings run so deep, intellectual and political issues assume a spiritual dimension. In the new age of globalization, prevailing views based on environmental ethics could become as "transformational" as any widespread beliefs that have arisen from the history of religions.

Understandably, huge dilemmas emerge over questions of moral philosophy regarding the extent of human responsibility for nature. Before the modern era, most moral philosophers shared the view that humanity is the moral center of the universe. Only a small minority were prepared to challenge the predominant opinion that humans are of exclusive or overriding moral significance. But there is now a rapidly growing literature on questions concerning the existence and nature of values in the non-human natural world.[90]

For professional environmental ethicists, the *minimal* position is that we all have environmental responsibilities, not only to one another, and to our children and grandchildren, but to all succeeding human generations everywhere, beyond the limits of our family, community and nation. From this minimalist perspective, it is conceded to be essential we develop a profound understanding of nature – of all species and ecological relationships – so as not to underestimate the impact of the human presence on the biosphere and thereby jeopardize the welfare of future human generations at projected demographic levels. But this position might be construed as anthropocentric, restricted by the vision of foreseeable future human needs: long-range sustainability to serve projected human interests.

Today many environmental ethicists believe that this minimal position is insufficient. Indeed some "deep ecology" spiritualists seem prepared to go quite far toward the extreme ideal of equality among all species that make up the ecosphere.[91] Yet most of the philosophic debate focuses on how to draw lines between "moderate" and "immoderate" claims. For example, can we distinguish "interaction" with nature from "interference" with nature?[92] Is "interference" necessarily non-beneficial to non-human components of nature? If not every component of the natural environment has an absolute "right to life," as maintained by some deep ecology theorists, how can humans make morally acceptable distinctions? Condemned to a relativist position, those seeking the middle ground may have to accept that public policy on obligation-to-nature issues will have to fluctuate uncomfortably between two competing, and essentially incompatible, value systems: the ethical and the institutional.

Such dilemmas are increasingly likely to be political rather than scientific or philosophical in nature, and public policy on such matters will rarely be entirely ethical. Yet in the context of ocean management, environmental politics has brought an ethical dimension to many issues that might otherwise have been regarded as technical or managerial problems. Nowhere has the obligation-to-nature ethic been more insistent than in the international politics of whaling and the harvesting of other marine mammals.[93] The extraordinary success of the anti-whaling and anti-sealing movements is well documented, and many pragmatists, who would personally prefer further experimentation with regulatory measures based on scientific evidence, are almost ready to acquiesce in the prohibitory approach through moratoria or otherwise, although the latter seems inappropriate in the case of "stocks" that are not seriously threatened.[94] Whales, dolphins and similar creatures attract so much human affection, perhaps as kindred species, that they have become beneficiaries of the animal rights movement.[95] Few, however, are prepared to cross the line and argue that all other marine species, including commercial stocks of fish, should be extended the same right to life. At the present stage of human development it is difficult to envisage a collective decision to forego ocean-based sources of animal protein. The primacy of human need is evident in most ethical systems, and only a small, though growing, minority are strict vegetarians.

Moderate ethicists, shunning absolutist concepts such as the "right to life" of all species, are unlikely to yield to it when it creates a threat to a supply of universally accepted human food, such as fish. But more difficult dilemmas arise when the non-human species has a limited consumptive appeal, as in the case of whales and seals, which are either a traditional food in certain aboriginal communities or a luxury food in certain affluent societies.[96] Ethicists strongly opposed to the hunting and eating of such species – as distinguished from conservationists concerned with resource sustainability – often have to contend with competing ethical considerations focused on such value concepts as cultural diversity, aboriginal prerogative and communal or local entitlement. Such ethical dilemmas have to be confronted in the diplomatic arena, and increasingly in arenas penetrated, or even "hi-jacked," by non-governmental participants representing civil society institutions.[97] So the taking of sides on these substantive issues is often influenced, if not determined, by how strongly one feels about the procedural or institutional implications of such "intrusions" or "participatory interventions."

Supporters of the obligation-to-nature ethic are likely to attract allies more easily over issues arising from "non-consumptive" uses of the ocean and its resources. Like the sport hunting (as distinct from conservationist culling) of foxes and grizzly bears on land,[98] the sport fishing of salmon and the sport hunting of polar bears and other marine mammals may seem more difficult to justify, although to most observers the issue of animal cruelty seems quite different in the case of salmon. In many ocean-side communities, sport fishing is a fairly important branch of the tourism industry, and coastal community development policy must take economic considerations into account. Similarly, in many other contexts of coastal recreation and tourism, public policy must accommodate economic interests without sacrificing the universally approved goal of marine biodiversity.[99]

Despite the general preference for moderate rather than immoderate public policy on obligation-to-nature issues, there is no doubting the growing intensity of minority sentiment in various contexts. At least in circumstances where there is evidence of animal cruelty, prohibition rather than regulation seems to be increasingly favored by a majority. Yet even in these cases it is clear that such concerns have not yet been reflected in public policy outside a limited number of (mostly Western) societies, and questions must be asked about the ethnocentric tendency of these countries to attempt to impose their cultural values on others.

Equity in ocean governance

Broad as the range of ethical issues in ocean governance is seen to be under the obligation-to-nature rubric, the range is further expanded if we incorporate the many correctives that have been advocated and negotiated in the arena of ocean diplomacy. The need for equity as well as effectiveness in ocean law, policy and management has been recognized since the late 1960s. An equitable redistribution of global wealth derivable from the ocean was at the heart of Arvid Pardo's vision for a new ocean order that resulted in the

UN decision to launch the UNCLOS III process of radical reform. Equitable correctives agreed upon at the conference include (among others):

1 the extension of coastal state jurisdiction, perceived as protection for vulnerable developing coastal states;
2 the common heritage ethic reflected in the International Seabed Authority (ISA), intended as a cooperative global system for generating revenues from deep ocean mining that would be allocated among the most deserving states on a need basis;
3 provisions for the transfer of ocean-based technologies from developed to developing economies; and
4 recognition of the special entitlements of landlocked and other geographically disadvantaged states.

However, in practice relatively little benefit has accrued to the intended beneficiaries from these and other correctives. The extension of coastal state jurisdiction has allowed many developing coastal states to develop their own, relatively viable, fishing industries, but most developing coastal states still lack the capacity to take advantage of their legal entitlements. The ISA was substantially weakened through the post-UNCLOS III "revisions" negotiated in the early 1990s,[100] and in any event the postponement of hopes for deep ocean mining has delayed the intended result of wealth redistribution under the common heritage regime. The technology transfer provisions have remained in place, but seem to have little, if any, operational vitality in an age dominated by market-driven ideologies such as the intellectual property movement supported by transnational corporations. Similarly, the rights vested in the world's 40 or more landlocked (and other geographically disadvantaged) countries have not yet been translated into the kind of beneficial cooperative arrangements envisaged in the LOS Convention.[101] In short, UNCLOS III has very largely failed to correct the inequities inherent in the inter-state system of ocean use and management.

In response, the Brundtland Commission placed an emphasis on the need for ocean equity in Chapter 17 of its Agenda 21; and more recently the Independent World Commission on the Oceans has addressed the same need with greater firmness.[102] In light of these two reports, priority should be given to a number of equity issues, of which the following five in particular might be singled out.

Geographically disadvantaged countries

The dilemma of how to implement the UNCLOS III provisions in favor of landlocked and other geographically disadvantaged states is institutional in nature. Even if this grouping is limited to the truly landlocked, there is no forum in existence, or in prospect, where these problems can be discussed on the basis of shared interest. It may be unrealistic to expect the elites of these countries, with so little expertise in ocean affairs, to mount an effective collective strategy for attracting the world community's attention to this issue of neglect.[103] Moreover, many of the new landlocked countries, formerly republics of the Soviet Union, have much more exciting possibilities of economic development through the production of their land-based hydrocarbon resources.

Small, developing island states

By contrast, this grouping of resource-deficient countries has an extremely high degree of ocean awareness and their new jurisdictional entitlements as a result of UNCLOS III, covering vast ocean spaces, have motivated a fairly vigorous diplomatic campaign before and after the Rio Summit of 1992.[104] Despite the lack of any provisions specifically directed at these states in the LOS Convention, several conferences have been organized and proposals developed on their behalf. Needless to say, the capacity-building problem for these semi-viable economies is paramount.

Coastal communities

Outside the inter-state framework of UNCLOS III, various equity issues can be articulated at the sub-national level of society. Coastal populations in most countries are peripheral to the power centers of central government and to the wealth centers of the corporate community, and often subject to the threat of natural disasters. In some nations, strong efforts have been made to involve local community institutions in "community-based coastal management" or "co-management" (or other multi-level approaches) to coastal management. Yet the equity argument for local engagement in such arrangements adds to the institutional complexity of the concept of integrated ocean (or coastal) management. Instead of replacing the politics of central government, local community participation often compounds the politicization process by contributing a further layer of political conflict.[105]

Traditional fishing communities

In most parts of the world, the pressure on fish stocks is intensified by the economically efficient but environmentally destructive methods of industrial fishing. These capital-intensive technologies are deployed mostly within corporate structures that are controlled outside the local community and designed essentially to maximize profits rather than to sustain the local economy and maintain its traditional lifestyle. Conservation measures taken by central governments in response to the phenomenon of industrial overfishing often result in prohibitions or strict regulations that cause distress in fishing communities that are almost totally dependent on traditional stocks and yet may have had only a minor role in causing the problem.[106]

Indigenous coastal fishing communities

All of these problems tend to be compounded when the local coastal community consists mostly of an indigenous people that is not strongly represented in state institutions and forms a minority with real or perceived grievances against the dominant ethnic sector of the population. Efforts to address the grievances of indigenous peoples and to promote the principle of aboriginal rights at the world community level resulted in the 1994 Draft Declaration on the Rights of Indigenous Peoples, but government resistance to such claims of special entitlement, in the form of independence, sovereignty, or self-government, is widespread. Yet at the national level, especially in Australia, Canada, New Zealand and the United States, public policy is evolving in support of traditional fishing rights of indigenous coastal fishing communities.[107] Also encouraging is the effort to develop "indigenous ecotourism," but here too both institutional and ethical dilemmas have to be confronted.[108]

The precautionary controversy

No account can be given of the ethical dilemmas of international ocean governance without some discussion of the long-running controversy over the "precautionary principle." The idea behind this concept can be considered a reaction to the perceived failures of conservationism based on scientific evidence and of ineffective regulatory regimes. Accordingly, for many proponents, the core meaning of the principle is the radical idea of a need to shift the burden of proof: from proof of unacceptable risk on the part of the accuser to that of minimal or tolerable risk on the part of the accused, the would-be-polluter.

However, the more extreme, onus-shifting formulations introduced at global conferences have encountered resistance, not only by developing countries concerned with their potential impact on economic growth, but also by the United States and other non-European developed countries. Invariably, the need to secure consensus in global negotiations has produced less stringent formulations. In some cases, the norm is characterized as an "approach," or "concept," not as a "principle," with a view to defeating the argument that the norm carries a binding obligation in international law. In other cases, the term "principle" may be permitted, but the formulation omits the reversal of onus of proof, which has caused particular consternation in the scientific community. In still others, of course, the legal impact of the terminology is reduced by reason of the non-binding nature of the instrument in question. Because of the diversity of terminology in global instruments, it has become difficult to demonstrate that any particular formulation has become established through state practice as a principle of customary international law.

As a result of this variance, it might be argued that the precautionary concept – not to beg the question – is at present an ethical norm that is still evolving in the world community, where deference must be made to the fact of economic disparity and the legitimacy of cultural diversity. The dilemma, which is at least partly ethical, is whether there is a "core" within the precautionary norm, an "essential" component that can be said to be an accepted part of world environmental ethic.

Ethics versus autonomy in ocean management

Perhaps the most conspicuous feature of international ocean affairs today is the struggle between the transnational ethical community and the interstate system. In the sectors discussed above, a sense of ethical commitment, held particularly strongly outside government service, is having an increasingly profound effect on governmental and intergovernmental decision making, threatening the statist tradition of national autonomy. Essentially, it is a struggle between ethical commitment and political legitimacy.

Virtually all ethical elements in international ocean governance come to the field through the global arena, where ethical principles and constructs are negotiated as important provisions of instruments, both binding and non-binding. Now, more than ever, the conventional international law of the ocean, in this expanded sense, represents a synthesis of the new idealism that is being brought to the arena in the form of ethical prescriptions and the pragmatism that has always resided there. Realistic expectations of the outcomes of international ocean diplomacy – and perhaps other "ethicized" sectors of international relations – require a balancing between the prescriptions contained in the negotiated texts and the various limiting factors and considerations that can be assessed only by contextual analysis.

These limiting factors and considerations, which may, but usually do not, become textually explicit, arise from a variety of economic, political, cultural, geographical and historical realities and perceptions that make up the milieu of international diplomacy. In the case of most ambitious global instruments, the milieu tends to reduce the real significance of the outcomes of conference diplomacy. Reluctance to implement or comply typically finds expression in invocations to sovereignty, sovereign rights and other modes of national prerogative. The classic dilemma of our age is how to reconcile the ethical imperatives produced in the diplomatic arena with the national entitlements still associated with the concept of national autonomy.

Conceptual dilemmas

"Sovereignty of science" tradition

Historically, the first sector of ocean management to evolve was fishery management, which has its roots in the 1880s, when North European zoologists and statisticians first came together to provide a scientific foundation for international efforts to regulate the fishing of overexploited stocks in the North Sea and adjacent waters. The next step was the formation in 1901 of the International Council for the Exploration of the Sea (ICES) to serve as a forum for the exchange and discussion of data. ICES is still the principal forum for such purposes despite the emergence elsewhere of other organizations in the field of ocean science, both at the global (e.g. the International Oceanographic Commission of UNESCO) and the regional level (e.g. the North Pacific Marine Science Organization – PICES, which serves as a counterpart to ICES for the North Pacific).[109]

For well over 100 years, then, a tradition has been building that places applied ocean science at the center of efforts to regulate ocean activities as rationally as possible, ideally within a framework of regulatory measures that can be adjusted periodically in light of updated research findings. Between the early 1920s, when the first international fishery commission of this kind was brought into existence, and the late 1950s, which saw the arrival of numerous FAO fishery commissions,[110] there was little doubt about the sovereign status of science. By then numerous international agreements contained language calling for the best available scientific knowledge as the rationale for regulating fishing and whaling. In the case of most Western countries, these instruments represented an extension of the "sovereignty of science" tradition in ocean affairs reflected in their domestic policy and legislation. Moreover, by the early 1960s this tradition was beginning to be exported from the fishery sector to marine pollution prevention and control and, of course, habitat protection and the protection of endangered species. In every sector of ocean management where a regulatory approach was seen to be the key to the control of overuse or misuse, the marine science community was given a pivotal role. Indeed, in some countries the scientific community has been the chief influence on the design of ocean management mechanisms.

The public policy dilemma is how much it is appropriate to expend on ocean science conceived as the essential foundation of ocean management. It would seem obvious that effective ocean management or governance depends on the availability of reliable information about the ocean: its structures, movements, organisms and processes. But the immensity of three-dimensional ocean space poses an enormous challenge to the scientific community, consisting as it does of 320 million cubic miles of water covering 140 million square miles of sea floor that encompasses seven-tenths of the surface area of Earth. Moreover, oceanography is a fairly recent convergence of four distinct fields of science: marine geology, ocean physics, marine chemistry and marine biology. The first systematic effort to integrate the scientific exploration of the ocean was as recent as the 1870s, when HMS *Challenger* embarked on its famous four-year voyage, the first truly global oceanographic expedition. Since then a great deal has been learnt about the ocean, the seabed and subsoil; and yet the marine sciences have only begun to scratch the surface.[111]

Despite its newness as a "master discipline," oceanography quickly became an important presence in the scientific community. Perhaps because of the exciting nature of so many of their discoveries in the deep, oceanographers acquired a rather prestigious status in the eyes of the general public. Compared with other "basic sciences," the field of marine science could be presented rather easily as an area of popular interest. Its obvious importance as a focus of research ensured a steady supply of public funds, and even now its total endowment may be second only to space science. The scale of public funds made available for ocean science has created expectations that the scientific-bureaucratic elite could bring rationality into the elusive "science" of ocean management, and even into the highly political domain of "marine policy."

Competing orders of rationality

The "sovereignty of science" tradition began to encounter challenges in the late 1950s. In the fishery sector, the first generation of fishery regulatory bodies had been governed by biological constructs, the most famous being the goal of "maximum sustainable yield." With the entry of economists into the field of fishery management and the introduction of such goals as "net economic yield," concepts of biological rationality had to compete with those of economic rationality.[112] Just as there seemed to be some hope for reconciliation between the two disciplines in the late 1960s, it became increasingly obvious that in practice, in the management of most major commercial fisheries, social and political realities (including commercial and economic development interests) were trumping both the biological and economic orders of rationality. Both the biologists and the economists were able to agree on pinning the blame for serious stock depletion on the problem of overcapacity, and on the need for limiting severely the right of entry. But successful examples of dramatic stock recovery due to strict entry controls are still relatively few. In many countries around the world, the normal situation is for fisheries to be managed, or more precisely mismanaged, chiefly on the basis of social and political "realities."

It can perhaps be argued that placing social and political considerations at the center of ocean management – or of any kind of management – is hardly "irrational." The concept of management applied to the ocean, or any other "natural" domain, is, after all, somewhat vague. The ocean itself, in its boundless vastness, cannot be managed; only users and institutions that impinge upon it. By this reasoning "ocean governance" is essentially an effort to provide a system of incentives and disincentives, backed up by an array of positive and negative sanctions, with a view to the larger, land-based purposes of social and economic development, limited but not controlled by "scientific realities" and "economic constraints." The most dramatic way of making this point is by noting that in Canada the *total* cost of fishery management is said to exceed the total revenue produced by the

Canadian fishing industry, a paradox explainable only by reference to the overriding social and political need to maintain peripheral coastal communities that cannot develop an alternative local economy.[113] So the "dilemma of rationality" is how to choose among these competing orders.

Lawyers – or, more properly, lawyer-diplomats – have also played a role in the competition among different orders of rationality in ocean management. By restructuring the law of the sea framework at UNCLOS III, the diplomatic community created the concept that coastal states have legal entitlements that cannot be abridged by reference to the traditional freedoms of the high seas. These coastal entitlements do not depend upon the coastal state's capacity or willingness to discharge its international responsibilities for the protection of the marine environment within its greatly extended limits of jurisdiction. So, in that sense, UNCLOS III created an ethos of coastal entitlement that is sufficiently robust to support the notion that a sophisticated level of effectiveness might eventually become attainable in the rational management of *coastal waters*.

The coastal zone management movement

The "goal concept" of comprehensive, cross-sectoral ocean management was conceived in the womb of the United States Marine Council, which was established in the early 1960s in response to constant prodding by Vice-President Hubert Humphrey.[114] One of the Council's first acts was to engage a blue-ribbon task force, the Stratton Commission, to study the need for a comprehensive cross-sectoral approach to the management of the ocean. Some emphasis was placed on offshore areas that had become areas of special interest to the United States since the famous Truman Proclamations of 1945, which revolutionized the concept of coastal state entitlement and responsibility far beyond the traditionally narrow belt of territorial waters.[115] The result of the Commission's work was *Our Nation and the Sea* (1969), a "visionary four-volume study describing all aspects of US interests in the oceans, complete with recommendations for policies and strategies for carrying them out."[116] Pre-eminent among the Commission's prescriptions was its call for federal government leadership in the establishment of a "coastal zone" around the territories of the United States, an area of overlap at the interface between the land and the ocean.

Although the area conceived by the Stratton Commission was narrowly defined, its concept of "coastal zone management" (CZM) carried the seed of modern thinking about techniques of coordinated management required in coastal areas, which are notorious for the multiplicity of agencies with related responsibilities. In short, the Stratton Commission provided the cradle for the nurturing of the modern (or "post-modern") concept of "integrated coastal management." The intellectual influence of *Our Nation and the Sea* was enormous, extending throughout most of the world, although few countries succeeded in matching the institutional energy expended in the

United States on the implementation of this overarching strategy of intra-governmental coordination.[117]

Despite inevitable disappointments in certain regions of the United States, most observers concede the relative success of the CZM experiment.[118] Yet those most skeptical about the attainability of the goals of integrated coastal or ocean management might feel entitled to blame this institutionally ambitious American model for the global trend to overly ambitious goals of ocean management and the associated risk of over-expectation.

The proliferation of "governing constructs"

Since the promulgation of the US *Coastal Zone Management Act*, we have become witness to an extraordinary proliferation of concepts in the field of ocean and coastal management. Most of these concepts might be regarded as available to guide, or even govern, decisions on the scope, focus, rationale or technique of ocean management.

Spatial constructs

These concepts are designed to serve legal or general administrative purposes. Some are of long standing, but many are new or redefined products of the global diplomatic arena. The most obvious examples are:

- Internal waters
- Archipelagic waters
- Coastal management zones
- Ports, harbors, roadsteads
- Territorial seas
- Bays (historic and non-historic)
- International straits
- Exclusive economic zones
- Enclosed or semi-enclosed waters
- Ice-covered waters
- Continental shelf (as legally defined)
- Regional seas
- International sea-floor ("International Area")
- High seas

Most of these 14 constructs define, or at least suggest, the scope of water areas that could be designated for ocean research, planning, regulation, monitoring, enforcement or other ocean management purposes. Even the high seas, the most extensive of all, has been used as the referent for certain kinds of management arrangements. Most are internationally defined legal regimes designed to serve general or multiple administrative purposes, suggestive of scope rather than focus.

Focal constructs

Like the foregoing, focal constructs are mostly areal in conception. But, by contrast with spatial constructs, these are created in order to focus national or sub-national administrative controls or regulatory arrangements on specific problems. They can be broken down into at least 13 categories:

- Wetlands
- Functional inshore zones
- Functional coastal zones
- Functional contiguous zones
- Specific types of fisheries (e.g. anadromous species, catadromous species, straddling stocks, high migratory species)
- Endangered or threatened species (e.g. marine mammals)
- Specific habitat areas
- Marine protected areas (e.g. marine parks)
- Special protected areas (under Art. 211 of UNCLOS)
- Particularly sensitive areas
- Functional offshore zones
- Large marine ecosystems
- Zones of peace (or non-nuclear zones)

These are all examples of actual or potential units of specific research, planning, management, regulation or policy application. The area in each category is determined by the nature of the problem for which it is designated, and it tends to be unifunctional in its rationale.

Choices among these focal constructs, as alternative levels of ocean planning and management, have to be made chiefly at the national level. In the case of Canada, for example, the choice cannot be made entirely by any one national agency. The Department of Fisheries and Oceans (DFO) envisages the establishment of "ecosystem-based management" objectives at the level of Large Ocean Management Areas (LOMAs), within which "integrated management" plans of various scales would be "nested."[119] However, it is recognized that the "identification of areas of interest for marine protected areas (MPAs), to be established by the Government of Canada" includes not only MPAs under the *Oceans Act*, which is administered by DFO, but also Marine Conservation Areas under Heritage Canada and Marine Wildlife Sanctuaries under Environment Canada.[120] The overlapping of agency mandates underline the virtue of "integrated management," but also the great difficulty of achieving such a goal.

Normative constructs

Unlike the spatial and focal constructs, normative concepts are non-areal, having a geographically general, if not universal, relevance. Essentially, they provide guidance for more specific goal setting or priority ordering in ocean management. The most familiar, and perhaps most influential on the development of ocean governance thinking, are six in number:

- Sustainability in ocean use
- Common heritage
- Integrated ocean/coastal management
- Biodiversity conservation
- Ecosystem management
- Precautionary principle or approach

Each of these goal concepts is "value-laden" and has its own relatively controversial history. Each might be conceded, even by its advocates, to be ambitious, or even idealistic, exhibiting its own order of rationality and ethic. Their practicality depends not least on how they are applied in relation to the spatial and focal constructs.

Most of these normative constructs are reflected in the language of national legislation around the world. In the case of Canada, chief emphasis is placed by DFO on the goal of integrated ocean management, but it is to be "guided" by seven "principles," including three of those listed above: (i) ecosystem-based management, (ii) sustainable development and (iii) the precautionary approach (in a modified, "erring on the side of caution," formulation). The four other principles are: (iv) conservation, (v) duty in shared responsibility, (vi) flexibility and (vii) inclusiveness.[121] Overall, primacy seems to be given to sustainable development, integrated management and the precautionary approach.[122]

Technical constructs

Within these frameworks of spatial, focal and normative constructs, there has emerged a host of ideas that represent alternative "techniques" of ocean management. It may be sufficient to list 14, although it would be possible to add many more, or break these approaches down into more specific categories:

- Data collection and synthesis
- Statistical analysis
- Population dynamics (e.g. maximum or optimum sustainable yield)
- Regulation of ocean activities
- Prohibition of certain ocean activities
- Biomass monitoring
- Community-based management
- Co-management of coastal waters
- Adaptive management
- Conflict resolution

- Sea use planning
- Regime building
- Climate change research and monitoring
- "Trade-and" linkage strategy

Like most taxonomies, this system of classification can be criticized on the ground that certain constructs could be included under two, if not three, of these headings. For example, the prohibition of certain ocean activities could be seen as the product of ethic as much as technique, and thus classifiable as normative. The same might be said for community-based management, conflict resolution and "trade-and" strategy. But the utilitarian test is whether all these constructs deserve to be distinguished with a view to the need for greater conceptual clarity.

The greening and glossing of UNCLOS

Since the end of UNCLOS III in 1982, there has been a steady succession of new global instruments with an oceanic orientation. Some of these instruments are orthodox, legally binding instruments of "treaty character;" others are legal in form but possessing provisions or norms of a "soft law" nature; and others again are not legal instruments in form yet have a fundamental "legal policy" significance – arguably a significance that seems to endow them with a potential legal effect that may be realized over a period of normative development. At the end of UNCLOS III the delegations were close to "burn out" after 15 exhausting years of immersion in ocean law and policy issues of unprecedented complexity. Collective fatigue ruled out any decision making on an intergovernmental apparatus that would be mandated to monitor subsequent treaty or related developments that might affect the interpretation and application of the LOS Convention's numerous, and often open-ended, provisions.

In the absence of a mechanism to provide studies and recommendations on changes taking place in the law of the sea, and especially in the environmental law of the sea, it has fallen to the "invisible college" of academic international lawyers to supply insights into the implications of this "greening" of the law of the sea. Those who are uncomfortable with the prospect of a plethora of unofficial, scholarly glosses of a "sacred text" such as UNCLOS may have to consider the alternative of finding a politically acceptable mode of "revision."

The LOS Convention itself provides for *amendments* through three distinct procedures under Articles 312–14.[123] In addition, Article 155 of the original text provided for a *review conference* to review Part XI on the International Area, but this provision has been transcended, or rendered virtually "dead-letter," by the replacement language of the 1994 Agreement Relating to the Implementation of Part XI of the UN Convention on the Law of the Sea of 10 December 1982.[124] This new language is much more flexible than what it replaced, leaving it to the discretion of the Assembly of the International Sea-Bed Authority whether this review should be conducted formally through a conference of the kind originally prescribed in Article 155, or in a less formal manner.

The 1994 instrument cited above is officially characterized as an "implementation agreement," but functions to serve the purpose of revision. The 1995 Agreement for the Implementation of the Provisions of the UN Convention on the Law of the Sea of 10 December 1982 Relating to Straddling Fish Stocks and Highly Migratory Fish Stocks[125] seems more appropriately described, although its addition of much more specific responsibilities might seem to carry out the function of revision beyond the original language of commitment. Moreover, there are other ways of "revising" existing treaties without resorting to the difficult and time-consuming diplomacy associated with formal amendment, and without resorting to the tried and politically proven, if juridically dubious, practice of "revision" through subsequent "implementation" agreements. Given the fundamental or constitutive nature of the LOS Convention, it might be suggested that at least part of the "revisionary" process could be executed through a group of non-official experts meeting in "scientific" or "quasi-glossatory" mode, by analogy with those scholars who compiled the influential Harvard Draft in the inter-war period.[126]

Conceptual balance in international ocean governance

The critical conceptual dilemma in contemporary ocean management lies in the selection or creation of a generally convincing frame of reference that is both progressive and realistic. Such a balanced framework would have to satisfy the following criteria of acceptability.

First, the LOS Convention retains "quasi-constitutional" status, especially on jurisdictional matters pertaining to the allocation of states' basic rights and obligations, and yet must be treated as a "living constitution" open to re-interpretation, development and revision in accordance with important changes in the mores of international society.

Second, there is no single "truth" in the conceptual apparatus of international ocean governance. The closest to a normative construct of universal validity is the first-order requirement that ocean uses must be as consistent as possible with the goal of sustainability. Sadly, the truth is that there is a compelling need to "breathe life and content into the notion of 'sustainability' to make it into a fundamental norm of the world ocean regime for the 21st century."[127] The other principal normative constructs (integrated ocean/coastal management, common heritage, biodiversity conservation, precaution and ecosystem management) are important, but not overarching, in the manner of sustainability. Their weight should be determined through contextual analysis and political interaction. For example, the principle of common heritage should continue to be applied to secure the goal of common benefit outside established limits of national jurisdiction, but interpreted restrictively inside these limits; otherwise the nation-building work of UNCLOS III would be undermined.[128] Ecosystem and biodiversity approaches to ocean management should be encouraged as innovative experiments in *ocean science*, where resources permit, and tested out as a foundation of *ocean management* operations in selected ocean areas and sectors, where appropriate institutional capacities are in place. Integrated ocean/coastal management likewise is experimental, deserving a special effort to raise the bar in intra-governmental collaboration, and also in cooperation between state and society. Precaution is generally accepted as a sensible guide to discourage overuse or misuse of ocean resources, but should not itself be overused or misused as a policy to justify the underuse of ocean resources or as an unduly costly burden on responsible use. The co-existence of various, incompatible formulations of the precautionary principle underlines the need to subject issues of how to apply that principle to the increasingly rigorous techniques of *risk analysis*.

Third, debates between science and ethics in ocean governance at the level of abstraction should be avoided since they encourage prejudgments and over-generalizations, generating heat rather than light. As rival "drivers" of international ocean governance, they have a political significance in the global context of inter-state disparities. Science, though inherently a difficult and expensive business, is culturally and politically neutral, to a large extent, whereas the dominance of ethics might put the cause of global ocean governance in jeopardy, seeming more attractive to cost-conscious affluent economies. The respective merits of science and ethics, when in collision, are best evaluated within the context of specific public policy issues, such as inshore versus offshore fishing policy, coastal pollution and tourism development, alternative ocean energy policy, offshore hydrocarbon development, the transportation and storage of ultra-hazardous wastes, and the regulation of land-based activities that contribute to contamination of the marine environment.[129]

Scope versus focus in ocean management

The proliferation of "governing constructs" in international ocean governance will make it very difficult to bring back the age when science was virtually sovereign in ocean management. The competition among different orders of rationality in the fishery management sector has spread to other sectors of ocean use. This kind of intellectual competition puts a public policy premium on breadth of vision rather than on depth of technical knowledge. The dilemma is how much "rigor" it may be safe to sacrifice in order to gain the highest possible level of "sophistication." Each of these manifestations of intellectual bias may be regarded as potentially harmful as a fully "rational" foundation of ocean management unless corrected by the other, equally legitimate, bias.

The search for intellectual balance between these biases is never-ending, fluctuating in its outcome from one context to the next. It would be unfortunate for the searchers for balance if mechanisms such as the Joint Group of Experts on the Scientific Aspects of Marine Environmental Protection (GESAMP) were to be dismantled for the lack of financial support. As matters now stand, it is difficult enough for funding agencies to understand the debates within the ocean-related disciplines and to make choices between "rigorous" and "sophisticated" proposals.

In recent years, the premises of marine ecology have become increasingly saleable. After enduring many years of resistance on the part of more "rigorous" disciplines (or siblings) within the oceanography family, marine ecologists have begun to win important "pro-scope" battles in the shaping of marine scientific research. With the prospect of NEPTUNE and other major oceanography programs based on the latest and highest technology, there is likely to be a very different kind of cross-disciplinary research methodology in oceanography that promises to yield results of a "sophisticated" kind without a damaging loss of "rigor." For example, from a scientific research (program design) perspective, the LME approach might make a great deal of sense, reflecting a balance between rigor and sophistication.

The difficulty is, however, that what makes intellectual sense as a unit of sophisticated scientific research does not necessarily make operational sense as a unit of international governance, regulation or management. Other things being equal, a smaller grouping of neighboring coastal states is more likely to be induced to participate effectively in cooperative ocean management, at the sub-regional level, than a larger grouping of states urged to cooperate at the LME level. The paradox is that the "scope" approach in the case of science and the "focus" approach in the case of institutional development are equally "rational," belonging to quite different orders of rationality.

The challenges

The regime-building role in ocean governance

In the present age of history we tend to emphasize the legitimacy of the global conference arena and to place our confidence in its outcomes: global prescriptions and institutions. As a result we have inherited a plethora of universally negotiated principles, concepts, goals, techniques and procedures. They are all well-intentioned, but the ocean management "framework" that has emerged from the global arena consists of highly disparate components, and analysis of the world "milieu" reveals an abundance of institutional, ethical and conceptual dilemmas.

In light of these dilemmas, the best prospect for progress toward effective ocean management seems to reside with bolder experiments in the formation and maintenance of maritime regimes.[130] By stressing the practical importance of international regimes in ocean governance, we have the best chance of circumventing at least some of these theoretical debates over the institutional, ethical and conceptual issues reviewed above.

The apparent advantages of the regime-building approach are at least six in number:

1. The sub-global level of diplomacy required for regional regime-building purposes reduces *consensus-building difficulties*. Smaller groupings can select from the global stock of ocean management principles, concepts, goals, techniques and procedures, choosing those that are found most appropriate for the target region.
2. The fear of *Western impositions* perceived to influence the outcomes of the global arena is unlikely to frustrate the formation of non-Western regional regimes. Admittedly it might be feared that Western funding in support of non-Western regimes may not be forthcoming for such regimes that are seen by Western observers to depart too radically from Western-favored outcomes of global diplomacy, but this may be a healthy fear that would work to achieve a balance between global orthodoxy and regional variation.
3. Regional maritime regimes can be designed specifically around the political, economic, geographical, oceanographic, social and cultural *realities of the region* they are intended to serve. Globally conceived "models" may still have suggestive value, but regional groupings should be encouraged to rely upon their own knowledge, experience and judgment.
4. Regionally negotiated regimes should generate a higher *degree of commitment* to the goals and principles of ocean management on the part of participating governments than they display in response to the outcomes of global diplomacy. Such evidence of serious regional or sub-regional commitment should have the effect of attracting substantial external support for ocean management programs of operational importance in the target region.
5. Regional regimes are essentially *experiments* and can be adjusted, as subsequent experience may dictate, much more easily than the outcomes of global conference diplomacy. For example, the initial version of such a regime may be tentative on matters of political or cultural sensitivity such as the role of non-governmental institutions, but become bolder as experience demonstrates the net benefits of a broader, societal, "governance" approach that would involve partnerships between "track-one" and "track-two" participants.
6. Developing coastal and island states will see the building and maintenance of regional ocean regimes as relatively *low-cost kinds of intergovernmental interaction*, compared with global interactions. The smaller the grouping, the more economically feasible it will be to integrate the domains of ocean diplomats and ocean managers.[131]

The effectiveness of maritime regimes

There is, of course, a large and constantly growing literature on international regimes. Recent studies concentrate less on their structure and process than on their apparent effectiveness. The most rigorous effort so far to confront theory with evidence has produced some interesting insights.[132] First, the researchers, concentrating chiefly on 15 case studies of environmental regimes, distinguish between two kinds of effects that can be compared for evaluative purposes: behavioral and functional. Behavioral effects ("outcomes") are revealed in changes in behavioral patterns, especially bureaucratic actions and interactions at national and regional levels. Functional effects ("impacts") are revealed in longer-term changes beyond human behavior, which in the case of environmental regimes means changes in the sector of the natural environment targeted by the regime. Second, this study shows, on the whole, a higher degree of effectiveness achieved by reference to behavioral effects than by reference to functional effects.[133] Third, much of the work is devoted to the various factors that contribute to regime effectiveness, and the regimes studied are distributed under three classifications: effective regimes, mixed-performance regimes and ineffective regimes. Three of the ocean-related regimes studied are judged to be effective (North Sea dumping,[134] ocean dumping of low-level radioactive wastes[135] and tuna management in the West Central and Southwest Pacific);[136] four ocean-related regimes are judged ineffective (the Mediterranean Action Plan,[137] oil pollution from ships at sea,[138] the International Whaling Commission[139] and CCAMLR);[140] and two are classified in the intermediate mixed-performance category (land-based marine pollution in the North Sea[141] and the management of high-seas salmon in the North Pacific).[142]

The conclusions reached by these authors are important, and deserve to be summarized:

1. Most of the regimes studied have succeeded in changing "actor behavior" in the directions intended. Regimes engaged in activities beyond mere standard setting – devoted to functions such as planning and implementation – tend to be more effective than those not so engaged.[143]
2. Even those regimes that achieve significant results have modest beginnings. New mechanisms need time and experimentation to find an appropriate mode of operation. The initial difficulties tend to be the most difficult to overcome.[144]
3. Most regimes fail to solve the problems they were designed to solve, and often fail by a wide margin. Nearly 60 percent of the cases studied scored low in terms of functional or problem-solving effectiveness. Moreover a fair degree of effectiveness is not at all the same thing as *efficiency*.[145]
4. Failure to deal effectively with substantive problems may, however, be

accompanied by success in the development of problem-solving capacity. Effective capacity-building measures may be expected to lead eventually to effective problem-solving measures. So, in the face of intractable substantive difficulties, the regime builders may be wise to focus their energies initially on the building of institutional and individual capacities.[146]

5 Entrepreneurial leadership is important in the design, development and implementation of international regimes. Leadership often needs to be cultivated through interaction within and between governments, delegations, transnational networks and small groups. Sometimes informal leadership is more important than the leadership provided by secretariats and delegates appointed to chair conferences and committees.[147]

6 Power may be overestimated as a factor in determining regime effectiveness. The evidence suggests that power is more potent in bringing about behavioral change than in promoting the common good, and at worst may be counter-productive in dealing with problems that are "politically benign."[148]

Levels of approach to maritime regime building

There are three principal levels of sub-global international regimes within the context of ocean management: the regional sea level, the LME level and the sub-regional (or neighborhood) level.

In its first two decades (1975–95), the UNEP Regional Seas Programme received acclaim from most observers. What seemed impressive was its innovativeness and good intentions, and its foundation on the concept of common interest in "shared waters," which appeared to promise a collective willingness to cooperate. Now, however, much of the original optimism invested in the Programme has declined, if not evaporated. Many have wished to believe that at least the prototype designed for the Mediterranean has been relatively effective, but the recent re-evaluation of the Mediterranean Action Plan scores it low as "ineffective."[149] Most of the success achieved in that region is attributed to the action of international bodies such as the European Community, not to the Plan itself.[150] Some advocate the need to revitalize the Regional Seas Programme on the ground that the conceptual framework exists, but it may be the wrong framework.

Much more recently, several initiatives have been taken to experiment in ocean regime building at the large marine ecosystem level. The LME approach has been institutionalized in the Southern Ocean,[151] and there are LME advocates for institutional experiments elsewhere. As an approach to *scientific program development*, the LME level has much to commend it. Arguably only at this level can hopes for innovative, systematic, cross-disciplinary "consilience" be realized. But the case for creating operationally viable ocean management regimes (as distinct from scientific programs) based on LMEs still has to be made. Diplomatic and bureaucratic energies and priorities focus very sharply on sovereign territories and national waters. There is very little evidence of governmental willingness to ignore man-made national boundaries in the sea, except in situations where the prospect of joint development opportunity transcends the normal preference for national autonomy. The larger the LME, the less likely it provides a realistic platform for effective ocean regimes.

Of the three approaches, the sub-regional level may be the most likely to support effective ocean regimes. Admittedly the scale of such endeavors is only one factor out of many, and almost certainly not the most important. Regime effectiveness turns on numerous variables that seem only indirectly related, if at all, to the geographical scale of the regime. But, other things being equal, a relatively small area of "neighborhood" waters seems the best candidate as a unit of cooperative ocean management, particularly if the co-users are only three or four in number.[152] The experience in the Gulf of Thailand, facilitated by the Southeast Asian Programme in Ocean Law, Policy and Management (SEAPOL), suggests that regime building on the part of previously uncooperative littoral states (viz Cambodia, Malaysia, Thailand and Vietnam) may become effective at least in the behavioral sense. Moreover, neighborhood waters such as the Gulf of Thailand may often qualify as semi-enclosed waters, where the littoral states "should" or "shall" cooperate in accordance with Article 123 of the LOS Convention.

Types of maritime regimes

Because the word "environment" tends to be all-encompassing, the literature on international regimes gives the impression that almost all regimes are environmental in nature. For the sake of greater clarity, it would be useful to apply criteria to the task of classification. Environmental terminology has become emotive in effect, both attracting and repelling at the same time. Regime classification should not be difficult, since all regimes, by definition, are functional settlements and arrangements.

It may be sufficient to suggest the existence of a growing family of international maritime regimes: some wholly environmental in purpose, some partly environmental, and some non-environmental. In a recent article on the future of the Arctic Ocean, five possible categories are distinguished: (i) law of the sea Arctic regimes, (ii) Arctic environmental regimes, (iii) integrated Arctic sustainable development regimes, (iv) Arctic science regimes and (v) strategic regimes.[153] Given the limitations inherent in the Arctic Ocean, it seems unlikely that other oceans would not benefit from an even wider range of categories and sub-categories of functionally defined international regimes.[154]

Conclusions

Our present era has been characterized as "an age of absurd expectations." Those of us in the ambitious field of international ocean governance may be wise to take stock of the dangers of "overexpectation." The above reflections give cause surely for a degree of pessimism, guarded against the dangers of despair. At least, it can do nothing but good to be clear about the impediments to significant progress and the different kinds of faith that must be embraced.

The *institutional* dilemmas discussed in this essay are difficult, but possible, to deal with. Some advances in cooperative ocean management (e.g. in the sector of high seas fishery management) will require, sooner or later, revisions of the relevant provisions of the LOS Convention. It will require a huge diplomatic effort to capture anything close to worldwide consensus on the necessary revisions within an inter-state system that still depends upon the consent of sovereignty-holders. The problem of treaty implementation will remain severe in most regions of the world as long as the basic tasks of capacity building receive inadequate support from the major sources of international development assistance. Even more urgently, the sustainability of fish stocks within the limits of national jurisdiction, which account for over 90 percent of protein ocean food captured in the wild, looks like an impossible goal in the absence of substantial investment in fishery education, at all levels of sophistication. Yet there is some evidence that the issue of ocean food security is beginning to receive closer attention outside the technical communities. Non-state institutions can play a crucial role in promoting realistic change on the part of state institutions through constructive partnerships. There are some encouraging examples of harmonious partnerships between track-one and track-two institutions in the context of international ocean governance, as indeed reflected in the success of ACORN. Moreover, we might be carefully optimistic that intergovernmental organizations like FAO now recognize the utility of proceeding jointly with hard-law and soft-law instruments instead of feeling an obligation always to choose between them.

The problem with *ethical* dilemmas is that ethicists are notoriously difficult to satisfy with compromises. It is difficult to see a convergence between the two distinct cultures that seem to make up the community of ocean management specialists. The ecological/ethical perspective on ocean affairs, which is justified by the interconnectedness of environmental concerns, has had a significant impact on the politics of marine affairs, especially in the more highly developed countries. Above all, the "transnational ethical community" has attracted more financial support for its "lobbying" purposes than the other culture, which might be characterized as reflecting a "technical" or "sectoral" perspective. Specialists in ocean science and technology sectors, by their own admission, are not usually adept in lobbying. They may be disadvantaged by the rigorous, quantitative nature of their own training and the relativist, comparativist cast of their judgment. It is not always easy to transact compromise judgments with those of another culture endowed with "moral clarity"! However, the pressures of the policy-making arena create the need for constant experiments in accommodation. Lawyers –

the merchants in words – are likely to find themselves increasingly engaged in supplying mediational language for articulating international commitments, even at the cost of precision.

The *conceptual* dilemmas arise from the problem of oversupply. The multiplicity of ways of characterizing the complex problems of ocean management arises chiefly from the number of disciplines that are needed to bring extensive "sophistication" as well as intensive "rigor" to the field. The best way of obtaining a balance among different kinds of expertise would be by investing more public resources in the development of specialized, graduate-level, university courses in marine affairs. Such interdisciplinary courses already exist in a few Western universities, but they are too few in number to produce the degree of educational roundedness that is so badly needed around the world in the crucial arena of ocean management. Educational correctives are unlikely to be found until we can all collaborate effectively in raising the public and political levels of maritime consciousness.

Notes

1 "Government" and "governance" both refer to "purposive behavior, to goal-oriented activities, to a system of rule; but government suggests activities that are backed by formal authority, by police powers, to ensure the implementation of duly constituted policies, whereas governance refers to activities backed by shared goals that may or may not derive from formally prescribed responsibilities and that do not necessarily rely on police powers to overcome defiance and attain compliance." James N. Roseanau "Governance, Order and Change in World Politics" in James N. Roseanau and Ernest-Otto Czempiel (eds) *Governance Without Government: Order and Change in World Politics* (Cambridge University Press, Cambridge: 1992) 4. Governance, the "more all-encompassing phenomenon," is a system of rule that works only if "it is accepted by the majority (or at least the most powerful of those it affects), whereas governments can function even in the face of widespread opposition to their policies." Ibid. Global governance implies "an absence of central authority and the need for collaboration or cooperation among governments and others who seek to encourage common practices and goals in addressing global issues." Leon Gordenker and Thomas G. Weiss "Pluralizing Global Governance: Analytical Approaches and Dimensions" in Thomas G. Weiss and Leon Gordenker (eds) *N.G.O.'s, the U.N., and Global Governance* (Lynne Rienner, Boulder: 1996) 17 at 17.

2 The classic work depicting the anarchic nature of international society is Hedley Bull *The Anarchical Society: A Study of Order in World Politics* (Macmillan, London: 1977). Those who perceive the nation-state to be "in retreat" may conclude that international society is becoming more, not less, anarchic. Susan Strange *The Retreat of the State: The Diffusion of Power in the World Economy* (Cambridge University Press, Cambridge: 1966). Advocates of global governance tend to be more optimistic, both those who place their faith in the efficiency (and ultimately the fairness) of freer trade, and those who welcome signs of an emerging "world culture." Even outside these categories, some scholars now argue, "the international system (in spite of its lack of an overarching regime or world government) is several stages beyond anarchy." Robert C. North *War, Peace, Survival: Global Politics and Conceptual Synthesis* (Westview Press, Boulder: 1990) at 136.

3 See Lawrence Juda and R.H. Burroughs "Prospects for comprehensive ocean management" (1990) 14 *Marine Policy* 23. The concept of "comprehensive" ocean management seems to call for an unprecedented degree of integration, arguably beyond that attainable in the "real world," especially in the case of developing countries. Gerard Peet "Ocean Management in Practice" in Paolo Fabbri (ed.) *Ocean Management in Global Change* (Elsevier Applied Science, London: 1992) 40.

4 The term "consilience," the "jumping together" of knowledge from diverse disciplines, was coined by Edmund O. Wilson in *Consilience: The Unity of Knowledge* (Alfred A. Knopf, New York: 1998). Many would support cross-disciplinary integration as an important component of the institutional goal of integrated ocean management.

5 James A. Caparaso "Changes in the Westphalian order: Territory, public authority, and sovereignty" (2000) 2(2) *International Studies Review* 1 at 2.

6 Ibid., at 1.

7 Ibid., at 2.

8 Ibid., at 5.

9 See Richard B. Bilder "Beyond Compliance: Helping Nations Cooperate" in Dinah Shelton (ed.) *Commitment and Compliance: The Role of Non-Binding Norms in the International Legal System* (Oxford University Press, Oxford: 2000) 65.

10 On the historical development and application of the term "sovereignty" see Helmut Steinberger "Sovereignty" in Rudolf Bernhardt (ed.) *Encyclopedia of Public International Law Volume IV* (North-Holland Publishing Co., New York: 2000) 500.

11 See Michael Reisman "Sovereignty and Human Rights in Contemporary International Law" in Gregory H. Fox and Brad R. Roth (eds) *Democratic Governance and International Law* (Cambridge University Press, Cambridge: 2000) 239 at 240.

12 See ibid., generally.

13 Early in 2002 the George W. Bush Administration's proposal to permit exploratory drilling in areas within Alaska's National Wildlife Refuge adjacent to the Yukon border failed to gather sufficient Congressional support. It was also controversial internationally, and apparently had not been officially discussed with the Canadian or Yukon government despite the responsibilities accepted by the United States under the 1987 Agreement on Conservation of the Porcupine Caribou Herd. Arguably these responsibilities are also reflected in emergent principles of international environmental law. It might be thought that the United States has a legal duty to consult with its neighbor in a situation such as this because of the threat to porcupine caribou and other migratory wildlife in the transboundary region posed by the contemplated action.

14 See Oran R. Young *International Cooperation: Building Regimes for Natural Resources and Environment* (Cornell University Press, Ithaca: 1989) at 187 with particular reference to shipping in the Arctic Ocean.

15 In the context of fishery management, this failure is particularly conspicuous. However, the fault lies not in the text itself; rather in the lack of political will on the part of almost all coastal states that acquired vastly extended areas of national jurisdiction under the EEZ and continental shelf regimes, but have failed to make serious efforts to achieve the goals of cooperative ocean management and sustainable ocean use implicit in the LOS Convention, and made more explicit in subsequent global instruments. See Edward L. Miles "The concept of ocean governance: Evolution toward the 21st century and the principle of sustainable ocean use" (1999) 27 *Ocean Management* 1.

16 See David L. Blaney and Naeem Inayatullah "The Westphalian deferral" (2000) 2(2) *International Studies Review* 29-64 at 29-30. On the idea of "shared sovereignty" applied to US possessions, see Donald C. Woodworth "The exclusive economic zone and the United States insular areas: A case for shared sovereignty" (1994) 25 *Ocean Development and International Law* 365.

17 Over 120 such references are made to international organizations or agencies, though not by name, in the "environmental" provisions of the LOS Convention. See chart in Lee A. Kimball, Douglas M. Johnston, Phillip M. Saunders and Peter Payoyo "Conservation and Management of the Marine Environment: Required Initiatives and Responsibilities under the 1982 United Nations Convention on the Law of the Sea" in IUCN *The Law of the Sea: Priorities and Responsibilities in Implementing the Convention* (IUCN, Bonn: 1995) 121 at 131–54. For an older but detailed study of the ocean management roles of regional organizations, see Douglas M. Johnston and Lawrence M.G. Enomoto "Regional Approaches to the Protection and Conservation of the Marine Environment" in Douglas M. Johnston (ed.) *The Environmental Law of the Sea* (IUCN, Bonn: 1981) 285.

18 For an evaluation, see Lee A. Kimball "Whither international institutional arrangements to support ocean law?" (1997) 36 *Columbia Journal of Transnational Law* 308. See also Barbara Kwiatkowska "Institutional marine affairs cooperation in developing state regions" (1990) 14 *Marine Policy* 385. For annual updates on institutional developments in the field, see *International Organizations and the Law of the Sea: Documentary Yearbook* (Netherlands Institute for the Law of the Sea, Utrecht).

19 As the organized world community has grown to over 180 nation-states, the problems of effective decision making and implementation have multiplied. The idea of "world government" seems to be becoming more utopian, despite the emergence of techniques such as consensus for facilitating the making of large-scale organizational decisions. Etzioni, note 1 at 597. Some argue that greater effectiveness depends on a membership's willingness to delegate more authority to the secretariat. For an early, but still instructive, view of the possibilities, see Thomas G. Weiss *International Bureaucracy: An Analysis of the Operation of Functional and Global International Secretariats* (Lexington Books, Lexington: 1975).

20 At the meeting of parties to the Kyoto Protocol, held at The Hague in November 2000, the delegates failed to agree on a compromise over emissions control strategy under the Framework Convention on Climate Change. This failure was due to a clash between the European Union and another group of highly developed nations including the United States, Canada and Australia. Canada subsequently ratified, but at the time of writing (November 2004) several key countries, including the United States and Australia, have remained outside. Subsequent compromises seem unlikely to win over the Bush Administration or the US Congress.

21 Peter A. Gourevitch "The Governance Problem in International Relations" in David A. Lake and Robert Powell (eds) *Strategic Choice and International Relations* (Princeton University Press, Princeton: 1999) 137 at 140.

22 "Institutions do things for members that they cannot obtain without them. Members acquire incentives to preserve institutions. The test of power of an institution is thus its utility . . ." Ibid., at 138.

23 On the prospect of mutual gain in the general context of globalization, see Lloyd Gruber *Ruling the World: Power Politics and the Rise of Supranational Institutions* (Princeton University Press, Princeton: 2000) 15–32.

24 The trend to more centralized control of development assistance is supported

by considerations of economic efficiency such as the reduction of duplicated effort. The counter-arguments focus on the need to avoid placing too much control in a single bureaucracy, such as the World Bank or GEF, with similar mind-sets.

25 For example, the Southeast Asian Programme in Ocean Law, Policy and Management (SEAPOL), which was funded by Canada (first by the International Development Research Centre and then the Canadian International Development Agency) for forum-building purposes in Southeast Asia, focused on the problems of implementing the LOS Convention within that region.

26 As a consequence, it became increasingly difficult after UNCLOS III, and the following debate over revision of the deep ocean mining provisions in the late 1980s and early 1990s, to sustain the Law of the Sea Institute, the largest non-governmental forum for the review of law of the sea issues. However, the Institute resumed operations in 2001 at its new home at Berkeley School of Law in California.

27 By far the most active training (as distinct from educational) institution was, and still is, the International Ocean Institute, whose cross-disciplinary summer training courses at Halifax in Canada and elsewhere continue to attract mid-career specialists in ocean management.

28 On the individual, institutional and national requirements that have to be met for successful capacity-building initiatives in the ocean management field, see Jan H. Stel "Marine capacity building in a changing global setting" (1998) 22 *Marine Policy* 175.

29 In Southeast Asia the largest indigenous institution in the ocean field is the Malaysian Institute of Marine Affairs (MIMA) in Kuala Lumpur.

30 United Nations *The Global Partnership for Environment and Development: A Guide to Agenda 21* (United Nations, New York: 1993). See, in particular, Kheng Lian Koh, Robert C. Beckman and Lin Sien Chia (eds) *Sustainable Development of Coastal and Ocean Areas in Southeast Asia: Post-Rio Perspectives* (National University of Singapore, Singapore: 1995).

31 Jens Sorensen "National and international efforts at integrated coastal management: Definitions, achievements and lessons" (1999) 25 *Coastal Management* 3 at 13.

32 Ibid., at 28.

33 See "The Role of Social Institutions," infra.

34 See the discussion in Chapter 2 by Chircop and Hildebrand in this volume.

35 For some regional organizations it may be necessary, at least initially, to engage international consultants to develop program or project proposals in consultation with the leading ocean scientists and other specialists of the member states. Securing large-scale funds for such a purpose would require a high degree of sophistication in the oceans sector and a flair for grant writing, a combination that is not necessarily available within the secretariats that serve these organizations.

36 In the late UNCLOS III period, when thoughts turned to problems of implementation, commentators tended to focus on "marine regions." They differed on whether to define potential units of cooperative ocean management in geographical, political or diplomatic terms. Hopes were placed in the UNEP Regional Seas Programme, which mostly cut across existing regional political organizations. But these efforts have proved disappointing. See "Levels of Approach to Maritime Regime Building," infra.

37 It is a matter of scientific controversy how broadly to apply the concept of ecosystem management so as to satisfy all marine ecologists, fishery scientists and oceanographers; see Chapters 8 and 9 in this volume.

38 Other marine ecologists argue for the division of the sea into "primary compartments" ("ecological biomes and provinces"), which are admitted to be fuzzy, and "secondary compartments." These components would be identified by reference to a wide range of oceanographic, hydrographic, ecological and other criteria. Alan Longhurst *Ecological Geography of the Sea* (Academic Press, San Diego: 1995) 59–85.

39 Cooperation of this kind is, of course, called for in Article 123 of the LOS Convention.

40 On the early period of the INGO movement, see John Boli and George M. Thomas *Constructing World Culture: International Nongovernmental Organizations Since 1875* (Stanford University Press, Stanford: 1999) 13–49.

41 See generally Grant J. Hewison "The role of environmental nongovernmental organizations in ocean governance" (1996) 12 *Ocean Yearbook* 32 at 32–5. In retrospect, the authors of the bold prediction of a world "Peoples' Assembly" in 2000 underestimated both the diffuseness and the daily robustness of the "transnational ethical community." See Ronald St. J. Macdonald, Gerald L. Morris and Douglas M. Johnston "International law and society in the year 2000" (1973) 51 *Canadian Bar Review* 316 at 322–5.

42 For a Greenpeace perspective, see Kevin Stairs and Peter Taylor "Non-governmental Organizations and the Legal Protection of the Oceans: A Case Study" in Andrew Hurrell and Benedict Kingsbury (eds) *The International Politics of the Environment: Actors, Interests and Institutions* (Clarendon Press, Oxford: 1992) 110.

43 A criticism of the science establishment to which most environmental NGOs would subscribe is that "adequate answers to problems requiring scientific information characteristically involve a synthesis of all relevant knowledge, to which science is seldom more than a partial contributor." Lynton K. Caldwell *Between Two Worlds: Science, the Environmental Movement, and Policy Choice* (Cambridge University Press, New York: 1990) at 21.

44 Article 71 of the UN Charter provides that UN Economic and Social Council (ECOSOC) "may make suitable arrangements for consultation with non-governmental organizations, which are concerned with matters within its competence . . ."

45 Douglas M. Johnston *Consent and Commitment in the World Community: The Classification and Analysis of International Instruments* (Transnational Publishers, Inc., Irvington-on-Hudson: 1997) at 264.

46 See "Ethical Dilemmas," infra.

47 It has been estimated that in some countries which are almost continuously engaged in transactional dealings with other countries – at the state, governmental and ministerial levels – for every formal, legally binding agreement negotiated (and accepted through signature or otherwise) the same party now enters into three, four or more informal, non-binding instruments. The gap may be even wider, but it is difficult to measure due to the lack of comprehensive data. Johnston, note 45 at 18–20.

48 For example, as argued below, the need for precautionary restraint might be regarded as a "soft law norm," but it appears both in formal, binding agreements such as the 1995 Fish Stocks Agreement and in informal, non-binding instruments such as Agenda 21, albeit in various formulations.

49 For a recent, comprehensive account, see Shelton, note 9. See also Kenneth W. Abbott and Duncan Smidal "Hard and soft law in international governance" (2000) 54 *International Organization* 421.

50 It is scarcely heretical to suggest that most of the UNCLOS III provisions on environmental protection are on the soft side, as contrasted with the provisions on navigation and jurisdiction. Hewison, note 41 at 39.

51 The report of the World Commission on Environment and Development, *Our Common Future* (Oxford University Press, Oxford: 1987), provided the blueprint for UNCED, held in Rio in 1992, especially for the massive Agenda 21 and the Rio Declaration on Environment and Development, which consisted of 27 "principles" including the four cited.

52 The generality of environmental norms is equally evident when they are procedural in character, underlining the need for cooperation, environmental impact assessment, notification, consultation, information exchange, monitoring and reporting. On the difficulty of distinguishing clearly between principles, policies and goals, see David M. Dzidzornu "Four principles in marine environmental protection: A comparative analysis" (1998) 29 *Ocean Development and International Law* 91.

53 See, for example, the 1995 Fish Stocks Agreement (the "Agreement for the Implementation of the Provisions of the United Nations Convention on the Law of the Sea of 10 December 1992 Relating to the Conservation and Management of Straddling Fish Stocks and Highly Migratory Fish Stocks" (1995) 34 *International Legal Materials* 1547).

54 Many Western international environmental lawyers now greet the emergence of the "polluter pays principle" as a norm of customary international law. However, most developing countries today still refuse to recognize it as a binding norm, preferring instead to leave questions of inter-state liability at the previous, more amorphous, level of normative generality. Alan Khee-Jin Tan "Forest fires of Indonesia: State responsibility and international liability" (1999) 48 *International and Comparative Law Quarterly* 826.

55 Crawford B. Macpherson *The Life and Times of Liberal Democracy* (Oxford University Press, Oxford: 1977) 23–43.

56 Ibid., at 44–76.

57 Ibid., at 77–92.

58 Ibid., at 93.

59 Ibid.

60 Peter Bachrach *The Theory of Democratic Elitism: A Critique* (Little Brown and Company, Boston: 1967) at 49.

61 Carole Pateman *Participation and Democratic Theory* (Cambridge University Press, Cambridge: 1970) at 104.

62 Terrence Cook and Patrick M. Morgan *Participatory Democracy* (Canfield Press, San Francisco: 1971) at 4. This kind of democracy, the authors argue, has two principal distinguishing features: (i) the decentralization or dispersion of authoritative decision making and (ii) the direct involvement of amateurs in the making of decisions. Ibid.

63 Ibid., at 5.

64 More recently, one writer coined the terms "discursive democracy" and "communicative rationality" to describe the bridge that needs to be built over the gap between state and society. John S. Dryzek *Discursive Democracy Politics, Policy and Political Science* (Cambridge University Press, Cambridge: 1990) at 218.

65 The original tract, entitled *An Essay on the Principle of Population as it Affects the Future Improvement of Society*, included commentary on the works of earlier writers such as the French revolutionary republican Marquis de Condorect (1743–94) and the English radical rationalist William Godwin (1756–94). Subsequent enlargements of the tract provided copious, illustrative material and, although always lacking cogent empirical weight, the Malthusian argument has always been influential, even in the domain of biology where Charles Darwin and Alfred Wallace were moved to develop the principle of natural selection in response to the theory of chronically scarce food supplies.

66 Bjorn Lomborg "The truth about the environment" *The Economist* (4–10 August 2001) at 63. As perhaps the most influential vehicle of "eco-skepticism," *The Economist* has become a target of environmental criticism, but in 2001 it became convinced by the scientific evidence that global warming has to be taken seriously. "Burning Bush" *The Economist* (16 June 2001) at 77. See also "How many planets? A survey of the global environment" *The Economist* (6 July 2002) (*Survey*).
67 Mauricio Schoijet "Limits of Growth and the Rise of Capitalism" (1999) 4 *Environmental History* 515 at 527.
68 See, for example, Bedrich Moldan, Suzanne Billharz and Robyn Matravers (eds) *Sustainability Indicators: A Report on the Project on Indicators of Sustainable Development* (John Wiley & Sons, Chichester: 1997).
69 See, for example, Wilfred Beckerman "Sustainable development: Is it a useful concept?" (1994) 3 *Environmental Values* 191; Rudi M. Verburg and Vincent Wiegel "On the compatibility of sustainability and economic growth" (1997) 19 *Environmental Ethics* 247.
70 Robert U. Ayres, Jeroen C.J.M. van den Berrgh and John M. Gowdy "Strong versus weak sustainability: Economics, natural sciences, and 'consilience'" (2001) 23 *Environmental Ethics* 155 at 156.
71 Ibid.
72 The argument that future generations have the moral right to inherit a healthy generation is a premise of the relatively new, or re-articulated, principle of "intergenerational equity." See, in particular, Edith Brown Weiss *In Fairness to Future Generations: International Law, Common Patrimony, and Intergenerational Equity* (Transnational Publishers, Dobbs Ferry: 1989).
73 In recent years increased attention has been given by human rights advocates to the "collective" rights enunciated in the 1966 International Covenant on Economic, Social and Cultural Rights, which in the Western world has been accorded a secondary status compared with the 1966 International Covenant on Civil and Political Rights. This relegation of collective rights is usually attributed to the political dominance of the Western liberal democracies, which traditionally reflect the primacy of individual freedoms and liberties.
74 Ayres *et al.*, note 70 at 156.
75 Ibid., at 157. See generally Kjell Arne Brekke *Economic Growth and the Environment: On the Measurement of Income and Welfare* (E. Elgar Publishers, Cheltenham: 1997).
76 Weiss, note 72. See also, Anthony D'Amato "Do we owe a duty to future generations to preserve the global environment?" (1990) 84 *American Journal of International Law* 190.
77 See, for example, Daniel W. Bromley "Searching for sustainability: The poverty of spontaneous order" (1998) 24 *Ecological Economics* 231.
78 FAO, Code of Conduct for Responsible Fisheries (1995). Online at: www.fao.org./fi/agreem/codecond/codecon.asp (accessed 26 January 2004).
79 See note 53.
80 The SDRS consists of both "target reference points" (i.e. the desired outcomes for the fisheries) and "limit reference points" (i.e. upper limits to the rate of fishing or fishing effort level that should not be exceeded). Lawrence Juda "Rio plus ten: The evolution of international marine fisheries governance" (2001) 33 *Ocean Development and International Law* 109 at 123. For a fuller account, see Anthony Charles *Sustainable Fishery Systems* (Blackwell Science, Osney Mead, Oxford: 2001) at 185.
81 Juda, ibid., at 118–21.

82 FAO "New international plan of action targets illegal, unregulated and unreported fishing" Press Release 01/11, cited in ibid., at 118.
83 Resort to port state enforcement through voluntary regional arrangements has become a relatively successful method of combating sub-standard vessels, which are the principal source of safety of life at sea and vessel-based pollution problems. For a discussion of current arrangements in the Asia-Pacific region, see Douglas M. Johnston and Ankana Sirivnatnanon (eds) *Maritime Transit and Port State Control: Trends in System Compliance* (SEAPOL, Bangkok: 2000) 95.
84 The Code of Conduct is a non-binding instrument, consisting essentially of ethical to "soft law" principles and guidelines, not of "hard law" rules that might be enforceable by state institutions. The Fish Stocks Agreement, on the other hand, though a formally binding treaty, has been slow in attracting ratifications. For example, it was eight years after adoption before it received the collective ratification of the European Union on 19 December 2003. It is likely that FAO in the coming years will pursue a policy of multiple strategies based on a "suite" of both hard-law and soft-law instruments in the hope of extracting the benefits of both approaches.
85 William T. Burke "UNCED and the oceans" (1993) 17 *Marine Policy* 519 at 522.
86 To the extent that UNCLOS III reinforced the concept of *sovereign state entitlement* to offshore resources, since 1982 it has been necessary for the diplomatic community to "strengthen" the environmental provisions of the UN Convention on the Law of the Sea without seeming to be "revising" the Convention.
87 The history of nature conservation can be traced back thousands of years. To search for nature's "primeval virginal base-line state" we would have to start with the early Holocene (i.e. post-glacial) epoch. Neil Roberts *The Holocene: An Environmental History* (B. Blackwell, Oxford: 1989) at 6. See also Donald Worster (ed.) *The Ends of the Earth: Perspectives on Modern Environmental History* (Cambridge University Press, Cambridge: 1988).
88 The vagueness of the concept of "nature," though often emotionally satisfying, has invited criticism, and even ridicule, as a focus of public policy. The modern tendency is to differentiate specific components of nature, such as species (of different kinds), resources, habitat and breeding/spawning areas, ecosystems, wilderness areas, and other units of potential management, so that broad macro-management principles and policies can be replaced with more specific arrangements. However, some of the new terminology, such as "biodiversity" and "large marine ecosystem," is extremely generic and seems to invite overgeneralization.
89 See, for example, Peter Singer *Animal Liberation: A New Ethics for Our Treatment of Animals* (New York Review, New York: 1975); Steve F. Sapontzis *Morals, Reason and Animals* (Temple University Press, Philadelphia: 1987); Tom Regan and Peter Singer (eds) *Animal Rights and Human Obligations* (Prentice Hall, Englewood Cliffs: 1989); Tom Regan *The Case for Animal Rights* (University of California Press, Berkeley: 1983); Mary Midgley *Animals and Why They Matter* (Penguin Books, Harmondsworth, Middlesex: 1983); David DeGracia *Taking Animals Seriously: Mental Life and Moral Status* (Cambridge University Press, Cambridge: 1996); Stephen R.L. Clark *Animals and their Moral Standing* (Routledge, London: 1997).
90 Most of those who advocate animal rights concede the difficulty of arguing the absolute equality of all species. Many might agree, for example, that since species vary so greatly in their physiological capacity to experience pleasure and pain, at greatly different levels of "consciousness," the concept of "humane treatment" must be applied differentially. Logically, it follows that those non-human species closest to the human level of physiological capacity, such as

other mammals, have an ethical status below that of humans but superior to that of other life forms. Only a few perhaps would insist that even the least developed species have exactly the same right to life and protection as the most developed, and yet that small-minority opinion seems to be well entrenched and the number of adherents may be growing.

91 "Deep ecology" advocates, embracing the ethic of "very strong" sustainability, believe that "every component or subordinate system of the natural environment, every species, and every physical stock, must be preserved." Ayres *et al.*, note 70 at 160. Yet it must surely be conceded that human "interaction" with nature, by any definition, is unavoidable. One writer has argued that introducing the concept of "interference" with nature "tends to muddy the water rather than shed light": as a derivative concept, "it cannot play any kind of fundamental role in an environmental ethic and one would be unwise to try to construct an environmental ethic around it." Mark A. Michael "How to interfere with nature" (2001) 23 *Environmental Ethics* 135 at 159.

92 Is the domestication of species an "interference"? Or the saving of injured or diseased species? Or the preservation of endangered species? Or the introduction of alien species into an ecosystem? Or the prevention of natural erosion? Ibid., at 137–8.

93 See, for example, Peter J. Stoett *The International Politics of Whaling* (University of British Columbia Press, Vancouver: 1977); Robert L. Friedheim (ed.) *Toward a Sustainable Whaling Regime* (University of Washington Press, Seattle: 2001) 183.

94 The minke whale population, for example, is abundant and responsible harvesting would present no conservation threat. See Christopher D. Stone "Legal and Moral Issues in the Taking of Minke Whales" in Robert L. Friedheim *Report: International Legal Workshop* (Sixth Annual Whaling Symposium, Institute of Cetacean Research, Tokyo: 1996) Appendix, xvii–xxii. One suggested compromise is to focus on the need for reform within the International Whaling Commission, for example, by restoring the legitimacy of the Revised Management Scheme before any resumption of whaling is authorized in the case of abundant stocks. Robert L. Friedheim "Fixing the Whaling Regime: A Proposal" in Friedheim, ibid., at 311–35.

95 The animal rights movement has been particularly active in the context of marine mammals. See, for example, Anthony D'Amato and Chopra Sudhir "Whales: Their emerging right to life" (1992) 85 *American Journal of International Law* 21. But some animals are more "charismatic" than other. Russel L. Barsh "Food Security, Food Hegemony, and Charismatic Animals" in Friedheim, note 93 at 147–79.

96 Milton M.R. Freeman "Is Money the Root of the Problem? Cultural Conflicts in the IWC" in Friedheim, note 93 at 123–46.

97 On the role of NGOs in the meetings of the International Whaling Commission, see Robert L. Friedheim "Negotiating in the IWC Environment" in Friedheim, note 93 at 200–34, specially 210–13. See also Arne Kalland "The Anti-Whaling Campaigns and Japanese Responses" in The Institute of Cetacean Research *Japanese Position on Whaling and Anti-Whaling Campaign* (Institute of Cetacean Research, Tokyo: 1998) 11.

98 It can, of course, be argued that permitting regulated recreational hunting of abundant wildlife populations is the most cost-effective way of culling "surplus stocks," since it generates income for local communities and also revenues that defray the costs of management. Yet on ethical grounds, many react emotionally to the prospect of commercial gain from the taking of life in the wilderness, where, it is felt, there should be minimal interference with nature.

99 The rise of *ecotourism* is perhaps the most encouraging development in efforts to reconcile nature conservation with economic needs. See, for example, Erlet Cater and Gwen Lowman (eds) *Ecotourism: A Sustainable Option?* (Wiley, Chichester: 1994). But even the possibility of sustainable ecotourism is questioned. Rosemary Burton "The Sustainability of Ecotourism" in Mike J. Stabler (ed.) *Tourism and Sustainability: Principles to Practice* (CAB International, Wallingford, Oxford: 1997) 357.

100 For a description of these modifications, see Moritaka Hayashi "The 1994 agreement for the universalization of the Law of the Sea Convention" (1996) 27 *Ocean Development and International Law* 31.

101 For a recent review of the future role of land-locked countries in the ocean management context, see Douglas M. Johnston "The Role of Land-locked Countries in Subregional Coastal Management" in Martin S. Glassner (ed.) *Resource Management and Transit to and from the Sea* (SEAPOL, Bangkok: 2002) 54.

102 Independent World Commission on the Oceans *The Ocean Our Future* (Cambridge University Press, Cambridge: 1998) at 55–73.

103 For a discussion of the various diplomatic arenas available to land-locked countries, see Glassner, note 101 at 49–73.

104 The campaign on behalf of small, developing island states has been led by the islands of the Caribbean and the South Pacific. The countries of these two regions are politically well-organized and have relatively strong regional institutions. Along with other island groups, they have some degree of influence at the global level, making up almost a quarter of the membership of the UN General Assembly. In all of these respects, they have much greater political clout than the developing land-locked countries.

105 Many countries are experimenting with programs intended to assist coastal communities to become more centrally involved in fishery management and other kinds of coastal management. Canada, Sweden, Norway, the Netherlands and Australia are prominent among the donor countries giving priority to community-based and co-management approaches to ocean management. These programs are of particular interest in such countries as Mexico, Cuba, Bangladesh, Thailand, the Philippines, Vietnam, Indonesia, Zimbabwe and various island states of the South Pacific. "Co-management is thought . . . to do away with what is seen as the distant, impersonal, insensitive, bureaucratic approach now characterizing the role of government in fisheries management." Svein Jentoft, Bonnie McCay and Douglas Wilson "Social theory and fisheries co-management" (1998) 22 *Marine Policy* 423 at 424.

106 After the gathering of global consensus on a 200-mile exclusive economic zone (EEZ) in the 1970s, this new regime of extended coastal state jurisdiction was perceived in many countries as an opportunity to expand the national fishing industry. In many cases this led to major investments in the technologies of industrial fishing through capital-intensive rather than labor-intensive strategies. Increased corporate revenues did not necessarily coincide with the interests of the inshore fishing communities, and often the larger catches in offshore waters were clearly detrimental to the species on which the local community was traditionally dependent. On the whole, the EEZ has not been beneficial to traditional fishing communities, and today it is feared that corrective measures may be too little and too late.

107 On current indigenous claims and developments in the context of Canadian ocean law, policy and management, see the chapters in Part VI of this volume, for example, Chapter 11 by Ginn.

108 Richard Butler and Thomas Hinch *Tourism and Indigenous Peoples* (International Thompson Business Press, London: 1996).

109 On ICES, see A.E.J. Went "Seventy years agrowing: A history of the International Council for the Exploration of the Sea 1902–1972" (1972) 165 *Rapports et Proces-Verbaux des Reunions* 1. See also Warren S. Wooster "On the evolution of international marine science institutions" (1993) 10 *Ocean Yearbook* 172.

110 The first modern international fishery agreement devoted to conservation proper was the Convention for the Preservation of the Halibut Fishery of the Northern Pacific Ocean signed by Canada and the United States in 1923. For background and summary, see Douglas M. Johnston *The International Law of Fisheries: A Framework for Policy-Oriented Inquiries* (Yale University Press, New Haven: 1965) 572–84.

111 Elisabeth M. Borgese "The crisis of knowledge" (2001) 15 *Ocean Yearbook* 1.

112 For a contemporary discussion of maximum sustainable yield (MSY) and other objectives of fishery management in the 1960s, see Francis T. Christy, Jr and Anthony Scott *The Commonwealth in Ocean Fisheries: Some Problems of Growth and Economic Allocation* (The John Hopkins Press, Baltimore: 1965) 215–30. For a succinct outline of the five alternative objectives of fishery management, as now understood, see Charles, note 80 at 85–8.

113 In order to compare costs and benefits of "fishery management" it is necessary to look at the "fishery system" as a whole, since the goals are as much social as economic and biological. Charles, note 80 at 69–84. On the "big picture" of Canadian fisheries, see Brad de Young, Randall M. Peterman, A. Rod Dobell, Evelyn Pinkerton, Yvan Breton, Anthony T. Charles, Michael J. Fogarty, Gordon R. Munro and Christopher Taggart *Canadian Marine Fisheries in a Changing and Uncertain World* (Report prepared for the Canadian Global Change Program of the Royal Society of Canada) (NRC Research Press, Ottawa: 1999).

114 Judith Kildow "The roots and context of the coastal zone movement" (1997) 25 *Coastal Management* 231 at 235.

115 Ibid., at 232.

116 Ibid., at 235.

117 For an international perspective, see OECD *Coastal Zone Management: Integrated Policies* (OECD Publications and Information Centre, Washington, DC: 1993), where the concept of "integrated coastal zone management" is explained and discussed. For a definition, see ibid., at 25.

118 Marc J. Hershman, James W. Good, Tina Bernd-Cohen, Virginia Lee and Pamela Pogue "The effectiveness of coastal zone management in the United States" (1999) 27 *Coastal Management* 113.

119 Government of Canada *Policy and Operational Framework for Integrated Management of Estuarine, Coastal and Marine Environments in Canada* (Fisheries and Oceans Canada, Ottawa: 2002) at 16.

120 Ibid., at 17–18.

121 Ibid., at 9–10.

122 Government of Canada *Canada's Oceans Strategy* (Fisheries and Oceans Canada, Ottawa: 2002) at 10–12. Comparison of this document with the companion document cited in note 119 suggests it was not easy for DFO to establish a clear policy rationale for granting paramountcy to any one of these principles.

123 Under Article 312, any party to the Convention may request the convening of a conference to consider proposals for specific formal amendments of any provisions other than those relating to deep ocean mining activities under Part XI. Under Article 313 proposals of the same kind may be submitted by any party for application of "simplified procedures," which would dispense with a conference and facilitate amendments to which no objection is lodged by any parties

within 12 months from the date of circulation. A third amendment method exclusively under Part XI is set out in Article 314.

124 33 *International Legal Materials* 1309 (1994).

125 34 *International Legal Materials* 1542 (1995).

126 A gloss is a scholarly commentary that explains or interprets early writings. It is most unlikely that a wholly "scientific" undertaking of that kind could be entertained today in the highly politicized context of the law of the sea, although it is not so long since it was deemed appropriate to use the quasi-scientific (juridical) services of the International Law Commission (ILC) of the United Nations to engage in the tasks of "codification" and "progressive development" of international law. The law of the sea is a specialized area, and only a few of the existing members of the ILC can be considered experts in that area. So the stronger argument would be to create a new mechanism, a Council of Experts, whose members would be chosen by virtue of their specialized knowledge in that field. This would be consistent with the "specialist" rationale for creation of the International Tribunal on the Law of the Sea.

127 Miles, note 15 at 3.

128 The common heritage principle under Article 136 of the LOS Convention is applicable only to the international sea-bed area and its resources beyond the 200 nautical mile limit of the EEZ (under Article 57) or beyond the outer limits of the continental shelf, if applicable, under Article 76. It is still highly uncertain what resources in those deep ocean areas can be captured or exploited under present technical and economic criteria of exploitability.

129 An excellent overview of the range of current ocean uses and their impact on the marine environment is provided in a recent study by GESAMP (the Joint Group of Experts on the Scientific Aspects of Marine Environmental Protection) *A Sea of Troubles* (Reports and Studies GESAMP No. 70, UNEP, Geneva: 2001).

130 The study of maritime regimes is a sub-set of the much larger field of international regimes. For recent examples of the latter, see Edward L. Miles, Arild Underdal, Steinar Andresen, Jorgen Wettestad, Jon B. Skjaerseth and Elaine M. Carlin *Environmental Regime Effectiveness: Confronting Theory with Evidence* (M.I.T. Press, Cambridge: 2002); Shelton, note 9; Andreas Hasenclever, Peter Mayer and Volker Rittberger *Theories of International Regimes* (Cambridge University Press, Cambridge: 1997); Marc A. Levy, Oran R. Young and Michael Zurn "The study of international regimes" (1995) 1 *European Journal of International Relations* 267. For a recent study restricted to maritime regimes, see Mark J. Valencia *Maritime Regime Building: Lessons Learned and Their Relevance for Northeast Asia* (Martinus Nijhoff Publishers, The Hague: 2001).

131 On the theory of maritime regime building, see Phillip Saunders "Maritime Regional Cooperation: Theory and Principles" in Valencia, note 130 at 1; Douglas M. Johnston "Southeast Asia: Lessons Learned," ibid., at 73–86; Mark J. Valencia "Conclusions and Lessons Learned," ibid., at 149–71.

132 Miles *et al.* note 130.

133 Arild Underdal "Conclusions: Patterns of Regime Effectiveness," ibid., at 433–65.

134 Jon Birger Skjaerseth "Toward the End of Dumping in the North Sea: The Case of the Oslo Commission," ibid., at 65–85.

135 Edward L. Miles "Sea Dumping of Low-Level Radioactive Waste, 1964–1982," ibid., at 87–116.

136 Edward L. Miles "The Management of Tuna Fisheries in the West Central and Southwest Pacific," ibid., at 117–48.

137 Jon Birger Skjaerseth "The Effectiveness of the Mediterranean Action Plan," ibid., at 311–30.

138 Elaine M. Carlin "Oil Pollution from Ships at Sea: The Ability of Nations to Protect a Blue Planet," ibid., at 331–56.
139 Steinar Andresen "The International Whaling Commission (IWC): More Failure Than Success?," ibid., at 379–403.
140 Steinar Andresen "The Convention for the Conservation of Antarctic Marine Living Resources (CCAMLR): Improving Procedures but Lacking Results," ibid., at 405–29.
141 Jon Birger Skjaerseth "Cleaning Up the North Sea: The Case of Land-Based Pollution Control," ibid., at 175–95.
142 Edward L. Miles "The Management of High-Seas Salmon in the North Pacific," ibid., at 249–68.
143 Underdal, note 133 at 435.
144 Miles *et al.* "Epilogue," note 130 at 468.
145 Ibid., at 468–9.
146 Ibid., at 470–1.
147 Ibid., at 472.
148 Ibid., at 472–3.
149 Skjaerseth, note 134.
150 Ibid., at 321–5.
151 See Donald R. Rothwell "The Southern Ocean: Environmental and Resource Management under International Law" in Hance D. Smith (ed.) *The Oceans: Key Issues in Marine Affairs* (Kluwer Academic Publishers, Dordrecht: 2004) 297–311.
152 The case for "neighborhood" maritime regimes can be built on the record of success in various regions. See, for example, Douglas M. Johnston and David L. VanderZwaag "Toward the Management of the Gulf of Thailand: Charting the Course of Cooperation" in Douglas M. Johnston (ed.) *SEAPOL Integrated Studies of the Gulf of Thailand* Vol. 1 (SEAPOL, Bangkok: 1998) at 69–135.
153 Douglas M. Johnston "The future of the Arctic Ocean: Competing domains of international public policy" (2003) 17 *Ocean Yearbook* 596–624.
154 The leading scholar in the field of regime theory has suggested an initial typology consisting of four functional categories: regulatory, procedural, programmatic and generative. Oran D. Young *Governance in World Affairs* (Cornell University Press, Ithaca: 1999) at 25.

16 Principled oceans governance agendas

Lessons learned and future challenges

David L. VanderZwaag and Donald R. Rothwell

Introduction

This volume has addressed principled oceans governance from the perspective of two large coastal states that, during the past decade, have actively engaged in domestic, regional and international ocean governance agendas. It would be wrong to think that between them Australia and Canada can set the international ocean governance agenda, though they have certainly attempted to influence that agenda both regionally and globally.[1] What can be said is that a review of the Australian and Canadian experiences assists in understanding the "lessons learned" from both their successes and failures. Given the vast extent of their respective maritime domains, these lessons have application for nearly all coastal states seeking to engage in an oceans governance process. In addition, because of their distinctive political and constitutional dynamics, it is possible to identify particular challenges not only for other federations, but also for States which have complex and multilayered domestic governance structures. In this final chapter a brief review will be undertaken to identify the lessons learned from Australian/Canadian experiences, before turning to assess the challenges at the global, regional and national levels.

Lessons learned

Perhaps the most significant of all the lessons to be learned from the Australian and Canadian experiences is that there is no single model for operationalizing oceans governance at the national level. How each state approaches marine use issues will very much depend on its legal, cultural, political, social and economic context. Australia and Canada have taken distinctive approaches to national ocean governance with no commonality in the legal and policy frameworks or in bureaucratic structures. A range of challenges has been faced including bureaucratic, constitutional, inter-governmental, political, institutional and structural. Both countries continue to face legislative and regulatory fragmentation at not only the national level but also at the provincial and state levels. This only serves to highlight an important general lesson: the adoption of a national oceans statute or a national oceans policy does not ensure that a country will actually get its "oceans act" together.

There is common ground in the Australian/Canadian responses to compliance and enforcement strategies. Reflective of their federal systems, central administration and management is the norm with some devolution of authority to provincial and state authorities. An increased emphasis upon maritime security and cooperation with neighbors, which for Canada is a particular challenge given the concerns of the United States, is also a common feature.

Specific lessons can be drawn from implementation experiences for each of the main principles emphasized in this volume. For integrated coastal/ocean management at least four instructive realities stand out. First, development of integrated management plans is not a "quick process" but involves numerous steps including the need to:

- identify existing and proposed ocean uses;
- assess marine resource distributions and ecosystem relationships;
- engage affected interests;
- get endorsement of the plan by planning participants and decision-making authorities; and
- ensure implementation of the plan through monitoring and evaluation.

After nearly eight years of implementation efforts under its *Oceans Act*, Canada has yet to finalize a plan for the initial Large Ocean Management Areas (LOMAs) chosen.[2] Australia has only completed one regional plan, the South-East Regional Marine Plan.

Second, defining management planning areas may be an "iterative process." Countries have to start somewhere, and rather restrictive boundaries may be initially chosen to allow practical launching of the integrated planning process. For example, Canada's Eastern Scotian Shelf Integrated Management (ESSIM) Initiative[3] has focused planning on marine areas outside the territorial sea[4] and has not adopted a large marine ecosystem approach. It has excluded areas of the Western Scotian Shelf including numerous transboundary management challenges involving fish stocks and other living marine resources shared with the United States.[5]

A third lesson from Australian and Canadian integrated planning experiences is that a federated system of governance certainly does complicate and even call into question the eventual effectiveness of planning exercises. Neither Canada nor Australia has ensured the "buy in" by provinces or states, for example through financial incentives to encourage provincial/state coastal planning initiatives or a guarantee that federal permitting powers will be exercised consistent with provincial/state planning priorities. Offshore claims by First Nations add a further complication in some areas.

In Canada, an especially complex layering of jurisdictional accommodations seems likely to evolve in relation to integrated coastal/ocean planning.

In September 2004, Canada entered into the first federal/provincial memorandum of understanding for the implementation of Canada's *Oceans Strategy*. The Canada-British Columbia Memorandum of Understanding (MOU) Respecting the Implementation of Canada's Oceans Strategy on the Pacific Coast pledges the development of further sub-agreements including on implementation measures for coastal planning and integrated oceans management planning.[6] The Government of Canada has noted its intent to explore similar MOUs with other jurisdictions.[7] Canada's Oceans Action Plan, receiving an initial budget of CDN$28 million over two years, has as one of its four pillars a commitment to expand integrated management and federal–provincial partnerships.[8]

A fourth learning dimension in relation to integrated management is the common Australian and Canadian collaborative approach to planning. Both States have opted, at least initially, for leaving integrated management plan implementation to existing permitting and authorization processes for the various sectoral areas, such as fisheries and oil and gas exploration/exploitation.[9] The cooperative implementation experiences promise to be test cases for how far various departments and agencies will be willing to use their powers and resources to achieve plan objectives.[10]

The precautionary principle/approach has been a driver for oceans governance reform for both Australia and Canada. As strong supporters of the Stockholm and Rio processes and associated principles underpinning environmental management in the law of the sea, plus having relatively sophisticated environmental law and management regimes, both countries have a track record of giving effect to precaution in various forms. However, while the record superficially is impressive, it is clear there has been much "wading and wandering in tricky currents." This seems to have been as a result of variable political will, economic and social pressures, bureaucratic inadequacies, and judicial reluctance to fully engage with the principles.[11]

Other than adapting a strong "reverse listing" approach to ocean dumping where only wastes on a "safe list" may be considered for disposal at sea,[12] both countries have favored quite weak versions of precaution. For example, Australia's *Environmental Protection and Biodiversity Conservation Act 1999* (EPBC) adopts a rather diluted form of precaution.[13] Canada's *Framework for the Application of Precaution in Science-based Decision Making about Risk*[14] avoids suggesting a burden of proof reversal to development proponents and does not discuss possible standards of proof. Instead, the Framework favors trade and economic agendas and suggests that precaution should only be triggered upon an adequate risk assessment.[15]

Judicial considerations of the precautionary principle in Australia and Canada are still quite limited and the content of the principle and implications for administrative review remains largely to be determined.[16] Complicating the interpretive context are the ongoing differences of opinions among academic writers as to whether the precautionary principle provides little or no guidance,[17] or whether the principle is in fact clear and powerful, for example, calling for the reversal in the legal burden of proof.[18]

Further lessons to be learned in relation to precaution can be drawn from the field of fisheries management. Australia stands out as an example of how environmental assessment can be applied to fisheries as provided in the progressive EPBC Act. Canada's discussions of the precautionary approach as applied to fisheries show how there may be a tendency for managers and scientists to get caught up with a "quantification mentality" of trying to model fisheries reference points for healthy zones, cautious zones and critical zones. Broader law and policy implications of precaution may be ignored or marginalized, such as the need to develop environmentally friendly fishing gears and techniques and the need to ensure a network of marine protected areas.[19]

While Australia and Canada have committed to promoting ecosystem-based management,[20] both countries are facing difficult implementation challenges that are common around the globe. How to measure marine ecosystem integrity and health is still being debated,[21] along with what should be the most appropriate objectives and indicators.[22] Lack of sufficient funding for marine ecological research continues to be a concern.[23] Lack of scientific data and understanding of marine environments has been openly acknowledged.[24] Limited knowledge in relation to marine species at risk is becoming an especially pressing problem in light of the need to ensure that survival and recovery of endangered/threatened species are not jeopardized.[25]

There is also the challenge of fostering understanding among scientists and managers that ecosystem-based management does not just hinge upon inputs from the natural sciences. Conserving ecosystem integrity and ecosystem health are, indeed, emerging as dominant concepts in the humanities and social sciences.[26]

Both Australia and Canada have experimented with community-based and co-management arrangements from other countries, and various cautions emerge from the experiences. Use of terms like "community-based management," "co-management" and "community consultation" should be carefully distinguished to avoid suspicion and cynicism on the part of the communities. The term "community" should also be used with care in light of the numerous interests that may make up a community beyond the narrow fisheries sector. Effective community involvement requires adequate institutional and financial bases and government departments need to be sensitive to downloading governance responsibilities upon what may be over-stretched communities.[27] Modernizing national legislation may also be needed in order to provide an adequate legal foundation for the establishment of community-based and co-management arrangements.[28]

One of the great challenges for both countries is their response to indigenous rights in the coastal and oceans context, and both commonalities and significant distinctions between Australian and Canadian approaches have emerged. In Australia, there is no real recognition of marine native title, and whilst the oceans policy process has sought to engage indigenous communities, the success of those approaches seems to have been marginal at best.[29] In Canada, roles of First Nations in oceans management and planning have been only partly defined (in relation to fish and wildlife) pursuant to existing treaties or agreements.[30] While Canada has constitutionally entrenched aboriginal rights[31] in contrast to Australia,[32] Canada has likewise taken a position to denying aboriginal title to ocean spaces.[33] Neither country has fully worked out the spectrum of possible indigenous rights including the possible human right to environmental quality for indigenous peoples.[34]

Overall the Australian and Canadian approaches at national oceans governance can be labeled a partial success to date. During their first decade significant advances have been made. However, both countries remain at the early stage of the process and there are challenges still to be confronted.

Future international challenges

The Australian and Canadian experiences have undoubtedly been influential in international oceans governance. As two of the largest coastal States with complex jurisdictional and varied marine environments, there was sure to be significant interest in how they tackled oceans governance. Because of the interactions of their respective marine ecosystems, there was also going to be attention given to how they addressed the complex web of bilateral issues raised in seeking to manage "joint" maritime areas whether it be the Tasman Sea, Torres Strait,[35] the Gulf of Maine, Georgia Basin/Puget Sound,[36] or the Arctic Ocean.[37] In addition, because of their presence and role on the global stage – Australia as a principal supporter of the 1982 United Nations Convention on the Law of the Sea (LOS Convention) framework and initiatives for high seas ecosystem management, and Canada as a major player in moves to reform international fisheries and also global environmental management – there was inevitably going to be interest in their domestic oceans governance experiments and global implications. The regional and global agenda for principled oceans governance is therefore one which Australia and Canada have already been shaping through their own national agendas and proactively influencing through their diplomacy and initiatives at both regional and global forums.

Regionally, both countries have been active supporters of the Asia-Pacific Economic Cooperation (APEC) forum, which, in addition to hosting annual leader's summits, has a range of forums involving ministers and government officials. APEC has had a strong marine focus over the past decade and several working groups have actively addressed a range of marine environmental issues ranging from fisheries, resources conservation and transport. Both countries have played an active role in several of these working groups promoting domestic oceans and coastal governance arrangements, including the protection of marine biodiversity and promotion of marine protected areas.[38] The Department of Environment and Heritage (Australia) and Fisheries and Ocean Canada have been the principal agencies pursuing this agenda with ministerial support. Working in collaboration with other APEC members, Australia and Canada were able to win endorsement for the adoption of the Seoul Oceans Declaration at the 2002 APEC Ocean-Related Ministerial Meeting.[39]

At the global level, in addition to their ongoing engagement with a range of multilateral fora with a marine environmental focus,[40] the processes established under the framework of the LOS Convention have also given Australia and Canada additional opportunities to promote global oceans governance. The work of the United Nations Informal Consultative Process on Oceans and the Law of the Sea (UNICPOLOS) in particular has proven to be an important avenue for individual and collective Australian/Canadian concerns to be expressed over enhanced cooperation and coordination on oceans issues, fisheries governance, mechanisms to combat illegal, unregulated and unreported (IUU) fishing, and sustainable uses of the oceans including conservation and management of biological diversity.[41] Australia has been particularly active in this area, having sponsored meetings on high seas biodiversity conservation and the deep sea.[42]

What then are the key international issues for the future? One is the need for a more integrated legal framework that encompasses the key concepts and principles and provides the "tools" for principled oceans governance at the regional and global level. The LOS Convention provides a sound foundation for the legal framework, but it contains several weaknesses reflective of its 1970s roots and insufficiently incorporates contemporary principles of oceans governance. In particular, the Convention's Part XII environmental framework fails to develop fully principles of marine environmental management that were beginning to be recognized in the 1970s following the Stockholm Conference and which have gained further acceptance following the 1992 Rio Conference. Likewise, whilst the fisheries management provisions of the Convention concerning the EEZ, high seas, and straddling and highly migratory stocks have all been subjects of further development, the overarching framework of the Convention now seems to be somewhat lacking in how it seeks to balance sustainable resource and environmental management. It could be argued, then, that the time has truly come for a "Constitution of the Oceans"[43] which builds upon the legal framework of the LOS Convention and takes into account the parallel developments in integrated and principled oceans governance arising from environmental law and policy, international human rights law and policy, and, increasingly, international security law and policy. Whilst such a proposal is not new, it is inevitably greeted with wearied dismissal from those who remember the battles associated with the LOS Convention. It is, however, a proposal with increasing force as states take oceans governance to new levels of sophisticated integration not imagined during the 1970s.

Management of the high seas also poses a particular challenge. As coastal state claims extend further offshore, the area of "true high seas" gradually diminishes, resulting in further competition for increasingly scarce resources. Whilst there is at least a regime in place for the management of the deep sea-bed,[44] the same cannot be said for the living resources and associated ecosystems of the high seas.[45] Global high seas fisheries management is one of the last true gaps in the law of the sea, and whilst mechanisms are gradually emerging to manage these fisheries, the enforcement of these new regimes will remain a challenge for some considerable time to come. Here international oceans governance runs up against a range of other regimes, including international trade regimes as managed by the World Trade Organization and long-established principles of maritime law, including the rights of a "flag state."[46] Canada's long-standing efforts to confront the issue of straddling and highly migratory fish stocks has brought some success in tackling this high seas issue following the adoption of the 1995 UN Fish Stocks Agreement, whilst Australia's initiatives through bodies such as UNICPOLOS have further raised the profile of high seas biodiversity conservation. Whilst these are welcome developments, much more remains to be achieved before the high seas are sustainably managed.

Related to high seas governance is the challenge and potential of regional oceans governance. At one level this presents as an opportunity. Several of the world's oceans and seas already fall under a form of regional management, some of which are relatively sophisticated and reflect high levels of interaction and cooperation amongst the states parties.[47] However, the forms of regionalized oceans governance structures vary and environmental considerations are not always paramount.[48] Whilst it is possible therefore to point to the apparent success of some forms of regional oceans governance – including the broad Asia Pacific initiatives of APEC[49] – these successes seem to mask the reality where notwithstanding very strong regional interests and apparent institutional structures, success is not always guaranteed. Regional oceans governance, as both Australia and Canada have discovered from their own experiences, provides not only opportunities but also presents considerable challenges, not the least of which is the size of the ecosystem but also the weakness of the legal frameworks.[50]

A further agenda item is the speedy resolution of outstanding maritime claims. The new law of the sea created by the LOS Convention was estimated to create a need for the resolution of more than 350 maritime boundaries.[51] In more than two decades since the Convention's conclusion, there has been insignificant resolution of the vast majority of maritime boundary disputes, although new ones have emerged,[52] or are in the process of emerging.[53] Oceans governance will succeed in a maritime environment free of tensions over sovereignty and based upon principles of cooperation, not competition. It is therefore essential that as many contested maritime claims as possible be settled expeditiously so as to avoid rising political and military tension.[54] Some disputes remain intractable and incapable of speedy resolution,[55] others are more straightforward and their finalization would greatly assist in moving forward some bilateral oceans management issues.[56]

Closely related to maritime boundaries is maritime security.[57] In the wake of the 2001 terrorist attacks upon the United States, there has been a global upsurge of initiatives responding to international terrorism. A number of these have had a maritime perspective,[58] including the amendment or adjustment of existing legal frameworks and new responses such as the 2003 "Proliferation Security Initiative" that has been promoted by several leading maritime states, including Australia.[59] Whilst national responses are clearly important, regional and global responses remain essential in addressing the security threat posed by terrorist organizations and the shipment of hazardous cargoes such as weapons of mass destruction. Here a philosophical challenge must be faced. The law and policy of international oceans governance has been predominately founded upon "the peaceful uses of the seas and oceans"[60] consistent with the original Grotian vision of the sea as being free for all to use.[61] Maritime security initiatives, however, have inevitably placed constraints upon the use of the oceans for certain activities and have sanctioned certain enforcement measures to give effect to this new regime. Whether these initiatives remain consistent with the freedom of navigation remains to be settled. Likewise, there is the potential for too much power over the oceans to be conferred on the United Nations Security Council – a body that has little interest in oceans management other than for purely security purposes.

Concluding remarks

Numerous challenges remain for the development of national agendas for principled oceans governance. The interrelationships of various principles such as precaution, the ecosystem approach, intergenerational equity and public participation have yet to be worked out.[62] The jurisprudential underpinnings of sustainability principles need to be considered, including whether the content of principles should be ascertained in the positivist tradition of focusing upon what States have agreed to in treaties, texts and practices or whether normative currents should be sought from other sources such as natural law derived from human reason, the laws of nature or divine inspiration.[63]

While it is easy to feel "lost at sea" in light of all the institutional, ethical and conceptual dilemmas surrounding ocean governance,[64] sustainability principles do set a course towards rethinking and redefining human–nature relationships.[65] Perhaps the greatest challenge of all is trying to obtain wider social understanding and consensus on the ethical perspective that should guide principled decision making. Various ecophilosophical viewpoints have been advocated.[66]

A growing cadre of writers in the ecosystem management and ecological sustainability areas may provide the epistemic groundswell needed to push states and individuals beyond traditional anthropocentric ethical theories and the utilitarian tendency to give market and commercial values prevalence in public policy decisions.[67] The emerging groundswell,[68] largely ecocentric in nature but recognizing that humans are an integral part of nature, urges that ecological integrity be maximized through the establishment of biodiversity reserves while ecosystem health is maintained through ensuring ecologically sustainable human uses.[69]

Both Australia and Canada are already riding the groundswell through recognition of marine ecological integrity and marine ecosystem health as important policy goals.[70] However, the goals have yet to be achieved and ocean law and policy voyages have hardly begun.[71] Lofty words remain to be translated into actions.[72]

Notes

1 For example, Australia has led international efforts to develop governance arrangements for high seas biodiversity, a major gap in international law, and, among other initiatives, hosted a global conference on the topic in June 2003. See Department of the Environment and Heritage "Workshop on the Governance of High Seas Biodiversity Conservation." Online at: www/deh.gov.au/coasts/international/highseas/index.html (accessed 11 April 2005). Canada has been active in trying to strengthen the 1995 UN Agreement on Straddling and Highly Migratory Fish Stocks and regional fisheries management organizations, particularly the Northwest Atlantic Fisheries Organization. As part of these activities, Canada hosted an international Conference on the Governance of High Seas Fisheries and the United Nations Fish Agreement: Moving from Words to Action in St. John's, Newfoundland, 1–5 May 2005.

2 LOMAs selected as priorities are: Placentia Bay/Grand Banks, the Scotian Shelf, the Gulf of St. Lawrence, Beaufort Sea and the Pacific North Coast. See Fisheries and Oceans Canada, *Report on Plans and Priorities for 2005–2006* at 31. Online at: www.tbs-sct.gc.ca/est-pre/20052006/F0-P0/F0-P0r56.1_e.asp? (accessed 8 April 2005).

3 For an overview, see R.J. Rutherford, C.J. Herbert and S.S. Coffen-Smout "Integrated ocean management and the collaborative planning process: The Eastern Scotian Shelf Integrated Management (ESSIM) Initiative" (2005) 29 *Marine Policy* 75–83.

4 Fisheries and Oceans Canada *Eastern Scotian Shelf Integrated Ocean Management Plan (2006–2011) (Draft for Discussion)* (Oceans and Coastal Management Report 2005–07: February 2005) at 10.

5 However, consideration is being given to expanding the planning area to include the entire Scotian Shelf. Ibid., at 11. For a review of ecosystem parameters for the Scotian Shelf, see K.C.T. Zwanenburg, D. Bowen, A. Bundy, K. Drinkwater, K. Frank, R. O'Boyle, D. Sameoto, and M. Sinclair "Decadal Changes in the Scotian Shelf Large Marine Ecosystem" in K. Sherman and H.R. Skjoldal (eds) *Large Marine Ecosystems of the North Atlantic: Changing States and Sustainability* (Elsevier Science, Amsterdam: 2002) 105–50.

6 Fisheries and Oceans Canada "Canada and British Columbia Join Forces to Implement Canada's Oceans Strategy" News Release NR-PR-04-053e (18 September 2004).

7 Ibid., at 2.

8 Fisheries and Oceans Canada, "Oceans Action Plan: Phase I" Backgrounder BG-PR-05-002e (Fisheries and Oceans Canada–Pacific Region, Vancouver: 2 March

2005). Online at: www-comm.pac.dfo-mpo.gc.ca/pages/release/bckgrnd/2005/bg002_e.htm (accessed 25 April 2005).

9 See Lawrence Juda "Changing national approaches to ocean governance: The United States, Canada, and Australia" (2003) 34 *Ocean Development and International Law* 161–87.

10 Canada has noted that regulations could be issued under the *Oceans Act* if enforcement or other regulatory gaps exist in relation to plan implementation. Government of Canada, *Policy and Operational Framework for Integrated Management of Estuarine Coastal and Marine Environments in Canada* (Fisheries and Oceans Canada, Ottawa: 2002) at 31.

11 See Chapter 6 by VanderZwaag, Fuller and Myers and Chapter 7 by Kriwoken, Fallon and Rothwell in this volume.

12 *Canadian Environmental Protection Act, 1999* S.C. 1999, c. 33, s. 125; *Environmental Protection (Sea Dumping) Act 1981* (Cth), s. 19.

13 Section 3A provides a definition of the precautionary principle as part of the principles of ecologically sustainable development: "if there are threats of serious or irreversible environmental damage, lack of full scientific certainty should not be used as a reason for postponing measures to prevent environmental degradation. . . ."

14 Online at: www.pco_bap.gc.ca/default.asp?Language=E&page=publications&doc=precaution/pr7 (accessed 3 February 2005).

15 Principle 4.3 provides: "Sound scientific information and its evaluation must be the basis for applying precaution; the scientific information base and responsibility for producing it may shift as knowledge evolves."

16 See Chapters 6 and 7 in this volume.

17 For example, see Cass R. Sunstein, "Beyond the precautionary principle" (2003) 151 *University of Pennsylvania. Law Review* 1003; Jaye Ellis and Alison FitzGerald, "The precautionary principle in international law: Lessons from Fuller's Internal Morality" (2004) 39 *McGill Law Journal* 779–800.

18 For example, see Karla Sperling "If caution really mattered" (1999) 16 *Environmental and Planning Law Journal* 425–40; Kathryn Chapman, "Case Comment 114957 Canada Ltée (Spray-Tech, Société d'arrrosage) v. Hudson (ville): Application of the precautionary principle in domestic law" (2002) 15 *Canadian Journal of Administrative Law Practice* 123–36.

19 See Fisheries and Oceans Canada *Proceedings of the National Meeting on Applying the Precautionary Approach to Fisheries Management* (Canadian Science Advisory Secretariat Proceedings Series 2004/03, DFO, Ottawa: December 2004).

20 For example, *Australia's Oceans Policy* includes notions of ecosystem-based management under its statement of principles for ecologically sustainable ocean use: "If the potential of impact of an action is of concern, priority should be given to maintaining ecosystem health and integrity." Commonwealth of Australia *Australia's Oceans Policy: Caring, Understanding, Using Wisely* Vol. 1 (1998) at 19. Canada's *Oceans Strategy* in the context of its commitment to the precautionary approach calls for promotion of an ecosystem-based management approach and urges priority be given to maintaining ecosystem health and integrity, especially in the case of uncertainty. Government of Canada *Canada's Oceans Strategy: Our Oceans, Our Future* (Fisheries and Oceans Canada, Ottawa: 2002) at 11–12.

21 See Chapter 8 by Hatcher and Bradbury in this volume.

22 For example, through a National Workshop on Objectives and Indicators for Ecosystem-based Management in 2001, Fisheries and Oceans Canada has adopted three conceptual ecosystem objectives and has identified ten components of those objectives. The conceptual ecosystem objectives are: to conserve

enough components (ecosystems, species, populations, etc.) so as to maintain the natural resilience of the ecosystem; to conserve each component of the ecosystem so that it can play its historic role in the food web; and to conserve the physical and chemical properties of the ecosystem. For a discussion, see Fisheries and Oceans Canada *Habitat Status Report on Ecosystem Objectives* (Habitat Status Report 2004/01, DFO, Ottawa: September 2004). Online at: www.dfo-mpo.gc.ca/csas/csas/status/2004/HSR2004_001_E.pdf (accessed 25 April 2005).

23 For example, concerns have been expressed about the Department of Fisheries and Oceans' lack of information on fish stocks and their habitats and the lack of adequate financial resources to conduct oceans-related science. See Senate of Canada *Fish Habitat: Interim Report of the Standing Committee on Fisheries and Oceans* (Senate of Canada, Ottawa: November 2003) at 33.

24 For example, Australia's South-east Regional Marine Plan highlights the limits of present knowledge about the marine environment and future resources. See National Oceans Office *South-east Regional Marine Plan: Implementing Australia's Oceans Policy in the South-east Marine Region* (National Oceans Office, Hobart: 2004) at xi.

25 For example, abundance data are available for a comparatively small number of species with an estimate that such data are only available for about 26 of the 750 species of marine fish on Canada's Atlantic Coast. See J.A. Hutchings and J.K. Baum, "Measuring marine fish biodiversity: Temporal changes in abundance, life history, and demography" (2005) 360 *Philosophical Transactions of the Royal Society* (315–38) as referred to in David L. VanderZwaag and Jeffery A. Hutchings "Canada's marine species at risk: Science and law at the helm, but a sea of uncertainties" (2005) 36 *Ocean Development and International Law* 219–59.

26 See, for example, J. Baird Callicott and Karen Mumford "Ecological Sustainability as a Conservation Concept" in John Lemons, Laura Westra and Robert Goodland (eds) *Ecological Sustainability and Integrity: Concepts and Approaches* (Kluwer Academic Publishers, Dordrecht: 1998); J.B. Ruhl "The myth of what is inevitable under ecosystem management: A response to Pardy" (2004) 21 *Pace Environmental Law Review* 315–23.

27 See Chapter 10 by Binkley, Gill, Saunders and Wescott in this volume.

28 As highlighted by a recent Canadian report on the future of Pacific Coast fisheries which urges revision of the 1867 *Fisheries Act* to provide a firm legal basis for co-management. See Donald M. McRae and Peter H. Pearse *Treaties and Transition: Towards a Sustainable Fishery on Canada's Pacific Coast* (Fisheries and Oceans Canada, Vancouver: April 2004) at 52.

29 See Chapter 13 by Clark and Chapter 14 by Dillon in this volume.

30 See Chapter 12 by Jones in this volume.

31 See Thomas Isaac and Anthony Knox "Canadian aboriginal law: Creating certainty in resource development" (2004) 53 *University of New Brunswick Law Journal* 3–42.

32 See Julie Cassidy "The stolen generation: Canadian and Australian approaches to fiduciary duties" (2002–03) 34 *Ottawa Law Review* 175–238 at 209–10.

33 For arguments in relation to aboriginal title in the offshore, see Chapter 11 by Ginn in this volume.

34 See Charie Metcalf "Indigenous rights and the environment: Evolving international law" (2003–04) 35 *Ottawa Law Review* 101–40 at 122–3.

35 See Dennis Renton "The Torres Strait Treaty after 15 Years: Some Observations from a Papua New Guinea Perspective" in James Crawford and Donald R. Rothwell (eds) *The Law of the Sea in the Asian Pacific Region* (Martinus Nijhoff, Dordrecht: 1995) 171–80.

36 See Lawrence P. Hildebrand, Victoria Pebbles and David A. Fraser "Cooperative ecosystem management across the Canada–US border: Approaches and experi-

ences of transboundary programs in the Gulf of Maine, Great Lakes and Georgia Basin/Puget Sound" (2002) 45 *Ocean & Coastal Management* 421–57.

37 See David VanderZwaag, Rob Huebert and Stacey Ferrara "The Arctic Environmental Protection Strategy; Arctic Council and multilateral environmental initiatives: Tinkering while the Arctic marine environment totters" (2002) 30 *Denver Journal of International Law and Policy* 131–71.

38 See generally the reports of the APEC Marine Resources Conservation Working Group. Online at: www.apec.org/content/apec/apec_groups/working_groups/marine_resource_conservation.html (accessed 27 April 2005).

39 Seoul Oceans Declaration, adopted by the 1st APEC Ocean-related Ministerial Meeting, Seoul, Korea, 22–26 April 2002. Online at: www.apec.org//content/apec/ministerial_statements/sectoral_ministerial/ocean-relatedO/ocean-related.html (accessed 27 April 2005).

40 Australia has been a particularly strong supporter of the International Whaling Commission and the prohibitions adopted under the Commission prohibiting commercial whaling. Both Australia and Canada are active contributors to the Convention on International Trade in Endangered Species of Wild Fauna and Flora (CITES).

41 See the *Report of the work of the United Nations Open-ended Informal Consultative Process on Oceans and the Law at the Sea at its fifth meeting* UNGA Doc A/59/122 (1 July 2004). A senior official of the Australian Department of Environment and Heritage, Mr Phil Burgess, is the current Co-Chairperson of the Consultative Process.

42 See "Letter dated 24 May 2004 from the Permanent Representative of Australia to the United Nations addressed to the Secretary-General" UNGA Doc A/AC.259/12 (24 May 2004).

43 See generally the discussion in Report of the Independent World Commission on the Oceans *The Ocean Our Future* (Cambridge University Press, Cambridge: 1998).

44 See Part XI of the 1982 UN Convention on the Law of the Sea.

45 See, for example, Kristina M. Gjerde and David Freestone "Unfinished business: Deep-sea fisheries and the conservation of marine biodiversity beyond national jurisdiction. Editors' Introduction" (2004) 19 *International Journal of Marine and Coastal Law* 209–22.

46 See Rosemary Gail Rayfuse *Non-Flag State Enforcement in High Seas Fisheries* (Martinus Nijhoff Publishers, Leiden: 2004).

47 The Southern Ocean falls under the 1959 Antarctic Treaty and the 1980 Convention for the Conservation of Antarctic Marine Living Resources. The Mediterranean Action Plan was launched in 1975 and has been responsible for numerous instruments and action programs, see A. Vallega "The Mediterranean and Black Seas" in Hance D. Smith (ed.) *The Oceans: Key Issues in Marine Affairs* (Kluwer Academic Publishers, Dordrecht: 2004) 231 at 244–9.

48 A clear contrast exists within the Asia Pacific, where in the Southwest Pacific there exist detailed marine environmental mechanisms under a framework developed by the United Nations Environment Programme (UNEP) and the South Pacific Regional Environment Programme (SPREP), whilst in the area of the South China Sea fronting Japan, Korea, China, Vietman and the Philippines there has been minimal cooperation on environmental and resource management issues, though emerging cooperation on security matters, see discussion in Donald R. Rothwell "Is there an Asia Pacific environmental law?" (2000) 5 *Asia Pacific Journal of Environmental Law* 307–17.

49 The Southern Ocean is often pointed to as a success, see Donald R. Rothwell "The Southern Ocean: Environmental and resource management under international law" in Hance D. Smith (ed.) *The Oceans: Key Issues in Marine Affairs* (Kluwer Academic Publishers, Dordrecht: 2004) 297–311.

50 For example, Canada is giving increasing priority to reforming the Northwest Atlantic Fisheries Organization (NAFO) with key issues including vessels fishing over quotas, fishing of stocks under a moratorium and under-reporting of catches, see Fisheries and Oceans Canada *2005–2010 Strategic Plan: Our Waters, Our Future* at 18. Online at: www.dfo-mpo.gc.ca/dfo-mpo/plan_c.htm (accessed 8 April 2005). The apparent success of APEC initiatives also need to be seen against a weak implementation framework based primarily on political cooperation and not hard or even soft law mechanisms.

51 See Robert W. Smith "A geographical primer to maritime boundary-making" (1982) 12 *Ocean Development and International Law* 1 at 3 who identified 376 potential maritime boundaries in need of settlement under the regime created by the 1982 UN Convention on the Law of the Sea.

52 The Australian/Timor Leste maritime boundary dispute in the Timor Sea is an example. After having initially been settled by ongoing negotiations between Australia and East Timor, this boundary was unravelled following Timor Leste gaining independence in 2002 and engaging in a fresh round of ongoing negotiations with Australia. See Stuart Kaye "Negotiation and Dispute Resolution: A Case Study in International Boundary-Making – The Australia-Indonesia Boundary" in Alex G. Oude Elferink and Donald R. Rothwell (eds) *Oceans Management in the 21st Century: Institutional Frameworks and Responses* (Martinus Nijohff, Leiden: 2004) 143–66.

53 Following claims by some coastal states to an extended continental shelf, there will be a need for the finalization of new maritime boundaries with adjacent or opposite coastal states where previously there was no overlapping continental shelf claim. This is an issue which Australia and New Zealand have already confronted in the Tasman Sea. See Alexander Downer and Philip Ruddock "Australia and New Zealand Agree Maritime Boundaries" Media Release FA112b (25 July 2004). Online at: www.foreignminister.gov.au/releases/2004/fa112b_04.html (accessed 27 April 2005).

54 In 2005, there was ongoing tension between Japan and South Korea over disputed sovereignty concerning the Dokdo islets and the management of surrounding marine resources, see "South Korean parliament passes bill on islands in dispute with Japan" *Associated Press Newswires* 26 April 2005. There has also been tension between Malaysia and Indonesia over disputed islands offshore Borneo, see "Indonesia not to boost presence in disputed area despite collision" *Agence France Presse* 10 April 2005.

55 The disputes over the Spratley Islands in the South China Sea remain an example.

56 Canada remains engaged in several maritime boundary disputes with the United States in the case of the Juan de Fuca Strait region, the Dixon Entrance between British Columbia and Alaska, the Beaufort Sea and Machias Seal Island in the Gulf of Maine. Australia has yet to resolve all its boundaries with Indonesia, Timor Leste and in the Antarctic. See Ted L. McDorman "Canada's Oceans Limits and Boundaries: An Overview" in Lorne K. Kriwoken, Marcus Haward, David VanderZwaag and Bruce David (eds) *Oceans Law and Policy in the Post-UNCED Era: Australian and Canadian Perspectives* (Kluwer Law International, London: 1996); Stuart Kaye, "Australian Ocean Boundaries: An Overview" in ibid.

57 Juan Luis Suárez de Vivero and Juan Carlos Rodríguez Mateos "New factors in ocean governance: From economic to security-based boundaries" (2004) 28 *Marine Policy* 185 at 185, 187 who note "current maritime boundaries are being linked more and more with the concept of *global security.*" This link is further developed in David Wilson and Dick Sherwood (eds) *Oceans Governance and Maritime Strategy* (Allen & Unwin, St Leonards, NSW: 2000).

58 Canada has been particularly mindful of these developments given its close trading relationship with the United States and the emphasis upon port security in places such as Halifax and Vancouver.
59 See Michael Byers "Policing the high seas: The proliferation security initiative" (2004) 98 *American Journal of International Law* 526–45.
60 1982 UN Convention on the Law of the Sea, Preamble.
61 See generally the discussion in R.R. Churchill and A.V. Lowe *The Law of the Sea* 3rd edition (Manchester University Press, Manchester: 1999) 4–5.
62 See, for example, David VanderZwaag, "The precautionary principle and marine environmental protection: Slippery shores, rough seas, and rising normative tides" (2002) 33 *Ocean Development and International Law* 165–88 at 174–5.
63 For a discussion of the tensions between positivistic and natural law viewpoints, see P.W. Birnie and A.E. Boyle *International Law and the Environment* 2nd edition (Oxford University Press, Oxford: 2002) at 18–20.
64 See Chapter 15 by Johnston in this volume.
65 See David VanderZwaag *Canada and Marine Environmental Protection: Charting a Legal Course Towards Sustainable Development* (Kluwer Law International, London: 1995).
66 Prue Taylor, referring to the work of Robyn Eckersley, identifies five major streams of ecophilosophical thinkings: resource conservation, human welfare ecology, preservationism, animal liberation and ecocentrism. She advocates the application of an ecocentric perspective to the development of international law in particular. See Prue Taylor *An Ecological Approach to International Law: Responding to Challenges of Climate Change* (Routledge, London: 1998) 31–2.
67 For a discussion of the dominance of traditional ethical theories, see John Lemons and Pamela Morgan "Conservation of Biodiversity and Sustainable Development" in John Lemons and Donald A. Brown (eds) *Sustainable Development: Science, Ethics, and Public Policy* (Kluwer Academic Publishers, Dordrecht: 1995) at 101–2.
68 For a discussion of the continuing growth in the "transnational ethical community" in its environmental and human rights components, see Douglas M. Johnston and David L. VanderZwaag "The ocean and international environmental law: Swimming, sinking and treading water at the millennium" (2000) 43 *Ocean & Coastal Management* 147–61 at 151–3 and 157.
69 See, for example, J. Baird Callicott and Karen Mumford "Ecological Sustainability as a Conservation Concept" in John Lemons, Laura Westra and Robert Goodland (eds) *Ecological Sustainability and Integrity: Concepts and Approaches* (Kluwer Academic Publishers, Dordrecht: 1998); Richard O. Brookes, Ross Jones and Ross A. Virginia *Law and Ecology: The Rise of the Ecosystem Regime* (Ashgate Publishing Company, Burlington, Vermont: 2002) at 269 (where the controversial endpoints of ecosystem management are discussed). On the need to transform the human–natural environment relationship from minimizing harm to maximizing harmony, see David R. Boyd "Sustainability law: (R) Evolutionary directions for the future of environmental law" (2004) 14 *Journal of Environmental Law & Practice* 357–85 at 364–5.
70 See note 20.
71 The "troubled waters" yet to be faced include deciding on what constitutes an adequate network of marine protected areas, and how to further define ecosystem health and what are acceptable impacts on interferences by human uses.
72 In Canada, how far the *Oceans Act* has been effectively implemented in practice is being audited and a Report of the Commissioner of the Environment and Sustainable Development to the House of Commons is expected in September 2005.

Index